Microbial Technology in the Developing World

Microbial Technology
in the
Developing World

Edited by

E. J. DaSilva
Division of Scientific Research and Higher Education, UNESCO

Y. R. Dommergues
BSSFT Laboratory, Nogent sur Marne

E. J. Nyns
Bioengineering Department, University of Louvain

and

C. Ratledge
Department of Biochemistry, University of Hull

OXFORD NEW YORK TOYKO
OXFORD UNIVERSITY PRESS
1987

Oxford University Press, Walton Street, Oxford OX2 6DP
Oxford New York Toronto
Delhi Bombay Calcutta Madras Karachi
Petaling Jaya Singapore Hong Kong Tokyo
Nairobi Dar es Salaam Cape Town
Melbourne Auckland
and associated companies in
Beirut Berlin Ibadan Nicosia

Oxford is a trade mark of Oxford University Press

Published in the United States
by Oxford University Press, New York

British Library Cataloguing in Publication Data
Microbial technology in the developing world.
1. Biotechnology—Developing countries
I. DaSilva, E. J.
660'.6'091724   TP248.2
ISBN 0–19–854719–6

Library of Congress Cataloging in Publication Data
Microbial technology in the developng world.
1. Biotechnology—Developing countries.
DaSilva, E. J.
TP248.195.D48M53   1987   660'.62'091724   87–7951
ISBN 0–19–854719–6

Set by Cambrian Typesetters, Frimley, Surrey
Printed in Great Britain by
St Edmundsbury Press,
Bury St Edmunds, Suffolk

Contents

Contributors

H. D. Burges, Glasshouse Crops Research Institute, Worthing Road, Littlehampton, West Sussex BN17 6LP, UK.

Rita R. Colwell, Department of Microbiology, University of Maryland, College Park, Maryland 20742, USA.

E. J. DaSilva, Division of Scientific Research and Higher Education, UNESCO, Place de Fontenoy, Paris 7e, France.

H. G. Diem, BSSFT (CTFT/ORSTOM/CNRS), 45 bis Avenue de la Belle Gabrielle, 94736 Nogent-sur-Marne, Cedex, France.

H. W. Doelle, Biotechnology Unit, Department of Microbiology, University of Queensland, St Lucia, Brisbane 4067, Australia.

Y. R. Dommergues, BSSFT (CTFT/ORSTOM/CNRS), 45 bis Avenue de la Belle Gabrielle, 94736 Nogent-sur-Marne, Cedex, France.

B. L. Dreyfus, ORSTOM, BP 1386, Dakar, Senegal.

A. S. El Nawawy, Agriculture Research Centre, Giza, Arab Republic of Egypt.

E. L. Foo, UNEP/UNESCO/ICRO, Microbiological Resources Center, Karolinska Institute, S 104 01 Stockholm, Sweden.

J. Freire, *Rhizobium* MIRCEN, IPAGRO, Caixa Postal 776, 90 000, Porto Alegre, Brazil.

H. G. Gyllenberg, Department of Microbiology, University of Helsinki, SF–00710 Helsinki, Finland.

C. G. Heden, UNEP/UNESCO/ICRO Microbiological Resources Center, Karolinska Institute, S 104 01 Stockholm, Sweden.

S. O. Keya, *Rhizobium* MIRCEN, University of Nairobi, POB 30 197, Nairobi, Kenya.

Sisko Knuth, Department of Microbiology, University of Helsinki, SF-00710 Helsinki, Finland. (Present address: Research Laboratories Alko Ltd., P.O.B. 350, SF-00101 Helsinki, Finland.)

M. I. Krichevsky, Microbial Systematics Section, National Institute of Dental Research, National Institutes of Health, Bethesda, Maryland, USA.

W. Kurylowicz, State Institute of Hygiene, Warsaw, Poland.

W. Kurzatkowski, State Institute of Hygiene, Warsaw, Poland.

J. Lamptey, Institute for Biotechnology Research, University of Waterloo, Ontario, Canada N2L 3G1.

C. W. Lewis, Energy Studies Unit, University of Strathclyde, 100 Montrose Street, Glasgow, UK.

D. Nianguo Li, Energy Projects Ltd, China.

M. Moo-Young, Institute for Biotechnology Research, University of Waterloo, Waterloo, Ontario, Canada N2L 3G1.

L. H. G. Morton, School of Applied Biology, Lancashire Polytechnic, Preston, Lancashire PR4 1RA, UK.

Eugenia J. Olguin, Instituto Mexicano Technologia Appropriades, Mexico City, Mexico.

H. W. Pearson, Department of Botany, University of Liverpool, PO Box 147, Liverpool L69 3EX, UK.

J. S. Pillai, Department of Microbiology, University of Otago Medical School, PO Box 56, Dunedin, New Zealand.

Poonsuk Prasertsan, Department of Agro-Industry, Prince of Songkla University, Haad Yai, Thailand.

K. J. Seal, Biodeterioration Division, Biotechnology Centre, Cranfield Institute of Technology, Cranfield, Bedford MK43 OAL, UK.

J. C. Senez, Laboratoire de Chimie Bacterium, CNRS, F–13402, Marseilles, Cedex, France.

F. Sineriz, PROIMI (Planta Piloto de Procesos Industriales Microbiologicos), Avda. Belgrano y Pje. Caseros, 4000 San Miguel de Tucuman, Argentina.

J. Solecka, State Institute of Hygiene, Warsaw, Poland.

K. H. Steinkraus, Institute of Food Science, Cornell University, Geneva, New York, 14456, USA.

†H. Taguchi, International Centre of Co-operative Research and Development in Biotechnology, University of Osaka, Suita-Shi, 565 Osaka, Japan.

Cynthia A. Walczak, Microbial Systematics Section, National Institute of Dental Research, National Institutes of Health, Bethesda, Maryland, USA.

† Deceased, June 1987.

Microbial technology and the developing countries—an introduction

Microbial technology, probably one of the oldest scientific practices, can be traced back some 6000 years to the time when beer and wine production apparently began in ancient Egypt. Then, artisanal in concept and traditional in practice, and long referred to and known as biotechnology, the applications of microbiology have gradually evolved into technologies that are used in industrial processes dealing with the production of vaccines, antibiotics, amino acids, organic acidulants, feed supplements, and fermented foods.

The current upsurge of interest in today's biotechnology has stemmed from recent developments in recombinant DNA technology arising from research carried out in the 1970s. Despite the shift in focus, be it from a commercial or expedient science policy standpoint, the importance of applied microbiology and the unique role of the invisible microbe in technological development cannot be belittled.

Most simply defined, biotechnology is the harnessing of micro-organisms—or plant and animal cells that are made to behave as if they were micro-organisms—either for the production of specific materials for man or as a service process for the improvement of the environment. Biotechnology, as perceived by several academic and governmental policy makers, is a frontier area for all countries, both developed and developing, and amongst which the research priorities and policies may vary greatly. In industrialized societies, research initiatives are generally directed towards the production of novel commodity products, to the reversal of economic stagnation, reduction of unemployment, and to the conservation of the environment. On the other hand, the developing world, beset by problems of population growth and inadequate standards of living, has embraced the biotechnological promise of applied microbiology as a means of complementing limited supplies of conventional fuels and utility products throught the use of non-conventional fuels and bio-materials, of enhancing agricultural productivity through primarily microbiological means, of using novel bioproducts to conserve human, plant, and animal resources, and of adopting new measures to counteract the biodeterioration of materials of economic, social, and cultural value.

In the foregoing context, the example of alcohol is an apt and distinctive one. Widely recognized as an echo of the oldest form of biotechnology, this commodity is now being used as a fuel in some developing countries rather than as a drink. Another significant example is the production of methane from resources once considered as waste, and ranging from agro-industrial residues to domestic refuse and sewage. Indeed, such is the knife-edge economy that many communities in the developing countries have to contend with, that even a small improvement in the efficiency of energy utilization or generation may have far-reaching effects on the process of rural development and, indeed, on the improvement of life and its quality in those communities. The late Jackson Foster, an applied microbiologist of repute, observed that:

There is an enormous gulf between the industrial nations, with their advanced microbial technology, and other nations of the world where it is almost lacking. If we could extend to other areas even a small fraction of the microbiological regulation we perform today, we could bring about something that could be likened to a diplomatic coup of major magnitude. The humanitarian and social consequences of such an achievement are immense. Countries could produce for themselves, microbiologically, a number of food, medicinal, agricultural, and industrial products which now either must be imported or done without altogether. Foreign exchange could be conserved for other essentials. The indirect benefits accruing from the establishment of any new manufacturing enterprise are well-known—the spawning of subsidiary industries and services which, however small, derive their justification from the primary process. This autocatalysis affecting economics, education, and social stability enhances greatly the effects of an industrial process, even a small-scale one.

Within this volume, an attempt has been made to bring together those aspects of microbial biotechnology, both of the traditional and modern, and perhaps also of the emerging types, that have a role to play in the economies of the developing countries. In several cases this is perceived not as a national contribution, but more as a contribution with a regional perspective or even at a particular village level. Several authors have discussed the effects and benefits of biotechnology on rural and developing economies. Such examples may, perhaps, serve as a beacon for others faced with similar problems and could be adopted in other regions. Of course, there are no easy, universal solutions. Account must be taken of the need for innovative adaptation in the use of biotechnology for technological progress and economic growth in the developing country concerned. Again, the two above-mentioned examples, dealt with in this volume, suffice to make the point. Whereas the exposé on biogas illustrates the national political will and committment in the provision of a non-conventional fuel to meet the domestic energy needs of the rural sector, the contribution on alcohol production focuses on the socio-cultural implications and the ancillary economic advantages associated with the fermentation process.

Biotechnology itself cannot be an answer to transforming the poor country into a rich one. The problem is much more complex than that. It does, however, represent a radical way of thought and is a subject capable of considerable modification and change. This is so because it involves both rural and urban sectors, the use of traditional household skills and modern sophisticated techniques, the elements of small-scale and large-scale production, and finally, political committment and public acceptance. It is a key area of science and technology that edges away from the more traditional sciences such as chemistry, which has provided so many fertilizers, herbicides, and pesticides that are used extensively in developing countries, but at considerable expense and with danger to the environment.

Nature has endowed several developing countries with vast coastlines abundant in littoral, estuarine, and marine natural resources so well suited to the development of marine microbiology. Marine biotechnology has come a long way from the early classical experiments of the 1940s. In spite of this, and the fact that it has been treated as a restricted area of research interest in comparison with others, biotechnology's latent economic promise needs to be more fully exploited by the developing world.

In adopting those biotechnological processes that will yield positive socio-economic returns on the investments made, judicious selection is needed. Nearly every developing country has either initiated or is planning country-wide programmes for tapping the biotechnological labour force of micro-organisms for national development, and it is important that the problems be realistically understood before the committment of scarce funds. Several options exist ranging from the development of alternative energy sources to the potential of non-conventional sources of food supplies. Some of these options are long-ranging in scope and economic returns.

Biological nitrogen-fixation (BNF) is a low-cost, non-polluting microbial technology that is of particular significance in tropical agriculture and forestry, especially in the developing countries. In the coming years, BNF, enriched by new lines of current research, will continue to be deployed strategically in nutrient and mineral recycling in land reclamation, in enhancing the fertility of arid and desert soils, in forestry and agro-forestry.

In the food sector, single-cell protein (SCP) is one striking example, amongst others, of the evolution of basic laboratory research into an industrial venture. Its introduction for the production of feed and food supplements in human and animal diets in the developing countries seems to be amply justified by the unceasing population growth in the developing world, the demand for animal protein to counteract malnutrition, the alteration of surpluses and shortages of food, and the abundance of renewable biomass resources. A traditional use of SCP is found amongst the many fermented foods that have been prepared and consumed by man through the millennia. These foods, for example, ontjom, tape, and tempe, have a striking correlation and nutritive value to western food technology research dealing with the production of meat analogues, meat-like nuggets, and other textured vegetable proteins.

The mid-1960s witnessed an increasing use of computers in the development and monitoring of fermentation control. With the advent and availability of personal computers, development of microbial data-bases and bio-information network systems appears to be more attractive. Another feature of the growing field of bioinformatics—the offshoot resulting from the interaction between informatics and biotechnology—is computer conferencing, a system that overcomes the slow pace of communication, that spurs joint research activities, that catalyses a more fruitful exchange of technological information across national borders, and that conserves investments and resources. However, in all this, the microbiologist, generally an amateur computer specialist, can in due course, break through the barriers encountered in accessibility to data-base resources on microbes and process technologies. The translation from concept to practice is discussed in this volume.

Research and development are meaningless without a dynamic infrastructure. To overcome several bottlenecks, such as the shortage of trained manpower, and to bridge the ever-widening gap in the microbiological capability of the industrialized and developing countries, an infrastructure facilitating global scientific interaction, and promoting regional and inter-regional co-operation has been established. This mechanism is designed to support, and twin research and training programmes carried out in various institutions belonging to Unesco's global network of Microbiological Resources Centres (MIRCENs), which function as agents of international technology transfer in the exploration of the existing know-how in microbial technology and its availability at present constrained by patent law.

As indicated earlier, microbial technology predates written history in concept and practice. Its potential as well as the promise of its several new offshoots, enriched by the techniques of enzyme and genetic engineering, are vast and undisputed. Sir Harold Hartley, acknowledged as the father of biochemical engineering, said nearly two decades ago:

> The national importance of the development of industrial microbiology is so great that it deserves the attention of the Minister of Technology who now carries responsibility for the welfare of the science-based industries of this country and for supporting new developments of the growth industries of the future, among which biochemical engineering must take a high place . . . Successful development . . . must take place at universities in order to attract and train the young men and women who will be needed in these industries in the future. We have the ability and the enthusiasts if they are given support on an adequate scale for the complex operations involved. What we need now are two or three strong centres of teaching and research directed specifically to industrial microbiology in order to safeguard the future of these industries in this country. Now is the time to retrieve the position that has been allowed to slip.

The case made for the promotion of microbial-based industries in the UK in 1967 still holds true for the developed world, and especially for the developing countries that have embarked on the biotechnological path to economic growth. In that context, we hope that the MIRCENs, with which we have all been associated in one way or another, will play a catalytic role in engendering inter-regional collaboration and in consolidating national development.

This volume should be seen as a complement to the new scientific periodical *The MIRCEN Journal* in accelerating the interaction and consolidation of collaboration between the existing technical expertise of the developed countries with the growth of qualified manpower resources in the developing countries.

It is our hope, too, that this assembly of chapters, written by experts from throughout the world, both of the developed and the developing countries, will illustrate the practices and potential applications of microbial technology. It is inevitable that in a book of this size and scope the lapse of time between its conception and final appearance in print will render some of the material slightly out of date. Readers are urged to consult current research literature for details of their own particular interests. The book can only serve as an illustration of world-wide activities; it is not meant to be a definitive text. Scientists in each country must therefore use it as a message of encouragement and example, rather than as a compilation of contemporary research or a recipe for direct action.

E. J. DaSilva
Y. R. Dommergues
E. J. Nyns
C. Ratledge

1

Nitrogen fixation in tropical agriculture and forestry

B. L. Dreyfus, H. G. Diem, J. Freire, S. O. Keya,
and Y. R. Dommergues

Nitrogen compounds comprise from 40 to 50 per cent of the dry matter of protoplasm, the living substance of plant cells. For this reason, nitrogen is required in large quantities by growing plants and is indeed the key to soil fertility. Non-nitrogen-fixing plants, for example cereals, obtain all the nitrogen they need from the soil. In Senegalese conditions this uptake was estimated to be as follows: 79–132 kg N ha/crop for pearl millet; 74–84 kg N ha/crop for rice; 134 kg N ha/crop for sorghum; and 121–139 kg N ha/crop for maize (Blondel 1971). Nitrogen-fixing plants, essentially legumes, take a part of the nitrogen they require from the atmosphere, the other part being provided by the soil.

When nitrogen fertilizers are available, soil nitrogen levels are maintained or improved by applying these industrially-fixed nitrogen sources. Such a technology, which allows continuous crop yields, is successfully used in intensive agricultural systems, but certain limitations have been progressively observed, i.e. increasing cost especially in developing countries, low yields from leaching and denitrification, especially in tropical conditions, and pollution of underground water by nitrates. The other alternative for maintaining or improving the nitrogen content in soil is to exploit nitrogen fixation. In this chapter, the microbial and the plant components of symbiotic nitrogen-fixing systems that are encountered in the tropics are dealt with. On the other hand, an attempt is made to explain how to improve the effectivity of these systems by assuring that the infective and effective strain of *Rhizobium* or *Frankia* is present in the soil or on the seed at the proper time (section 3), by reducing the impact of the limiting factors that can be controlled (section 4), and by using plant cultivars that have been bred for improved symbiotic performance. This last approach has not yet been really exploited, but it will probably lead to major applications in the short term, at least in the case of nitrogen-fixing trees.

1. The symbionts

1.1 Tropical rhizobia

1.1.1. Taxonomy

It is now recognized that bacteria which form nodules on legumes, and which are known under the general name of rhizobia, belong to two genera: *Rhizobium* and *Bradyrhizobium* (Jordan 1982, 1984; Elkan 1984). All strains of the *Rhizobium* genus are fast-growing. The genus *Rhizobium* comprises four species: *Rhizobium leguminosarum*, *Rhizobium meliloti*, *Rhizobium loti*, and *Rhizobium fredii*. The species *R. leguminosarum* results from the fusion of three species *Rhizobium trifolii*, *Rhizobium phaseoli*, and *R. leguminosarum* (in Jarvis *et al.* 1986). *R. fredii* is a new species of *Rhizobium* which nodulates soybean (Scholla and Elkan 1984).

The genus *Bradyrhizobium* comprises all the slow-growing strains of the former *Rhizobium* genus, i.e. the cowpea miscellany, and *Bradyrhizobium japonicum* (formerly known as *Rhizobium japonicum*).

Until now, it was generally recognized that most tropical nitrogen-fixing bacteria nodulating legumes were part of the cowpea miscellany. In fact, these bacteria may belong to either genus, *Bradyrhizobium* or *Rhizobium*. Many fast-growing strains have already been isolated from tropical trees belonging to the genera *Acacia*, *Leucaena*, and *Sesbania*. A taxonomic study under way at the ORSTOM laboratory in Dakar has shown that these strains are closely related to temperate fast-growing ones and consequently should be included in the *Rhizobium* group.

1.1.2. Specificity in connection with nodulation and effectiveness

It is relevant to make a distinction between the specificity related to the nodulation process and the specifity which concerns the nitrogen-fixing potential of the system.

The first type of specificity has been recognized for many years. It is well known that host plants exhibit different degrees of specificity, the species that nodulate with many types of strains being designated as promiscuous, in contrast to the specific ones. Probably, the most promiscuous species are to be found in group 1 of the legume trees defined later (section 1.1.3.2.).

The concept of specificity related to effectiveness comes from studies in *Stylosanthes* sp. (Date and Halliday 1980). This concept was found to apply to other legumes including tree legumes, such as *Acacia seyal*. This tree species nodulates both with *Rhizobium* and *Bradyrhizobium* strains, but in general it fixes nitrogen actively only with *Rhizobium* strains.

1.1.3. Examples of host relationships

1.1.3.1. Soybean

As indicated earlier, soybean which generally nodulates with *Bradyrhizobium japonicum*, also nodulates with fast-growing strains of the genus *Rhizobium*, that are now designated as *Rhizobium fredii*, and with *Bradyrhizobium* of the cowpea miscellany. However, it should be noted here that there are differences in the host-relatedness of soybean cultivars, the host range of Asian cultivars being broader than that of American cultivars. The former cultivars can nodulate with *Bradyrhizobium* of the cowpea miscellany, whereas the latter ones cannot (Roughley *et al.* 1980).

1.1.3.2. Tropical trees

Tropical tree legumes can be classified into three groups according to effective nodulation patterns with *Rhizobium* and *Bradyrhizobium* (Dreyfus and Dommergues, 1981).

1. Trees of group 1 nodulate effectively with *Bradyrhizobium*, e.g. *Acacia albida*, *Dalbergia melanoxylon*, *Dalbergia sisso*, *Prosopis africana*, *Pterocarpus erinaceus*, *Gliricidia sepium*.

2. Trees of group 2 nodulate effectively with *Rhizobium*, e.g. *Acacia nilotica* (var. *adansoni* and *tomento*), *Acacia raddiana*, *Acacia senegal*, *Prosopis juliflora*, *Sesbania grandiflora*.

3. Trees of group 3 nodulate with both genera, *Rhizobium* and *Bradyrhizobium*, e.g. *Acacia seyal*, *A. sieberana*, *Erythrophleum guineense*.

This classification is liable to revision since it has already been observed that host plants known to nodulate with strains of *Rhizobium* alone or of *Bradyrhizobium* alone were later found to nodulate with both genera. This has been already reported in *Acacia senegal* (section 1.1.2.), and is also well known in *Leucaena leucocephala*, a species that generally nodulates with *Rhizobium*, but which occasionally nodulates with *Bradyrhizobium* (Dreyfus and Dommergues 1981; Sanginga *et al.* 1986).

1.1.3.3. Sesbania rostrata

Stems of *S. rostrata* nodulate with strains which, in spite of their fast-growing characteristics, are closely related to *Bradyrhizobium*. These strains, which can nodulate the roots, are unique since they can grow on molecular nitrogen as the sole nitrogen in pure culture (Dreyfus *et al.* 1983). They could probably be incorporated in a new species or even a new genus of nitrogen-fixing bacteria (Jarvis *et al.* 1986). Roots of *S. rostrata* nodulate with *Rhizobium* strains that also nodulate the roots of non-stem nodulated *Sesbania* spp. Few of these strains are also able to nodulate the stems of stem-nodulating *S. rostrata*, but they are unable to grow on molecular nitrogen as the sole nitrogen source *in vitro*.

1.1.3.4. Parasponia

Parasponia sp. (Ulmaceae) is the only non-legume genus known to nodulate with *Bradyrhizobium* (Trinick 1981). Whereas most *Bradyrhizobium* strains are rather promiscuous, *Bradyrhizobium* strains isolated from *Parasponia* sp. fail to nodulate most of the tropical legumes that usually nodulate with *Bradyrhizobium* strains (Trinick and Galbraith 1980).

1.1.4. Selection of strains for inoculation

In this paragraph, we shall not deal with the methodology involved in the selection of strains, a topic that has been properly addressed in specialized books (e.g. Somasegaran and Hoben 1985) and that is briefly presented hereafter (section 3.2), but shall call attention to the following considerations.

Since many tropical soils have a low pH and often a high aluminium content, it is advisable to screen strains for their tolerance to acidity and high aluminium concentration. However, one should be well aware of the fact that the host plant is usually less tolerant to these soil constraints than the symbiotic bacterium. Thus, it is necessary not only to improve the tolerance of the micro-organisms, but also that of the host plant. This point is well illustrated by the behaviour of *Leucaeoa leucocephala* (2.2.2.2). Moreover, it is recognized that the process of symbiosis, notably nodulation, is about ten times more sensitive to acidity than either bacterial or root growth alone (Evans *et al.* 1985).

In contrast with the established dogma that locally isolated strains are best adapted to the environmental conditions, and consequently the most effective in these conditions, it was found that some introduced strains could markedly improve nitrogen fixation when used for inoculating. Thus, strain TAL 651 isolated from *Psophocarpus* sp. and maintained in NifTAL MIRCEN collection at Hawaii, was found to be the most efficient strain for *Acacia holosericea* grown in Senegal (Cornet and Diem 1982).

1.2. Frankia

Apart from legumes (and plants of the genus *Parasponia*), which are nodulated by rhizobia, about 200 plant species covering 19 genera and 8 families are nodulated by nitrogen-fixing micro-organisms known as *Frankia* (section 2.4). The actinomycetal nature of *Frankia* was detected using cytological techniques for many years (Becking 1975; Gardner 1976) and confirmed in 1978 when Torrey's group isolated a strain of *Frankia* and cultivated it *in vitro* for the first time (Callaham *et al.* 1978).

1.2.1. Actinorhizal nodules

Morphologically and anatomically different from legume nodules, actinorhizal nodules are formed from modifications of lateral roots that produce lobes characterized by successive dichotomous branching and, in the next stage, a coralloid mass composed of a cluster of lobes. The nodule continues to grow for several years until it reaches the size of a tennis ball. Larger nodules, up to 20 cm in diameter, have been found on the roots of 20-year-old *Allocasuarina stricta* in Tunisia. The lobes at the outer part of the nodular mass, which are the most recently formed, fix nitrogen actively. Depending on the plant species and/or the environmental conditions, the lobe may or may not develop a determinate nodule root growing more or less vertically upward.

The symbiotic micro-organism, *Frankia*, occupies specific cortical cell layers outside the central vascular cylinder unlike legume nodules which have the symbiotic bacterium *Rhizobium* or *Bradyrhizobium* located within the peripheral vascular bundles.

Cross-sections of actinorhizal nodules usually show two structures: (i) the encapsulated filamentous hyphae which ramifies profusely passing from one cortical cell to another; and (ii) vesicles which exhibit different characteristics according to the host-plant species. Vesicles can be septate (e.g. *Alnus glutinosa* or *Colletia spinosa*) or non-septate (e.g. *Purshia*), spherical (e.g. *Alnus, Colletia, Ceanothus*, or *Hippophaë*), pear, club-shaped, or filamentous (e.g. *Comptonia, Coriaria, Myrica*, or *Datisca*). Vesicles are considered as the sites of nitrogen fixation in the nodules (Torrey 1985). However, in some plants, namely plants belonging to the genus *Casuarina*, no vesicles have been detected, which raises the problem of the site of nitrogen fixation in these nodules. A third structure, called sporangia, may exist in actinorhizal nodules. Sporangia have been found in effective nodules of *Alnus glutinosa*, *Myrica gale* and in ineffective nodules of *Elaeagnus umbellata* (Newcomb 1981). Whereas all strains of *Frankia* are able to produce sporangia in most culture media, only some strains can produce sporangia in the nodules of certain host-plants. It is well established that *Alnus* has both spore-positive and spore-negative nodules, the sporulating or non-sporulating character of the nodules being particular to certain strains of *Frankia*. In a recent study, Normand and Lalonde (1982) showed that the most effective strains of *Frankia* of *Alnus* were of the spore-negative type, whereas the spore-positive type strains had an effectiveness that was on the average 70 per cent less than that of the most effective spore-negative strains.

The haemoglobin content in nodules of legumes is correlated with their nitrogen-fixing ability. The haemoglobin of legumes, which is called leg-haemoglobin, serves to facilitate the oxygen flux to rhizobia respiring at extremely low, non-toxic, free oxygen concentration. In actinorhizal systems, the presence of a haemoglobin entity had been disputed for many years. Recently, Tjepkema (1983) through the use of sensitive spectrophotometry confirmed the presence of high concentrations of haemoglobin in nodule slices of *Casuarina cunninghamiana* and *Myrica gale*. Lower concentrations of haemoglobin were observed in the nodules of *Comptonia peregrina*, and *Elaeagnus angustifolia*, and negligible to trace amounts were found in *Ceanothus americanus* and *Datisca glomerata* (Tjepkema 1984; Appleby 1984).

1.2.2. Isolation and culture of *Frankia*

The difficulty of isolating *Frankia* from actinorhizal nodules and its subsequent cultivation explains why it took until 1978 to perform the first successful isolation (Callaham *et al*. 1978). The problem of isolating *Frankia* is not dealt with here as it has already been addressed in detail elsewhere (Lalonde and Calvert 1979; Baker and Torrey 1979; Diem *et al*. 1982a, 1983; Carpenter and Robertson 1983)

Frankia are slow-growing actinomycetes; even when cultured in the best adapted media currently known, the doubling time often exceeds 48 hours. Furthermore, the growth requirements of the different strains of *Frankia* are not yet well known.

Frankia cultured *in vitro* usually exhibit structures, i.e. hyphae, vesicles, and sporangia. Vesicles are generally formed in nitrogen-deficient media. Nitrogen fixation, as demonstrated by the acetylene reduction assay, or $^{15}N_2$ incorporation, has been shown to accompany the formation of vesicles, indicating that these structures are the site of nitrogen fixation. A fourth structure has recently been reported in cultures of *Frankia* of *Casuarina equisetifolia* (Diem and Dommergues 1985). This occurs when vegetative hyphae develop into wide torulose hyphae, which are called reproductive torulose hyphae (RTH), on account of their seemingly major role in the propagation of *Frankia* strains of *C. equisetifolia*.

1.2.3. Infectivity and effectivity

Attempting to classify the known *Frankia* isolates according to host specificity groups should be avoided since some strains of *Frankia* can infect two or more specificity groups, such as *Alnus* and *Elaeagnus* (Lalonde and Simon 1985), *Alnus* and *Comptonia* (Becking 1982), or *Casuarina* and *Hippophaë* (Diem *et al*. 1983; Zhang *et al*. 1984), and also because strains isolated from a given host are unable to infect this host, but can nodulate plants of other specificity groups (Diem *et al*. 1982b; Zhang *et al*. 1984). Some strains are known to be highly specific; *Frankia* strain ORS021001, for instance, nodulates only true *Casuarina* and not *Allocasuarina* species (Gauthier *et al*. 1984), whereas other strains have a wider host range, e.g. ORS 022602 (Puppo *et al*. 1885).

After an infective strain has been isolated, its effectiveness has to be evaluated. The procedure adopted for the selection of *Frankia* is similar to that used in the case of rhizobia, but screening strains of *Frankia* takes longer because nodulation is usually slower than in legumes (up to 2 weeks) and again, because the growth of some host plants is also slow at the plantlet stage, e.g. *Casuarina* spp. As in the legume symbiosis, some infective strains may be ineffective (e.g. Baker *et al.* 1980); furthermore, there may be some interaction between the host plant and *Frankia*, a fact that should be considered when selecting strains for their effectivity.

1.2.4. Influence of inoculation

It is well known that leguminous crops introduced in soils without specific *Rhizobium* strains benefit readily from inoculation, provided that the conditions for nodulation and nitrogen fixation are favourable. The situation is the same for actinorhizal plants: both pot and field experiments show that inoculation with the proper strain of *Frankia* usually enhance growth and nodulation of actinorhizal plants grown on soils that previously had a low or zero *Frankia* population.

As part of a large reclamation project in northern Quebec, more than seven million seedlings of actinorhizal plants, especially *Alnus crispa*, were successfully inoculated on an industrial scale with *Frankia* inoculum (Périnet *et al.* 1985).

Similar experiments performed in Senegal have demonstrated that inoculating *Casuarina equisetifolia* not only improved tree growth in the coastal sandy soils, but also contributed efficiently to stabilizing the moving dunes. The effects of inoculation were much greater than those obtained by applying relatively large amounts of nitrogen fertilizer (Table 1.1). Unpublished results by ORSTOM group indicate that inoculated *C. equisetifolia* produced more biomass than trees fed with nitrogen fertilizer. These findings concur with observations made by Sellstedt and Huss-Danell (1985) on *Alnus incana*, indicating that the energy required in nitrogen fixation is probably not great

Table 1.1

Influence of inoculation with *Frankia* ORS021001 on height, dry weight, and N_2 fixation of 11-month-old *C. equisetifolia* (Gauthier *et al.* 1985)

Treatments		Height (cm)	Dry weight (g/tree)	N_2 fixed (g N_2/tree)
Inoculation	N addition (g/tree)			
0	0.5	170 a	295 a	0
0	2.5	192 a	409 a	0
+	0.5	216 b	525 b	3.3–2.3[1]

Figures in same columns followed by same letter do not differ significantly, P = 0.05 (Duncan test).
1. First figure calculated from direct isotope dilution method, second figure from *A* value method.

enough to slow down biomass production in alders. One hypothesis suggests that, at least in the case of some actinorhizal plants, the symbiosis with *Frankia* favourably alters the hormonal balance of the plant, thus enhancing its growth.

The slow growth and poor yield of *Frankia* are considered a major impediment in the preparation of inocula for actinorhizal plants. However, technological progress will probably make it possible to produce commercial inocula in the near future (Righetti and Hannaway 1985).

2. The host plants

2.1. Legumes grown for grain, forage, green manuring, and mixed cropping

Leguminous plants other than trees are widely grown in the tropics, providing people and cattle with a protein-rich diet; they are also often used as green manure crops (e.g. Chee 1982) or introduced in mixed cropping systems (e.g. Sprent 1983; Simon 1986). All the plants presented hereafter belong to the subfamily Papilionoideae, whereas nitrogen-fixing trees are found in the three subfamilies of leguminous plants (section 2.2).

2.1.1. Grain legumes

Arachis hypogaea (peanut, groundnut) nodulates freely with *Bradyrhizobium* of the cowpea miscellany. Peanut cultivars exhibit clearcut variations in the nodulation ability, which affects the amount of nitrogen fixed. Nodulation is most sensitive to environmental constraints. In unfavourable conditions, nitrogen fixation may be totally inhibited. On the other hand, when soil has been properly managed, nitrogen fixation can reach 140 kg N_2/ha per crop (Ganry *et al.* 1985).

Cajanus cajan (pigeon pea) nodulates with *Bradyrhizobium* of the cowpea miscellany. Fixation rates of 90–150 kg N_2/ha/year have been reported (Dobereiner and Campelo 1977).

Cicer arietinum (chick-pea) has specific requirements for *Rhizobium*, which explains why inoculation is essential in soils that have never been planted with this plant before.

Glycine max (soybean), like chick-pea, has specific requirements for *Rhizobium* and *Bradyrhizobium* (section 1.1.3.1), so that inoculation is mandatory in areas where this legume is newly introduced. In favourable sites the amount of nitrogen fixed is in the range of 100–200 kg N_2/ha per crop (Gibson *et al.* 1982).

Phaseolus vulgaris has specific requirements so that inoculation is often beneficial. Acidity or nitrogen excess in soil are definitely harmful for the symbiosis. Other species of *Phaseolus*, namely *P. mungo* and *P. aureus*, are perfectly adapted to acid soils, where nitrogen fixation is satisfactory. In *P. aureus*, nitrogen fixation rates of 224 kg N_2 fixed/ha per crop have been reported (Dobereiner and Campelo 1977). Climbing cultivars are consistently superior in nitrogen fixation than most of the bush types. Three factors contribute to the variability encountered, and these are the supply of carbohydrates to the nodules, relative rates of N uptake from soil and time of flowering (Graham 1981).

Vigna unguiculata (cowpea) nodulates readily with *Bradyrhizobium* strains of the cowpea miscellany, which are present in most tropical soils, so that inoculation is not necessary. The amount of nitrogen fixed varies widely with the cultivar used: comparing the nitrogen fixation rates of four cultivars of cowpea grown in identical conditions, Eaglesham *et al.* (1982) found that these rates covered a span from 49 to 101 kg N_2 fixed/ha. Exceptionally high rates of 354 kg N_2 fixed/ha have been reported (Dobereiner and Campelo 1977).

Voandzeia subterranea (bambara groundnut) nodulates readily in many soils, which suggests that this legume is associated with strains of *Bradyrhizobium* of the cowpea miscellany. However, specific strains have been reported to be exceptionally effective (National Research Council 1979). Investigations are under way to screen the cultivars which exhibit the best nitrogen fixing ability.

2.1.2. Forage legumes

Centrosema pubescens does not seem to have very specific rhizobia requirements, so that inoculation is not essential except in some situations.

Desmodium spp has probably moderately specific rhizobia requirements. There are reports of exceptionally high nitrogen fixation rates (up to 360 kh N_2 fixed/ha/year) by *Desmodium intortum* (Dobereiner and Compelo 1977).

Lablab purpureus, ex-*Dolichos lablab* (lablab bean), often mentioned as a grain crop, is more frequently used as a forage legume. It nodulates easily with *Bradyrhizobium* of the cowpea miscellany, so that inoculation is generally not required.

Macroptilium atropurpureum, ex-*Phaseolus atropurpureus*, cv siratro, nodulates with most strains of *Bradyrhizobium*; thus it does not require inoculation. Nitrogen fixation rates are high (Dobereiner and Campelo 1977).

Neonotonia wightii, ex-*Glycine wightii*, nodulates with *Bradyrhizobium*, but, according to Dobereiner and Campelo (1977), this species is seldom well nodulated because it does not seem to tolerate unfavourable environmental factors, such as acidity or P deficiency.

Pueraria phaseoloides (tropical kudzu) generally nodulates profusely and spontaneously.

Stylosanthes spp. (stylo) fall into three groups according to their effectiveness response with rhizobia strains: (1) group PE (promiscuous and effective) comprises species nodulating with a wide range of strains of the cowpea miscellany (like *Arachis*, *Lablab*, *Macroptilium*, *Pueraria*, *Vigna*), (ii) group PI (promiscuous, but often ineffective) comprises species that frequently nodulate with a wide range of rhizobia, but are often ineffective in nitrogen fixation, and (iii) group S (specific) includes species that nodulate with a narrow range of rhizobia strains (Date and Halliday 1980). *Stylosanthes humilis*, which belongs to group PI, was reported to fix 90 kg N_2/ha/year (Dobereiner and Campelo 1977).

2.1.3. green manure and mixed cropping

Canavalia ensiformis (horse bean or jackbean), and *Canavalia gladiata* (swordbean) are not only valuable green manure and cover crops, but also highly productive pulse crops. *Canavalia ensiformis* was reported to fix ca 60 kg N_2/ha/year (Dobereiner and Campelo 1977).

Crotalaria juncea (sunhemp) is cultivated as a source of bast fibre in India. Since it nodulates with strains of *Bradyrhizobium* of the cowpea type, which are present in most soils, inoculation is seldom necessary. If the plant is given adequate phosphate, 1 ha of *C. juncea* can add up to 300 kg nitrogen to the soil (National Research Council 1979). Green manuring with *C. juncea* was reported to increase signficantly the yield of different crops in South America (Dobereiner and Campelo 1977).

Phaseolus javanica and *Pueraria phaseoloides* are successfully used as cover crops, providing nitrogen and humus, and shielding the soil from water erosion. To date, nitrogen fixation rates have not been estimated with precision.

Tephrosia spp. are often used for green manuring, e.g. *Tephrosia candida* in Brazil (Dobereiner and Campelo 1977).

2.2. Tree legumes

2.2.1 Introduction

In many tropical regions gradual loss of soil fertility, insufficient firewood, and shortage of fodder are becoming serious problems especially in the many areas where the human populations are growing. Increasing the use of fertilizers or importing fuel are theoretical, but not realistic solutions to these problems. Alternatives must be identified that are low in cost and in energy demand (Roskoski *et al.* 1982). The introduction or reintroduction of trees in the agro-systems and the development of the use of woody plants with low nutrient requirements could successfully help solve the problems outlined above. Legume trees, together with actinorhizal plants, are good candidates because many fix nitrogen from the atmosphere, and thus can grow in nitrogen-deficient soils. Furthermore, the common association of root systems with mycorrhizal fungi enables nitrogen-fixing trees to extract various crucial nutrients from infertile soils. The problem is much more complicated than generally thought, not only because of the vast range of variation in the nitrogen-fixing potential of the different species or even individual trees, but also because of the doubtful validity of figures published by experimenters who did not make allowances for the spacial and temporal variations that inevitably occur during biological nitrogen fixation. One fact is clear: the nitrogen-fixing potential of legume trees indeed covers a wide range, viz. from less than 1 to 300 kg N_2 fixed/ha/yr and over. Table 1.2. gives examples of reliable estimates of nitrogen fixation by a few tropical legume trees. Many of them have been obtained using the acetylene reduction method. This method has serious limitations, but since no simple, inexpensive alternatives exist, its widespread use will probably continue (Righetti and Hannaway 1985).

Table 1.2

Nitrogen fixation by trees in the tropics

Species	Method of estimation	N_2 fixed (kg/ha/yr)
Acacia mearnsii	B	200
	I	4–11
Acacia holosericea	A	6
Acacia pennatula	A	34
Gliricidia sepium	A	13
Inga jinicuil	A	35
Leucaena leucocephala	A	110
Prosopis glandulosa	C	25–36
Erythrina poeppigiana	D	57–66
Casuarina equisetifolia	B	58

A: acetylene reduction assay; B: nitrogen balance studies; C: by ^{15}N abundance studies indicating that 50% of total plant N comes from nitrogen fixation; I: isotope method (A value).

2.2.2. Host plants

The family Leguminosae is divided into three sub-families, Mimosoideae, Caesalpinoideae, and Papilionoideae, the best known genera of nitrogen-fixing trees being as follows (National Research Council 1979, 1980; Domingo 1983; Dobereiner 1984; Brewbaker *et al.* 1984; Halliday 1984 *a, b*). Caesalpinioideae: *Acrocarpus. Cassia, Cordeauxia, Schizolobium*; Mimosoideae: *Acacia, Albizia, Calliandra, Desmanthus, Enterolobium, Inga,*

Leucaena, Lysiloma, Mimosa, Parkia, Pithecellobium, Prosopis, Samanea;
Papilionoideae: *Dalbergia, Erythrina, Flemingia, Gliricidia, Sesbania*. Data
on nodulation ability are available for only some of the species belonging to
these different genera (Allen and Allen 1981; Kirkbride 1984).

As a general rule, members of the Caesalpiniodeae subfamily, i.e. mainly
trees found in tropic regions, do not nodulate, as is the case with six *Cassia*
species: *Cassia fistula, Cassia grandis, Cassia javanica, Cassia leiandra, Cassia
nodosa*, and *Cassia siamea* (Halliday 1984b). However, there are exceptions,
and these include *Acrocarpus fraxinifolius, Cordeauxia edulis*, and *Schizolobium parahyba*.

The majority of the Mimosoideae, which are also trees or shrubs growing in
the humid and dry tropics, nodulate, e.g. many species of the genus *Parkia*.
Parkia biglobosa, however, does not nodulate.

Nearly all the Papilionoideae, which are mainly shrubs and herbs, found
around the world, nodulate.

Table 1.3

Tentative classification of some native and introduced West African woody legumes
according to nodulation response patterns (after Dreyfus and Dommergues 1981)

Groups	Species	Specificity[1]
Group 1		
Nodulating with fast-	*Acacia farnesiana*	S
growing rhizobia	*Albizia lebbeck*	S
(Rhizobium)	*Acacia nilotica*	S
	Acacia raddiana	S
	Acacia senegal	S
	Leucaena leucocephala	S
	Prosopis juliflora	S
	Sesbania sp.	S
Group 2		
Nodulating with fast-		
and slow-growing		
rhizobia	*Acacia seyal*	S
(Rhizobium and		
Bradyrhizobium)		
Group 3		
Nodulating and slow-	*Acacia albida*	P
growing rhizobia	*Acacia holosericea*	P
(Bradyrhizobium)	*Acacia sieberana*	S
	Prosopis africana P	

1. S: specific; P: promiscuous.

Table 1.3. is an attempt to classify some native and introduced African nitrogen-fixing trees according to their nodulation patterns. The impression that tropical tree legumes are much more promiscuous than temperate ones, i.e. they nodulate with a wide range of tropical rhizobia, is generally accepted (Halliday 1985). In fact, the promiscuous character of nitrogen-fixing trees varies according to the host plant, some of them being definitely promiscuous (e.g. *Acacia albida* or *A. mearnsii*), whereas others are rather specific like *Sesbania* spp. Usually, hosts nodulating with fast-growing strains of *Rhizobium* appear to be more specific than those nodulating with slow-growers or simultaneously with fast- and slow-growers. Some of the most promising species of nitrogen-fixing trees classified according to the two main climatic zones of the tropics are dealt below.

2.2.2.1. Dry tropics

As a preliminary remark, it should be noted that, though nodules are seldom found on nitrogen-fixing trees in the field in the dry tropics, nodulation does effectively exist on the same species grown in greenhouses or nurseries. One probable explanation is that water stress severely inhibits nodulation (Felker 1984).

Acacia albida, *A. raddiana*, and *A. senegal* These *Acacia*, native to Africa, are usually considered as highly valuable legume trees serving especially as soil improvers in agroforestry, and sources of fuel-wood, forage, and gum (*A. senegal*). Their nitrogen potential seems to be rather poor, but could possibly be improved by capitalizing on the great variability of the host plant. The subject has not yet been adequately studied.

Acacia holosericea Native to Australia, this species nodulates readily with *Bradyrhizobium* strains of the cowpea miscellany, but its nitrogen-fixing potential is rather low (Cornet *et al.* 1985). In addition *A. holosericea* is most often attacked by root-knot nematodes, which precludes the use of this species in agroforestry.

Albizia lebbeck Native to Bangladesh, Burma and Pakistan, this tree has been propagated in many tropical and subtropical regions (National Research Council 1980). Large perennial nodules of large size have been observed on adult trees in Senegal, but the potential nitrogen fixation of this species has not yet been evaluated.

Cordeauxia edulis This small bush, native to the semi-desert region bordering Somalia and Ethiopia is remarkable for its unique tolerance to drought (National Research Council 1979). It has been reported to be nodulated.

Prosopis spp. The nitrogen-fixing characteristics of this genus, which is currently receiving the pride of place in semi-arid regions and saline soils, were recently reviewed (Felker *et al.* 1981). Measuring the nitrogen compartmentalization of the biomass, productivity of leaves, branches, trunk and reproductive tissues, Rundel *et al.* (1982) estimated a nitrogen-fixation rate of 23–36 kg N_2/ha/yr with a 33 per cent stand cover. From determinations of natural abundance $^{15}N/^{14}N$ ratios, Shearer *et al.* (1983) estimated that *Prosopis* fixed approximately 43–61 per cent of its nitrogen, this range of figures applying to six of the seven sites they studied in the Sonoran desert, California. *Prosopis juliflora*, a tree native to Central America and northern South America, has been introduced in many arid zones of the world; it grows fast, even on soils very low in nutrients, probably thanks to its good nitrogen-fixing potential. In some places, *P. juliflora* is an agressive invader and thus is considered a hindrance (National Research Council 1980). Other species of *Prosopis*, namely *Prosopis cinerea* and *Prosopis tamarugo* are widely utilized in arid countries; their nitrogen-fixing potential has never been evaluated. An experimental study of the nodulating ability of *Prosopis chilensis*, *P. tamarugo* and *P. alba* has been initiated recently. Among these species, *P. chilense* seems to respond best to inoculation. The effectiveness of the nitrogen fixation process appears to vary greatly with the different *Prosopis* populations (Torres 1985).

2.2.2.2. Humid tropics

Acacia auriculiformis   Native to Papua New Guinea and northern Australia, *A. auriculiformis* has been successfully introduced into Indonesia, Malaysia, and the Philippines. It produces profuse bundles of nodules and can thrive on soils deficient in nitrogen and organic matter, which suggests a good nitrogen-fixing capacity (Domingo 1983).

Acacia mangium   A native of Australia, and southern Papua New Guinea, *A. mangium* is now being tried in several places in South-East Asia (Domingo 1983) and Africa, namely southern Senegal, Benin, Congo, and the Ivory Coast. In the three latter countries it grows exceedingly well (National Research Council 1983).

Acacia mearnsii   A highland tree from southern Australia, *A. mearnsii* was introduced in India and Natal more than 100 years ago (Boland, *et al.* 1984), and successfully grown on tropical and subtropical plateaux, for example in Madagascar and Kenya. It has a remarkable nitrogen-fixing potential. However, in some soils, for example highland soils of Burundi in Eastern Africa, acidity and aluminium toxicity are so pronounced that nodulation is severely restricted, thus depriving the host plant of the benefice of nitrogen fixation.

Albizia falcataria   This species, native to the eastern islands of the Indonesian archipelago and the west of Irian, has been spread throughout South-East Asia. Domingo (1983) considered it to be one of the fastest growing trees in the world. It is reported to nodulate, but its nitrogen-fixing potential is probably lower than that of other species of the same genus since fertilization may be needed (National Research Council 1979).

Calliandra calothyrsus   Native to Central America, this small legume tree was introduced into Java in 1936. Its profuse nodulation, suggesting an active nitrogen fixation potential, and high litter production make it a first class soil improver, and as such, it is often used in rotation schedules and in intercropped systems (Domingo 1983). *Calliandra calothyrsus* has been successfully introduced beneath stands of non-nitrogen fixing trees, such as *Eucalyptus deglupta* and *Pinus merkusii* (National Research Council 1983).

Erythrina spp.   More than 100 species are planted as shade trees, windbreaks, living fences, support plant and for alley cropping, wood, food, and medicinal purposes. In Rwanda, *Erythrina abyssinica* nodulates profusely, which suggests that it is associated with *Bradyrhizobium* of the cowpea miscellany. Nitrogen fixation rates of 12–40 kg N_2 fixed/ha/yr have been reported from *Erythrina poeppigiana* in coffee and cacao plantations (Anon. 1986)

Gliricidia sepium   This is one of the most common trees of Mexico, Central America, and northern America. Estimates of nitrogen fixation based on nodule biomass and rates of nitrogenase activity are c. 13 kg N_2 fixed/ha/yr in the conditions prevailing in Mexico (Roskoski *et al.* 1982). *Gliricidia sepium* thrives well in the Amazon region. *G. sepium* has been introduced in Western Africa, but its nitrogen-fixing activity is impeded by attacks of root nematodes.

Leucaena leucocephala   This tree has been the focus of a great deal of research during the past decade (National Research Council 1977; IDRC 1983). Native to Central America, it has been planted in many tropical countries including South-East Asia (Domingo 1983), Africa (Okigbo 1984), and South America (Dobereiner 1984). *L. leucocephala* is often used in agroforestry. Kang *et al.* (1981*a*, *b*) found that five or six annual prunings of the *Leucaena* hedge rows yielded between 5 and 8 tons of dry tops/ha/yr with N-yield of between 180 and 250 kg/ha/yr and were able to sustain maize grain yield at about 3.8 tons/ha/yr for two consecutive years with no N addition. Another alternative to utilizing *Leucaena* prunings as a N source is to apply them as mulch or to incorporate them in the soil before planting. The prunings as a N source appeared to be more effective when incorporated in the soil than when applied as mulch. High maize grain yield was obtained with application of 10 tons fresh prunings or a combination of 5 tons fresh pruning and N at 50 kg/ha.

Leucaena leucocephala nodulates with fast-growing *Rhizobium*, probably closely related to *R. loti*. *Leucaena leucocephala* was credited with a very high nitrogen-fixing potential, ranging from 600 to 1000 kg N_2 fixed/ha/yr (Guevarra *et al.* 1978). A more realistic figure of 110 kg N_2 fixed/ha/yr has been calculated by Högberg and Kvarnström (1982). Using the isotope technique, K. Mulongoy and N. Sanginga (personal communication) found that *L. leucocephala* inoculated with its specific strain could fix up to 200 kg N_2/ha/yr in the conditions prevailing at the IITA station, Ibadan, Nigeria. Early efforts to grow *L. leucocephala* in acid soils were unsuccessful. Inoculating the host plant with acid-tolerant strains of rhizobia induced nodulation, but did not increase the production of the tree (Halliday 1985). Recently, genotypes of *L. leucocephala* tolerant to acidity were identified to acid Amazonian soils (Hutton 1984), which means that trees tolerant to acidity can probably be selected and planted in acid soils, where they could effectively fix nitrogen, if used in association with acid-tolerant *Rhizobium* strains.

Mimosa scabrella   Native to the Parana region of south-eastern Brazil, this species can be recommended for the mid-elevation cool tropics and subtropics (National Research Council 1980). It has been reported to respond positively to inoculation (Dobereiner 1984), but its nitrogen-fixing potential is unknown.

Pierocarpus indicus   In many countries of south-eastern Asia (e.g. Papua New Guinea, Indonesia, Malaya, the Philippines), *P. indicus* is reputed as choice timber for furniture. It nodulates, but its nitrogen-fixing potential has not yet been studied (Domingo 1983). Since many genotypes have been identified, it would be worthwhile exploiting the related variability and thus improving tree productivity.

Sesbania bispinosa and *S. grandiflora*   Native to many Asian countries, these tree legumes are widespread throughout South-East Asia, where they have a number of uses, including green manuring. Both species nodulate profusely, and are probably active nitrogen fixers, which could explain their extraordinary ability to restore soil fertility (Domingo 1983).

Parasponia spp.   The Parasponia genus is the only non-legume genus known to form nitrogen-fixing nodules with rhizobia (Trinick 1975, 1982; Becking 1982; Akkermans and Houvers 1983). Only the *Parasponia* species from the tropical Malay archipelago have been described. Like some legume trees listed above, *Parasponia* can be nodulated both by rhizobia and *Bradyrhizobium*, but it generally prefers specific strains of *Bradyrhizobium* (section

1.1.3.4). Very high nitrogen fixation rates, up to 850 kg N_2 fixed/ha/yr have been reported (Trinick 1981), but are open to question although relatively high figures can be expected.

2.2.3 Improvement of nitrogen fixation

The two approaches recommended for actinorhizal plants—one related to the symbiotic micro-organism, the other to the host plant—should be used concomitantly to improve nitrogen fixation by legume trees.

2.2.3.1. Inoculation with rhizobia

As in annual crops, inoculations on trees can only be expected to have beneficial effects if specific rhizobia are absent or scarce, or if the strain to be introduced is both more competitive and more effective than the native strains. The difference between annual crops and trees is that seedlings are, or should be, grown in a sterile substrate in a nursery to avoid attacks by root pathogens. In this situation, inoculation with specific rhizobia is systematically recommended and the beneficial effect is proportionally greater if the soil's nitrogen content is low. After the seedlings have been transplanted to the field, there are two possibilities. (i) if specific and efficient rhizobia are already present in plantation soils, the difference between uninoculated and inoculated plants tends to decrease within a few years; (ii) if the soil totally lacks the specific rhizobia, the beneficial effect of inoculation may be spectacular. An example of absence of response to inoculation was noted in a trial established in Senegal; 17 months after transplantation uninoculated *Acacia holosericea* were 1.05 m tall, while the inoculated trees were 1.19 m, which is not significantly different (Cornet *et al.* 1985). The reason for this virtually ineffective inoculation was that the soil already contained the *Bradyrhizobium* strains that nodulate *A. holosericea*.

An example of successful inoculation comes from an experiment conducted in Australia, in which, 15 months after transplantation, *Leucaena leucocephala* inoculated with *Rhizobium* strain C381 had reached a mean height of 2.10 m, whereas uninoculated trees only grew to 0.33 m (Diatloff 1973). In the Australian case, the inoculation generated a positive response because the soil did not naturally contain any of the *Rhizobium* strain specific to *L. leucocephala*.

2.2.3.2. Genetic improvement of the host

In symbiotic nitrogen-fixing systems, the host genotype is considered to interact with the infective strain of *Rhizobium* in conditioning the nitrogen-fixing potential. Exploiting the large differences that exist between the genotypes of a given plant species provides us with a promising approach that, for actinorhizal plants, will be discussed later (section 2.4.2). Except for

Leaucaena leucocephala and *Prosopis* spp., genetic variation for tree legumes has not yet been well explored (Koslowski and Huxley 1983).

2.2.3.3. Controlling the impact of limiting factors

Though this problem is discussed later in section 4, it is necessary to stress here that management practices such as application of fertilizers, may be good agriculturally, but may not always be suitable for forestry or agroforestry. Some alternatives exist. The best known is inoculation with mycorrhizal fungi. The double inoculation of trees with rhizobia and a vesicular-arbuscular mycorrhizal fungi at the nursery level is more effective than inoculation with rhizobia alone, the remarkable feature of inoculation with a vesicular-arbuscular mycorrhizal fungi being that it improves the uniformity of tree growth after transplantation (Cornet *et al.* 1985).

2.2.4. Discussion

Despite the existence of some very successful systems involving nitrogen-fixing trees in both forestry or in agroforestry, some authors have questioned the economic validity of these systems, since the application of nitrogen fertilizers is very easy, and their acquisition relatively cheap in comparison to the management of nitrogen-fixing systems (Turvey and Smethurst 1983). This outlook is probably correct in many cases occurring in temperate countries. On the other hand, in the tropics, active denitrification and intense leaching often decrease the effectiveness of nitrogen fertilizers to well below the level obtained in temperate climates. This fact, coupled with the high price of nitrogen fertilizers in many developing countries, challenge the wisdom of using the chemical approach in tropical forestry and agroforestry, which means that further efforts to improve biological nitrogen fixation are more than justified. Highly active nitrogen-fixing systems that can withstand the diverse and often harsh conditions of the tropics are not yet available, but by carefully selecting the genotypes of the symbiotic micro-organism and those of the host plant, the nitrogen-fixing potential of trees can be improved. Not so long ago this was a dream. Now it is a tangible objective that can be achieved in the near future.

2.3. Stem-nodulated legumes

2.3.1. Introduction

Most nitrogen-fixing legumes bear nodules on their root system. However, some species also form stem nodules and are called stem-nodulated legumes. Three genera have been reported to comprise stem-nodulated species: *Sesbania*, *Aeschynomene*, and *Neptunia*. So far, only two *Sesbania* species have been reported to produce stem nodules: *S. rostrata*, which is native to

Western Africa, and *Sesbania punctata*, probably native to Madagascar. The latter is presently being studied at the ORSTOM laboratory, Dakar. The *Aeschynomene* species known to have stem nodules are: *Aeschynomone afraspera*, *Aeschyomene denticulata*, *Aeschynomene elaphroxylon*, *Aeschynomene evenia*, *Aeschynomene filosa*, *Aeschynomene indica*, *Aeschynomene paniculata*, *Aeschynomene pfundii*, *Aeschynomene pratensis*, *Aeschynomene rudis*, *Aeschynomene schimperi*, *Aeschynomene scabra*, *Aeschynomene sensitiva*. The only *Neptunia* known to be stem-nodulated is *Neptunia oleracea*. In view of a recent comprehensive review (Dreyfus *et al.* 1984), attention is drawn mainly to the specific characteristics of these annual nitrogen-fixing systems and their implications for tropical agriculture.

2.3.2. Attributes of stem-nodulated legumes

2.3.2.1. Nodulation sites and nodule initiation

Nodulation occurs at predetermined sites on the stems. These sites have been identified as incipient root primordia, and remain dormant as long as they are not infected by specific rhizobia. In some species the root primordia protrude more or less through the stem epidermis; in other species they form a subepidermal dome hidden under the epidermis. In *Sesbania rostrata*, the root primordia always pierce the stem epidermis, forming a fissure that facilitates the penetration of the rhizobia. The genus *Aeschynomene* includes species with nodulation sites ranging from the *Sesbania* type with protruding root primordia (e.g. *Aeschynomene afraspera*), which nodulates readily, to the hidden root primordia type (e.g. *A. crassicaulis* and *A. elaphroxylon*), which is definitely less susceptible to rhizobial infection (Alazard 1985).

In *Sesbania rostrata*, the infection, which starts at the level of the fissure encircling the root primordia, spreads inward as narrow, branched intercellular threads to penetrate the host plant cells, where the rhizobia are finally released (Tsien *et al.* 1983; Duhoux 1984). In other words, the infection does not develop via the root hairs, like in most temperate legumes, but directly between the cells at the base of the root primordia, so that the nodule genesis seems to consist of three phases: intercellular infection, development of infection threads and intracellular infection. This mode of infection is not systematically observed in stem-nodulated legumes. In several *Aeschynomene* species, for instance, the mode of infection does not involve the development of infection threads. Following the stimulation of one or more meristematic zones at the base or within the root primordia by the presence of rhizobia, the nodule develops and become visible 2 (*S. rostrata*) to 8 days (*A. afraspera*) after stem inoculation (D. Alazard and E. Duhoux, personal communications). Like root nodules, stem nodules contain leghaemoglobin. This haemoglobin is present in large quantities, and furthermore has an oxygen affinity which is greater than that of the soybean (Appleby 1984), a plant known for its high nitrogen-fixing potential (Wittenberg *et al.* 1985).

2.3.2.2. Associated rhizobia

Strains of rhizobia-infecting stem-nodulated legumes have been classified into three main groups, according to their host relatedness.

The *Sesbania rostrata* group   All the strains in this group (type ORS570) are fast-growers. Their unique characteristics have already been presented (section 1.1.3.3). In continuous culture 'under optimum conditions of oxygen supply, nitrogen fixation rates are 300–400 nmol N_2/mg dry weight and nitrogenase activity reaches 2000 nmol/mg dry weight, the highest values yet recorded for any rhizobia species (Gebhardt *et l*. 1984). All rhizobia strains of infecting the stems of *S. rostrata* are highly specific.

The *Aeschynomene* group   Three subgroups have been recognized (Alazard 1985); subgroup α comprising fast-growing strains that have a narrow host range (e.g. *Rhizobium* of *A. afraspera*); subgroup ß, intermediate between subgroups α and γ (e.g. *Bradyrhizobium* of *A. indica*), and subgroup γ comprising slow-growing strains typically of the cowpea group (e.g. *Bradyrhizobium of A. elaphroxylon*).

The *Neptunia* group   The strains of this last group are fast-growers and are probably closely related to *R. meliloti*.

2.3.2.3. Stem inoculation

Spontaneous stem inoculation has often been observed, but is always irregular. The rhizobia are dust-borne, water-borne (by water on lowland soils and rainwater), and also conveyed by other vectors such as insects. Seed inoculation by specific rhizobia will induce excellent root nodulation, but only partial stem nodulation. Thus, the shoots have to be inoculated to ensure adequate infection of all the nodulation sites on the plant. The recommended method consists of spraying a liquid culture of the specific rhizobia on the stems using a standard sprayer. This procedure, together with a more sophisticated one, are described in recently published note by Dreyfus *et al.* (1985).

Considering the differences in the anatomy of nodulation sites (section 2.3.2.1), it is easy to understand that plants with the most highly evolved nodulation sites (e.g. *Sesbania rostrata*) nodulate readily, whereas those with hidden root primordia (e.g. *Aeschynomene elaphroxylon*) only nodulate if the stems are heavily inoculated.

2.3.2.4. The nitrogen-fixing potential

Various methods of estimation (isotopic, balance, and difference) used, and more recent unpublished data showed that *Sesbania rostrata* fixes between 100 and 300 kg N_2/ha in a period of *c*. 7 weeks (Rinaudo *et al.* 1983), thus indicating that it has one of the most effective nitrogen-fixing potential known to date. Such high performances can be explained by three sets of characteristics: (i) the unique properties of the associated rhizobia (section 2.3.2.2.) and the leghaemoglobin of *S. rostrata* (section 2.3.2.1); (ii) the fact that the stem nodule is a combination of functionally well integrated photosynthetic and nitrogen-fixing tissues, and as a result, satisfies part of its internal energy requirements (Eardly and Eaglesham 1985); and finally (iii) the tolerance of the whole system to combined nitrogen (section 2.3.2.5).

Stem-nodulated legumes other than *S. rostrata* do not exhibit such a high potential, although some, such as *Aeschynomene afraspera*, can be considered as good nitrogen fixers.

2.3.2.5. Tolerance to soil combined nitrogen

One of the dreams of the rhizobiologist has always been to obtain nitrogen-fixing systems that cannot be repressed by combined soil nitrogen. Root-nodulated legumes are repressed, although progress has recently been made through genetic improvement of the host plant (Gresshoff *et al.* 1985; Herridge and Betts 1985). On the other hand, stem-nodulated legumes are beautifully evolved systems, and are therefore capable of nodulating and sustaining nitrogen-fixing activity in the presence of relatively large amounts of mineral nitrogen in the rhizosphere. Thus, for *A. afraspera* grown in hydroponic conditions, the threshold of combined nitrogen is c. 10–15 mM for nodulation and c. 6 mM for nitrogen fixation. In soil, the threshold is ca 200 kg N/ha for nodulation and c. 100 kg N/ha for nitrogen fixation (D. Alazard, personal communication). These results confirm earlier findings with *Sesbania rostrata* (Dreyfus and Dommergues 1980). Exploiting this remarkable tolerance to combined nitrogen should probably be rewarding in the future (section 4).

2.3.3. Implications for agricultural practices

2.3.3.1. Soil fertility

Several field-simulation trials carried out in Senegal indicated that, when *S. rostrata* was applied as green manure in rice fields, the first following crop yielded 100 per cent more than the control plot, and the subsequent crop yielded 50 per cent more (Rinaudo *et al.* 1983). Other experiments using *Aeschynomene afraspera* as green manure showed similar beneficial effect on the rice yields (D. Alazard, personal communication). Such results evidence that stem-nodulated legumes can significantly contribute to the maintenance and the restoration of soil fertility in paddy soils.

2.3.3.2. Current limitations

We saw earlier that ploughing under stem-nodulated legumes as green manure can improve the soils used in tropical agriculture. Recent investigations carried out in Senegal indicate that these legumes can also be used as forage, which may be most interesting for the future. Consequently, we recommend developing the use of the stem-nodulated species mentioned previously. However, pending further genetic improvement, the culture of those legumes is subject to certain constraints. The main limitations that have been detected are: the high water requirements of most species, the photoperiodic response of some of them, especially *S. rostrata*, and this latter's sensitivity to root nematodes in well drained soils.

2.4. Actinorhizal plants

There are far fewer nitrogen-fixing actinorhizal plants than nitrogen-fixing legumes, but their numeric inferiority is offset by their generally excellent capacity to regenerate poor soils and, at the same time, produce not only timber and wood for fuel, but also shelter for cattle and crops.

2.4.1. The host range of *Frankia* systems

Because of their ability to form nodules with *Frankia*, which is an actinomycete, in 1978 the non-leguminous nitrogen-fixing plants became known as 'actinorhizal plants,' (Torrey and Tjepkema 1979), a name now used world-wide. Table 1.4 gives an updated synopsis of the tropical actinorhizal genera. The list will probably be expanded and/or revised as new species or genera are discovered. Recently, the nodulating ability of *Rubus ellipticus*, a plant hitherto recognized as actinorhizal, was questioned by Stowers (1985), in contradiction to earlier reports (Becking 1982) and this now casts doubt on the actinorhizal character of some species of the *Rubus* genus. New actinorhizal plants will probably be engineered in the future (section 5). It is clear that, despite the taxonomic affinities between certain of the actinorhizal genera, overall disparities are far too great to even consider that the nodulating ability may have evolved from a common ancestor (Bond 1983). The wide taxonomic host range of actinorhizal plants gives us reason for optimism (Righetti and Hannaway 1985), because this characteristic suggests the possibility of introducing nitrogen fixation into additional hosts.

2.4.2. Inter- and intra-specific variability in the nitrogen-fixing potential of actinorhizal plants

Lack of reliable estimates makes it difficult to rank the nitrogen-fixing potential of actinorhizal plants correctly. Some preliminary data indicate the existence of a rather wide range of potentials from good to rather poor. Consequently, the nitrogen-fixing potential of each individual species must always be carefully considered when choosing species for reforestation programmes. In addition, attention should be given to intra-specific variability within the nitrogen-fixing potential; it may be impressive within a single

Table 1.4

Genera and distribution of tropical and subtropical species of actinorrhizal plants

Family	Genus	Main known species	Distribution
Casuarinaceae	*Casuarina* *Allocasuarina* *Gymnostoma*		Tropical and subtropical Australia; Indo-Pacific area from India to Polynesia
Myricaceae	*Myrica*	*M. javanica*	Mountains of Indonesia and Philippines
		M. cacuminis	Mountains of Cuba
		M. punctata	Mountains of Cuba
		M. shaferi	Mountains of Cuba
Betulaceae	*Alnus*	*A. jorullensis*	South America
		A. nepalensis	Nepal
		A. japonica	Mountains, Asia
		A. maritima	Mountains, Asia
Elaeagnaceae	*Elaeagnus*	*E. latifolia*	Indonesia
		E. conferta	Indonesia
		E. philippensis	Philippines
Rhamnaceae	*Ceanothus*	*C. prostratus*	Pacific North American Coast to Mexico
	Discaria	*D. americana*	South America
		D. serratifolia	Argentina
		D. trinervis	
		D. nana	
	Colletia	*C. paradoxa*	South America
		C. spinosissima	
		C. armata *(= spinosa)*	
	Trevoa	*T. trinervis*	Chile
	Talguenese		Chile
	Kentrothamus		Chile
Coriariaceae	*Coriaria*	*C. japonica*	Mountains of Asia
		C. nepalensis	Mountains of Asia
		C. sinica	China (Hunan)
Rosaceae	*Rubus*	*R. ellipticus*	Continental Asia, Sri Lanka, Luzon, Indonesia
	Cowania	*C. mexicana*	California, Mexico
	Dryas	*Dryas* sp.	Mountains of Asia
	Cercocarpus	*C. ledifolius*	California
Datiscaceae	*Datisca*	*D. cannabina*	Mediterranean area to Central Asia
		D. glomerata	SW and NW Mexico

Table 1.5

Comparison of two clones of *Casuarina equisetifolia*[1] (Sougoufara *et al*, 1987)

	Shoot			Nodule		
	d.wt.		N total	d.wt.	ARA[2]	
Treatment	(mg per plant)	N (%)	(mg per plant)	(mg per plant)	per plant	per g nodule
Uninoculated						
Clone α	130 a	0.73 a	0.95 a	0 a	0 a	NA[4]
Clone ß	90 a	1.02 a	0.92 a	0 a	0 a	NA
Inoculated with *Frankia*[3]						
Clone α	660 b	1.71 b	11.29 b	54 b	2.88 b	54 b
Clone ß	1730 b	2.02 c	34.93 c	88 c	4.58 c	56 b

[1] 7 month-old cuttings, nine replicates.
[2] Acetylene reducing activity, expressed as μmol C_2H_4 per plant or per g nodule, dry weight (d.wt.).
[3] Each cutting was inoculated with 2 ml (20 μg of proteins) of a 4-week-old culture of *Frankia* strain ORS021001. Values in columns followed by the same letter are not significantly different, P = 0.05 (nodule dry weight; ARA), P = 0.01 (shoot dry weight, total N and N%).
[4] NA—not applicable.

species. This is illustrated in Table 1.5 which reports the results of a preliminary experiment conducted out at the ORSTOM Research Station of Bel Air, Dakar, Senegal. Two clones of *C. equisetifolia* (α and ß) characterized by different nitrogen-fixing potentials were grown in a nitrogen-deficient sandy soil; one set of plants was not inoculated, the other was inoculated with *Frankia* strain ORS021001 (Diem *et al.* 1983). In the former, both clones followed a very poor but not significantly different growth pattern. By contrast, in the inoculated group clone ß produced 2.6 times more biomass (expressed in terms of dry weight and total nitrogen) than clone α. Concomitantly, the nodule weight and the nitrogen-fixing activity of the whole plantlets (measured using the acetylene reduction method) of clone ß were 1.6 times significantly higher than the same characteristics of clone α. Interestingly, the specific nitrogen-fixing activity (that is the activity expressed on the nodule weight basis) of both clones did not differ significantly. Since the characteristic is probably directly conditioned by the genotype of the *Frankia* strain used, the result should not be surprising.

2.4.3. Exploiting the variability of the host plant

The above experiment clearly implies that the nitrogen-fixing potential of *Frankia* systems could be improved not only by selecting the best strains of *Frankia*, but also by selecting the most suitable host plant. In the preceding case, selection was based on the nodulating ability of the plant, since the goal was to produce hypernodulating clones.

The host's role in the symbiosis of crop legumes was recognized many years ago (Caldwell and Vest 1977). However, this characteristic has only been fully exploited by a few research groups (e.g. Attewell and Bliss 1985; Phillips and Teuber 1985; Gresshoff *et al*. 1985). In the case of woody plants, the problem is somewhat easier to solve for two reasons. First, wood plants constitute wild material, which has not yet been submitted to breeding manipulations and which often exhibits an extensive genetic variability. Secondly, perennial plants can be selected through vegetative propagation techniques; it is economically feasible to use these techniques for trees, but not for annual crops with a plant density of 100 000 or more individuals per ha.

In the case of *Alnus* spp. host selection via micropropagation techniques was successfully used by Lalonde's group (Lalonde and Simon 1985). Another example of this approach has already been presented (Table 1.5), which clearly shows that, by screening the clones of *Casuarina equisetifolia* which have the best nodulation, hence the highest nitrogen-fixing potential, nitrogen fixation rates can be markedly increased. The only problem encountered with *Casuarina* in the past was that vegetative propagation based on the use of cuttings was unsatisfactory because of frequent plagiotropism. This problem has been recently solved by Duhoux *et al*. (1986), who devised a most reliable micropropagation technique based on the use of immature female inflorescences. Field experiments are currently under way to check the behaviour of clones selected for their high nitrogen fixation potential and micropropagated to the method just referred to.

3. Inoculation

The purpose of inoculating a legume or an actinorhizal plant is to ensure that an adequate number of cells of the associated strain of rhizobia or *Frankia* is present on the seed of the host plant or in the soil at sowing or planting time, so that a quick infection of the root system occurs, leading to effective nitrogen fixation. In this section, we focus our attention mostly on the often overlooked problem of the need for inoculation, the other aspects of the topic being only briefly reviewed since they have been developed in detail in many publications (e.g. Brockwell 1977; Date and Roughley 1977; Burton 1979; Meisner and Cross 1980; Thompson 1980; Williams 1984).

3.1 The need for inoculation

The following rules could serve as a general guide:

1. When dealing with nursery soils, which should systematically be sterilized to eliminate pathogens, inoculation with the nitrogen-fixing micro-organism is always required (in this case, rhizobia or *Frankia* inoculation should be complemented by the introduction of mycorrhizal fungi).

2. When planning the introduction of a new species of nitrogen-fixing plant in a site where the species has never been grown before, one should make a distinction between promiscuous and specific host plants. If the legume plant is promiscuous, that is capable to nodulate with many types of rhizobia, inoculation is generally unnecessary. By contrast, if the legume has specific rhizobia requirements, inoculation should usually be recommended. In the latter situation, a beneficial effect of inoculation would be expected, provided no limiting factor interfere. Actinorhizal plants are rather specific, so that inoculation is mandatory.

3. When growing a nitrogen-fixing plant in a site where it had already been cultivated before, inoculation is often useless, provided that the associated strains of rhizobia or *Frankia* have survived. One should be well aware of the fact that there are many exceptions to these general rules so that, before starting any inoculation programme, it is necessary to determine whether the associated symbiotic strains of rhizobia or *Frankia* are present or not in the soil to be planted, and establish inoculation need and requisite management practices.

3.1.1. Simple methods of diagnosis

The occurrence of rhizobia or *Frankia* in the field can be simply observed by examining the legumes or actinorhizal plants whose inoculation group is similar to that of the species to be introduced at the site. The presence of nodules is at least *prima facie* evidence of the presence of the symbiont. In the case of rhizobia, it is possible to have an idea of the effectiveness of the symbiosis by splitting the nodules open and examining the colour of the section. Only effective nodules have a pink or red interior, whereas green or black nodules usually indicate ineffective nodulation. The physical appearance of the plant is a good indicator of an effective symbiosis for legumes (Meisner and Gross 1980) and actinorhizal plants. When host plants are green and healthy, the nodules are probably effective, a conclusion that could be erroneous when the nitrogen content of the soil is high enough to sustain the plant growth.

The presence, abundance and effectiveness of naturally occurring rhizobia or *Frankia* can also be assessed in the greenhouse, using different types of devices such a tubes or Leonard jars (Williams 1984).

3.1.2. Field experiments

A precise evaluation of a plant's response to inoculation should be based on the comparison of inoculated and uninoculated plants grown in the field. Date (1977) devised a simple experimental device with three treatments: (i) no inoculation, (ii) inoculation with the associated strain, and (iii) addition of nitrogen fertilizer (applied at the rate of 100–150 N/ha). Table 1.6 shows the possible responses of this type of trial. Since responses to inoculation are highly site-specific, it is advisable to test a maximum number of locations.

More refined trials have been suggested (Meisner and Gross 1980; FAO 1983; Halliday 1984 *a*, *b*), but they are complex and time-consuming. The most elaborate approaches involve the use of isotope techniques (LaRue and Patterson 1981). These last methods have the advantage of indicating the amount of nitrogen that has been effectively fixed by the inoculated plants. This information is most important, especially when the yield of inoculated and uninoculated plants are similar, but when part of the total nitrogen of the inoculated and hence nitrogen-fixing plant is derived from nitrogen fixation. In such a situation, the beneficial effect of inoculation can be detected only when using the isotope method.

3.2. Strain selection and maintenance of cultures

The selection of strains for inoculants should be based on two main criteria: (i) infectivity and effectiveness in nitrogen fixation, and (ii) tolerance to environmental constraints (e.g. acidity, soil toxicity, salinity, combined nitrogen, drought, pesticides). Other characteristics may also be desirable: (i) competitive ability, a property required when a superior strain is to be introduced in a soil where native, but less effective, strains are already present; (ii) ability to nodulate a large number of host plants (a trait which

Table 1.6

Possible outcomes of simple three-treatment inoculation trial for determining need to inoculate legume seed (from Date 1977)

	Uninoculated		Inoculated		Inoculated + nitrogen	
Outcome	Nodulation	Plant growth	Nodulation	Plant growth	Nodulation	Plant growth
1	—	poor	—	poor	—	poor
2	—	poor	—	poor	—	good
3	—	poor	+ E	good	+ or −	good
4	—	poor	+ I	poor	+ or −	good
5	+ I	poor	+ E	good	+ or −	good
6	+ I	poor	+ I	poor	+ or −	good
7	+ E	good	+ E	good	+ or −	good

− = Plants not nodulated.

+ = Plants nodulated.

+ or − = Plants may or may not be nodulated depending on whether the applied nitrogen inhibits nodule formation.

E = Effective in nitrogen fixation.

I = Ineffective in nitrogen fixation.

has the advantage of simplifying the distribution of inoculant); and (iii) ability to survive during distribution, storage, and use by farmers.

Multistrain inoculants have been successfully used despite findings suggesting that they should be avoided because of possible antagonistic and competitive effects (Williams 1984).

Methods to select superior strains of rhizobia in the laboratory and in the field have been described elsewhere (Date and Roughley 1977; Stowers and Elkan 1980; Halliday 1984*a*). These methods can be used to select *Frankia* strains. 'Even though the screening procedure is lengthy, attempts to shorten the sequence are ill-advised' (Halliday 1984).

Several methods of culture maintenance are successful: agar cultures in screw-cap tubes or bottles, dried soil, freeze-drying, and drying on porcelaine beads (Date and Roughley 1977; Stowers and Elkan 1980). In addition to freeze-drying the best method is to keep the liquid culture mixed with the same amount of glycerol in a freezer at −80°C. Preference should always be given to methods that minimize subculturing of cultures, thus reducing chances of variation by mutation.

3.3. Types of inoculant and methods of application

The quality of inoculants partly depends upon the suitability of the carrier, with best results usually obtained with peat. Many substitutes for this carrier have been proposed, e.g. bentonite, lignite, cellulose powder, various powdered crop residues. Peat has proved generally superior to other carriers. However, the composition of peat is highly variable and thus there are advantages in using a synthetic carrier of constant quality. This is achieved with inoculants obtained by entrapping rhizobia in a polyacrylamide gel or alginate instead of polyacrylamide. Field trials have proved that this last type of inoculant could be successfully used at the farm level (Ganry *et al.* 1985). When possible, peat is sterilized by heating, autoclaving or γ-radiation, and care should be taken to adjust both the moisture level and pH if high rhizobia populations are to be maintained during prolonged storage (Gibson *et al.* 1982). Further details on the preparation of inoculants are given in different reviews, such as those of Brockwell (1977) or Williams (1984).

Three procedures for inoculating are currently used: (i) mixing the inoculant with the seed immediately before sowing: (ii) suspending the inoculant in an adhesive, e.g. gum arabic or substituted methylethyl cellulose, moistening the seed with this and then pelleting the seeds by rolling them in finely ground coating materials at near neutral pH values, such as lime or rock phosphate; and (iii) inoculating the soil itself with the inoculant, by placing it below or alongside the seed bed.

When introducing rhizobia into soil by either the seed or soil inoculation technique, consideration should be given to the factors relevant to the establishment of an organism in an alien environment.

Detrimental factors, in many cases can be overcome through direct control measures, i.e. liming or application of fertilizers, use of inoculants with a high content of rhizobia or *Frankia* cells, protection of the micro-organisms used, and by introduction of the inoculant well before the legumes are planted.

The beneficial effect of high inoculum rates to overcome stress factors has been observed by many authors (Freire 1977; Selbach *et al.* 1978), mainly at the first planting of a legume in an area or as a way of introducing a new rhizobia strain into soil with a native population. A high number of cells around the germinating seed signifies a high number of nodules at the primary roots and inoculation success considering that the native rhizobia are dispersed in the soil. However, success in introducing a strain is dependent on the rate between the number of cells in the inoculum, in the native population, and in the competitive ability of the introduced strain in relation to the native ones. An important point is that nearly all of the experimental work for the introduction of rhizobia strains into a populated soil has been performed for one planting or for a short period of time. Long-term experiments with repeated introduction of the desired rhizobia strain need to be carried out.

As indicated earlier, protection of the symbiotic micro-organisms can be achieved through pelleting of the seed with calcium carbonate or rock phosphate, which have high beneficial effects on nodulation and yield. This technique developed in Australia and disseminated throughout the world has indeed proved to be of high value for the introduction and establishment of forage legumes in pastures in acid soils with no Al and Mn toxicities. The process also serves as an alternative to liming.

Introduction of rhizobia prior to the legume has been applied with success in the 'cerrado' or savannah soils of central Brazil as a means of overcoming the problem of deficient nodulation in the first planting of soybeans (Vargas *et al.* 1982). These areas after opening are usually seeded with upland (not irrigated) rice for 2–3 years. Now farmers are inoculating the rice seed, and when soybean is planted, already, find an established rhizobial population. This technique is much more economical than using large amounts of inoculant. Five or more repetitions are recommended for the first planting when not preceded by the inoculated rice.

Since *Frankia* is probably not motile in the soil, like vesicular-arbuscular mycorrhizae, we suggest to mix thoroughly the *Frankia* inoculant with the soil of the container where the plant is grown.

In the case of stem-nodulated legumes, inoculation is easily achieved by spraying the stem with a liquid culture of specific rhizobia, when the plants are 20–30 days old; inoculating the roots is achieved using any of the procedures described just above.

4. Limitations

Assuming that the plant selected is capable of a high level of production in a particular environment and that the inoculant strain is capable of forming an effective symbiosis with the host, it is then necessary to ensure that all other major limitations are minimized by the adoption of appropriate management practices. Here we shall briefly present some of the major limiting factors that can be encountered in the tropics.

4.1. Soil pH

Most legumes show reduced nodulation with a pH less than 5.0, not only because of the high hydrogen-ion concentration, but also because of the resulting toxicity of aluminium and manganese. Aluminium ions damage the roots of the host plant, which restricts nodulation. Liming is often proposed to alleviate the direct and indirect effects of acidity, but caution must be exercised in the use of this amendment. Ill-managed liming can induce deficiencies in magnesium, copper, zinc, and boron. Mulching has been recommended to stimulate nodulation as it controls manganese toxicity (Granhall 1986).

Some tropical legumes are most tolerant to acidity, nodulating well at pH as low as 4.5–4.7, e.g. *Arachis hypogaea*, *Macroptilium lathyroides*, *Desmodium uncinatum*, *Vigna sinensis*, and *Stylosanthes guyanensis*. *Stylosanthes capitata*, a legume with high tolerance to soil acidity, can nodulate in Leonard jars only when the growth medium is acidified to a pH lower than 5.0, and the calcium and phosphorus levels are lowered 10-fold (Halliday 1984*a*). In contrast with what was claimed earlier, the optimum pH of *Glycine max* is not in the 6.5–7.0 range. Actually, soil acidity *per se* is not the limiting factor for this legume, but if acidity induces aluminium and/or manganese toxicity, nodulation together with nitrogen fixation and yield may be seriously affected (Freire 1984). The effect of Al seems to be indirect on the root growth, on nodulation initiation and on the physiology of the host in the uptake and transport of calcium. A close interrelation exists between Ca, Al, and Mn. Tolerance to Al is related to the uptake and utilization of phosphorus. There is also some indication that uptake of Mn is higher in a soil with higher exchangeable Al (Cabeda *et al.* Cabeda and Freire 1986) than at soil lower in Al.

Legumes show large inter- and intraspecific differences in tolerance to soil acidity. Similarly, rhizobia strains also demonstrate wide variation in acid tolerance. This genetic diversity of the host plants and rhizobia should be exploited for better plant growth and nitrogen fixation (Graham and Chatel 1983).

Plant growth, nodulation, and nitrogen fixation are impaired at high pH by induced deficiencies in manganese, iron, boron, soil salinity, and anionic imbalances (Graham and Chatel 1983). In alkaline or sodic soils, recommended management practices are: addition of gypsum or organic matter. As with soil acidity, an appropriate strategy could be the selection of both host plants and associated *Rhizobium* but in this situation, the goal is to obtain plants and bacteria that are able to tolerate alkaline conditions.

4.2. Mineral influences

After nitrogen, phosphorus is most generally the limiting nutrient in tropical soils. Among tropical legumes, some species are more efficient in using phosphorus than the others, e.g. *Acacia holosericea* (Cornet *et al.* 1985), *Stylosanthes* and *Lupinus* (Gibson *et al.* 1982). Differences also occur between cultivars.

Both calcium and magnesium are essential for the host plant and the symbiotic bacterium. A change in the Ca/Mg ratio can reduce the accessibility of either nutrient. An imbalance is generally induced by high applications of lime or potassium fertilizers (Bergersen 1977; Munns 1977; Andrew 1978).

The effect of potassium on N_2 fixation is indirect, through the host physiology (Andrew 1977). In soils, low in productivity in the tropics, it is common to find high levels of available K and no response to fertilization. This may mislead the farmer for not using the fertilizer. When, however, productivity is increased, extraction of K is also increased and deficiency may appear. Potassium may also be detrimental to nodulation and N_2 fixation. In non-limed soils high in available Mn, the application of K as KCl increases the availability of Mn in the soil and in plant tissue and reduces nodulation (Borkert, *et al.* 1974).

Sulphur and some micronutrients are common limiting factors to N_2 fixation and legume productivity in the poor acid soils of the tropics and subtropics e.g. at the 'cerrado' or savannah soils of South America (Franco 1977). In high pH or limed soils zinc, and boron deficiencies may be a problem specially in high levels of productivity. For alfalfa, boron is a common limiting factor. Molybdenum as part of the nitrogenase molecule plays an essential function in the N_2 fixation. High expectations were witnessed some years ago on the 'miracle' effects of molybdenum. Deficiency in Mo may occur in some acid soils and disappear when soil is limed in most of the cases. In soils with toxic Al and/or Mn, the isolated application of Mo has no beneficial effect.

4.3. Combined nitrogen

The effect of application of nitrogen fertilizers on nodulation and nitrogen fixation is complex, and vary with the form and rate of application, the plant species or cultivar, the strain of rhizobia, the environmental conditions, and the amount of available nitrogen already present in the soil.

Mineral nitrogen in small amounts is beneficial to starting or nodulation and N_2 fixation specially in severely deficient soils (Franco 1977; Freire 1984). This is clearly shown in Leonard jars with sand and nutrient solution. However, in field conditions severe deficiencies very seldom occur.

It has been the classical recommendation of use as 'starter' nitrogen for the growth of legumes. However, field evidence of beneficial effects are not common at least for grain legumes. In the absolute majority of the experimental work conducted with soybeans in Brazil there was initially no economical responses to the use of mineral nitrogen, either at low or high application rates at different times of the growth cycle given that the plants were well nodulated (Barni et al. 1978; Freire 1984). In some experiments or crop fields the observer may be confused by taller and greener plants where N is applied at planting time. However, well nodulated plants without N will rapidly compensate and at harvest time the yield will be the same. It must, however, be mentioned that positive responses may be obtained in fields where nodulation and/or N_2 fixation has been inhibited, e.g. inefficient rhizobia strains, temperature, or moisture stress, etc. From a practical point of view it would be an uneconomical general recommendation for farmers to apply N fertilizers for soybeans. For some time in Brazil there was a general practice in the use of a 'starter' of 10–20 kg N for soybeans. Campo and Sfredo (1981) estimated that there was a wasting of 87 000 tons of N at the fertilizers formulas commonly used in the country. For other legumes similar results have been obtained (Kolling et al. 1985). For the short cycle with field beans (P. vulgaris) it seems that the N_2 fixation cannot supply the needs of the plants and that supplementary N would be beneficial.

In sandy soils, when moisture is adequate nitrification may yield high levels of nitrates that in some cases could even inhibit nodule formation.

It is generally recognized that large additions of nitrogen fertilizer almost always inhibit nodulation and nitrogen fixation. A promising approach to overcome this inhibition is to select lines of plants for good nodulation in the presence of nitrate nitrogen (Gresshoff et al. 1985), and then to inoculate such plants with rhizobia strains effective in nitrogen fixation in the presence of combined nitrogen. One should note here that stem-nodulated legumes possess the ability to simultaneously use combined nitrogen from the soil and atmospheric nitrogen (section 2.3.). Other interesting approaches include deep-banding or the use of slow-release fertilizers. Caution should be adopted in the use of nitrification inhibitors (Gibson et al. 1982).

4.4. Pathogens

Viral, fungal, and bacterial pathogens, together with insects and nematodes, may interfere with the plant growth and its ability to fix nitrogen. The deleterious effect of nematodes on nodulation and nitrogen fixation by legumes has been reported by many authors (Gibson 1977; Gibson *et al.* 1982). Controlling nematode attacks by a nematicide was shown to significantly improve nitrogen fixation of peanut (Germani *et al.* 1980).

4.5. Importance of management practices

Availability of inoculants and research capability are not sufficient for the diffusion of the practice of inoculation and for obtaining higher productivity from N_2 fixation. The use of rhizobia is just one of the items in a package of practices for legume growth and production. Thus, it must be promoted together with a long series of other inputs and agronomic factors. In many places it is of little use to promote the isolated application of rhizobia inoculant if, for example, lime and/or phosphorus are not applied on account of a lack of knowledge or because prices are too high. Here comes the role of the extension services, showing to the farmers that they can obtain more from their land. After the farmers are convinced, they will make the necessary efforts for the availability of the necessary inputs, credit, etc. This is what happened in southern Brazil 20 years ago. A research/extension integrated programme for the improvement of soil fertility changed completely the poverty situation of the farmers in millions of hectares of poor acidic soils high in exchangeable Al and Mn (Beaty *et al.* 1972). The expansion of soybeans had started some years behind and the effect of inoculation was usually poor. After the diffusion of new management practices for high applications of lime and fertilizers then the farmers started to see the beneficial effects of the rhizobia N_2 fixation. If it is assumed that N_2 fixation now provides an average of 50 kg/N/ha/year the contribution of the rhizobia-soybean symbiosis in Brazil is about 600 thousand tons of N for the 12 million hectares cropped to this legume in 1984.

5. Future

In the last decades a tremendous amount of work has been devoted to the study of biological nitrogen fixation. Exciting results have been achieved at the level of genetics, physiology, and ecology of symbiotic systems, but agriculture and forestry have only recently fully benefited from these remarkable advances in our basic knowledge. Fortunately, some approaches are progressively emerging that will probably give way to practical applications in the near- or mid-term (e.g. host selection via micropropagation techniques). Other research strategies will involve much more effort and time before fundamental results be transferred to the field.

5.1. Approaches for increasing the nitrogen-fixing potential of existing systems

Increased nitrogen fixation can be achieved by acting upon one, or better, both components of the symbiosis. Rhizobia and *Frankia* strains can now be genetically engineered by molecular techniques. Thus, one can foresee, in the short term, that the remarkably high nitrogenase activity of *Rhizobium* strain ORS 571 (from *Sesbania rostrata*) will be transferred to strains of *Rhizobium* nodulating other legumes, probably endowing the latter plants with an increased nitrogen-fixing potential.

The improvement of the host plant should be carried out in parallel to any genetic improvement of rhizobia or *Frankia* strains. Adapted technologies are now at hand for exploiting the variability of the host plant. Exploiting the spontaneous genetic heterogeneity of trees is an elegant means to improve nitrogen fixation by these plants, provided that vegetative propagation techniques are available that would allow mass production of selected clones. The variability of the host plant can be broadened via tissue culture, which offers new exciting possibilities (Lalonde and Simon 1985).

5.2. Improving tolerance to environmental constraints

Very actively nitrogen fixing systems would be useless if they were not tolerant to environmental constraints that limit their activity (section 4). Consequently, it is mandatory to select or engineer nitrogen-fixing systems whose activity is not, or only slightly, reduced by limiting factors. On the one hand, one can use the techniques of genetic manipulation of micro-organisms, which have improved so dramatically in the past few years. On the other hand, plant cell and tissue culture techniques are advanced enough to be successfully used to improve the host performances in hostile environments. The following examples will illustrate the benefit that could result from the manipulation of the host. One major limiting factor is soil salinity, which generally affects the host plant more than the symbiont. Salt-tolerant variant plants can now be produced using somaclonal variation. This approach is based on the fact that by culturing a callus on a salt-enriched medium, some cells of the callus may exhibit salt tolerance. Through continuous subculture on salt-enriched medium, the callus appear to 'gain' a substantial degree of tolerance (Nabors *et al.* 1980). The next step is to regenerate the plant from cultured calluses, a goal that is sometimes difficult to achieve. Nitrate and, to a lesser extent, other forms of combined nitrogen, retard nodulation and nitrogen fixation. Such an inhibition is obviously most detrimental for the economy of nitrogen especially in intensive agriculture. To circumvent this inhibition the best approach is to use nitrogen-fixing systems that are capable of continuing to fix nitrogen even in the presence of large amounts of combined nitrogen. Nitrate tolerant soybeans have been isolated from a soybean cultivar by Gresshoff's group in Australia (Delves *et al.* 1985). As indicated earlier, *Sesbania rostrata* , a stem-nodulated legume, has the unique ability to absorb combined nitrogen with roots and simultaneously fix atmospheric nitrogen with its stem-nodules, even when the amount of soil mineral nitrogen is as high as 200 kg nitrogen/ha. It has been suggested that transferring the stem nodulation characteristic from this plant to non-stem nodulated legumes would be a way to develop new, uninhibited nitrogen-fixing systems (Dreyfus and Dommergues 1980).

5.3. Transfer of nitrogen-fixing ability to non-nitrogen-fixing plants

The following strategies have been envisaged or are already at hand.

5.3.1. Genetic transfer of nitrogen fixation ability to the plant cells

The transfer of cloned genes between nitrogen-fixing micro-organisms and plants should be considered as a long-term project since the stability of the genes required for nitrogen fixation following their incorporation, expression and inheritance by the whole plant after their introduction into the plant cells is not yet known. However, if one take into account the development of our knowledge in plant molecular biology (Evans *et al.* 1985), several possibilities are offered by this strategy.

5.3.2. Hybridization of nitrogen-fixing and non-nitrogen-fixing plants

Gene transfer by wide hybridization has not yet been exploited to obtain new nitrogen-fixing systems. Fortunately, sophisticated technologies such somatic hybridization by protoplast fusion or embryo rescue (National Research Council 1982) are now available, which could hopefully be most helpful in this unexplored field of research.

5.3.3. Micro-grafting

Recently, Kyle and Righetti (1985) showed that the possibility exists of introducing nitrogen-fixing capability of actinorhizal Rosaceae into non-nitrogen-fixing plants of the same family by grafting the former plants onto nitrogen-fixing root stocks. Adopting a similar approach Lalonde's group, at Laval University, is exploring the possibility of grafting *Betulus* sp., a non-nitrogen-fixing tree, on *Alnus* sp., an actively nitrogen-fixing actinorhizal plant.

5.3.4. Indirect transfer of nitrogen-fixing capability through mycorrhizal fungi

The transfer of the whole set of genes required to fix nitrogen from a bacterium to a mycorrhizal fungus is probably feasible and easier than the transfer to a plant. However, a number of difficulties lies ahead, one prerequisite to genetical engineering of nitrogen-fixing endomycorrhizal fungi being the availability of a reliable method to grow this fungus *in vitro*.

5.4. Conclusion

To conclude we would like to stress two points. The first one, which has been already expressed by J. Postgate at the Corvallis meeting (Evans *et al.* 1985), is that 'for a strategic research topic such as the fixation of nitrogen, an open interdisciplinary approach is mandatory', uniting the efforts of chemists, biochemists, geneticists, physiologists, microbiologists, ecologists, and plant molecular biologists. The second message is that the more logical current and future application of biological nitrogen fixation is in land reclamation, forestry, and agroforestry. In such situations, biological nitrogen fixation by itself can meet totally or, at least, largely the nitrogen requirements of the plants. By contrast, in intensive agriculture, maintenance of crop yields cannot be obtained solely through nitrogen fixation. Then it is tempting to use exclusively nitrogen fertilizers. In fact, it would be much wiser to develop management practices based on the integrated use of industrial and biological nitrogen.

References

Akkermans, A. D. L. and Houvers, H. (1983). Morphology of nitrogen fixers in forest ecosystems. In *Biological nitrogen fixation in forest ecosystems: foundations and applications* (eds. J. C. Gordon and C. T. Wheeler) pp. 7–53, Nijhoff/Junk, The Hague

Alazard, D. (1985). Stem and root nodulation of *Aeschynomene* spp. Appl. Env. Microbiol. 50, 732–4.

Allen, O. N. and Allen, E. K. (1981), *The Leguminosae. A source book of characteristics, uses, and nodulation.* University of Wisconsin Press, Madison.

Andrew, C. S. (1977). Nutritional restraints on legume symbiosis. In *Exploiting the legume-Rhizobium symbiosis in tropical agriculture* (eds. J. M. Vincent, A. S. Whitney and J. Bose) pp. 253–74. College of Tropical Agriculture, University of Hawaii.

—— (1978). Legumes and acid soils. In *Limitations and potentials for biological nitrogen fixation in the tropics* (eds J. Dobereiner, R. H. Burris and A. E. Hollaender) pp. 135–60. Plenum, New York.

Anon. (1986). Erythrinas provide beauty and more. *NFT Highlights*, 86–02.

Appleby, C. A. (1984). Leghemoglobin and *Rhizobium*. *Ann. Rev. Plant Physiol.* 35, 443–78.

Attewell, J. and Bliss, F. A. (1985). Host plant characteristics of common bean lines selected using indirect measures of N_2 fixation. In *Nitrogen fixation research progress* (eds H. J. Evans, P. J. Bottomley and W. E. Newton) pp. 3–9. Nijhoff/ Junk, The Hague.

Baker, D., Newcomb, W. and Torrey J. G. (1980) Characterization of an ineffective actinorhizal microsymbiont, *Frankia sp. Eu Il Can. J. Microbiol.* 26, 1072–89.

—— and Torrey, J. G. (1979). The isolation and cultivation of actinomycetous root nodule endophytes. In *Symbiotic nitrogen fixation in the management of temperate forests* (eds J. C. Gordon, C. T. Wheeler, and D. A. Perry) pp. 36–56. Forest Research Laboratory, Corvallis.

Barni, N. A., Gomes, J. E. S., and Gonçalves N. A. (1978). Resposta da soja *Glycine max* (L.) Merrill à adubação nitrogenada no florescimento *Agron. Sulriograndense* 14, 243–50.

Beaty, M. *et al.* (1972). An integrated extension program for promoting rapid change in traditional agriculture on oxisols and ultisols in sub-tropical climate. *J. Agron. Education* 1, 37–40.

Becking, J. H. (1975). Root nodules in non-legumes. In *The development and function of roots* (eds. J. G. Torrey and D. T. Clarkson) pp. 507–66. Academic Press, London.

—— (1982). N_2-fixing tropical non-legumes. In *Microbiology of tropical soils and plant productivity* (eds Y. R. Dommergues and H. G. Diem) pp. 109–46. Nijhoff/ Junk, The Hague.

Bergersen, F. J. (1977). Factors controlling nitrogen fixation by rhizobia. In *Biological nitrogen fixation in farming systems of the tropics* (eds A. Ayanaba and P. J. Dart) pp. 153–65. John Wiley, Chichester.

Blondel, D. (1971). Contribution à l'étude de la croissance-matière sèche et de l'alimentation azotée des céréales de culture sèche au Sénégal. *Aron. Trop.* 26, 707–20.

Boland, D. J., Brooker, M. I. H., Chippendale, G. M., Hall, N., Hyland, B. P. M., Johnston, R. D., Kleining, D. A., and Turner, J. D. (1984) *Trees of Australia*. Nelson, CSIRO, East Melbourne.

Bond, G. (1983). Taxonomy and distribution of non-legume nitrogen-fixing systems. In *Biological nitrogen fixation in forest ecosystems: foundations and applications* (eds J. C. Gordon and C. T. Wheeler) pp. 55–87.

Borkert, C., Freire, J. R. J., Vidor, C., Grimm, S. S. (1974) Efeito do calcário e do cloreto de potássio sobre as concentrações de manganés nos oxissolos Santo Ángelo e Passo Fundo e absorção de manganés por duas cultivares de *Glycine max* Merr. *Agron Sulriograndense.* 11, 45–52.

Brewbaker, J. L., Van den Beldt, R. and Macdicken, K. (1984). Fuelwood uses and properties of nitrogen-fixing trees. *Pesq. agropec. bras. Brasilia* 19, 193–204.

Brockwell, J. (1977). Applications of legume seed and inoculants. In *A treatise on dinitrogen fixation. IV. Agronomy and ecoloy* (eds R. W. F. Hardy and A. H. Gibson) pp. 277–309. John Wiley, New York.

Burton, J. (1979). New developments in inoculating legumes. In *Recent advances in biological nitrogen fixation* (ed. N. S. Subba Rao) pp. 380–405. Oxford and IBH publishers, New Delhi.

Cabeda, M. S. and Freire, J. R. J. (1968). Informe preliminar sobre os efeitos da toxidez de manganês e alumínio sobre a nodulação e fixação do nitrogênio em soja

(*Glycine max* Merr.) em solos ácidos do Rio Grande do Sul. In *Anais da IV Reunião Latinoamericana sobre Inoculantes para Leguminosas*, pp. 282–94. Porto Alegre.

——, —— and Ludwick, A. E. (1968). Informe preliminar sobre a toxidez de manganês e alumínio em soluçao nutritiva sobre a soja (*Glycine max* Merr.). In *Anais da IV Reunião Latinoamericana sobre Inoculantes para Leguminosas*, pp. 168–78. Porto Alegre.

Caldwell, B. E. and Vest, H. G. (1977). Genetics aspects of nodulation and dinitrogen fixation by legumes: the macrosymbiont. In *A treatise on dinitrogen fixation* (eds R. W. F. Hardy and W. S. Silver) pp. 557–76. John Wiley, New York.

Callaham, D., Del Tredici, P., and Torrey, J. G. (1978). Isolation and cultivation *in vitro* of the actinomycete causing root nodulation in *Comptonia*. *Science* 199, 899–902.

Campo, R. and Sfredo, G. (1981). O nitrogêno na cultura da Soja. EMBRAPA, Informe Técnico.

Carpenter, C. V. and Robertson, L. R. (1983). Isolation and culture of nitrogen-fixing organisms. In *Biological nitrogen fixation in forest ecosystems: foundations and applications* (eds J. C. Gordon and C. T. Wheeler) pp. 89–106. Nijhoff/Junk, The Hague.

Chee, Y. K. (1982). The importance of legume cover crop establishment for cultivation of rubber (*Hevea brasiliensis*) in Malaysia. In *Biological nitrogen fixation technology for tropical agriculture* (ed. P. H. Graham) pp. 369–77. CIAT, Calif.

Cornet, F. and Diem, H. G. (1982). Etude comparative de l'efficacité des souches de *Rhizobium* d'*Acacia* isolées de sols du Sénégal et effet de la double symbiose *Rhizobium-Glomus mosseae* sur la croissance de *Acacia holosericea* et *A. raddiana*. *Bois For. Trop.* 198, 3–15.

——, Otto, C., Rinaudo, G., Diem, H. G. and Dommergues, Y. (1985). Nitrogen fixation by *Acacia holocericea* grown in field-simulating conditions. *Oecol. Plant.* 6, 211–18.

Date, R. A. (1977). Inoculation of tropical pasture legumes. In *Exploiting the legume-Rhizobium symbiosis in tropical agriculture* (eds. J. M. Vincent, A. S. Whitney, and J. Bose) pp. 293–311. College of Tropical Agriculture, University of Hawaii.

—— and Halliday, J. (1980). Relationships between Rhizobium and tropical forage legumes. In *Advances in legume science* (eds R. J. Summerfield and A. H. Bunting) pp. 597–601. Royal Botanical Gardens, Kew.

—— and Roughley, R. J. (1977). Preparation of legume seed inoculants. In *A. treatise on dinitrogen fixation. IV. Agronomy and ecology* (eds R. W. F. Hardy and A. H. Gibson) pp. 243–76. John Wiley, New York.

Delves, A. C., Day, D. A., Price, G. D., Carroll, B. J., and Gresshoff, P. M. (1985). Regulation of nodulation and nitrogen fixation in nitrate tolerant, supernodulating soybeans. In *Nitrogen fixation research progress* (eds H. J. Evans, P. J. Bottomley, and W. E. Newton). p. 41. Nijhoff/Junk, The Hague.

Diatloff, A. (1973). *Leucaena* needs inoculation. *Qld Agric. J.* 99, 642–4.

Diem, H. G. and Dommergues, Y. R. (1985). *In vitro* production of specialized reproductive hyphae by *Frankia* strain ORS021001 isolated from *Casuarina junghuhniana* root nodules. *Plant Soil* 87, 17–29.

——, ——, and —— (1982a). Isolation of *Frankia* from nodules of *Casuarina equisetifolia*. *Can. J. Microbiol*, 28, 526–30.

——, ——, and —— (1982b). Isolement et culture *in vitro* d'une souche infective et effective de *Frankia* isolée de nodules de *Casuarina sp*. *C. R. Acad. Sci. Paris* 295, 759–63.

——, ——, and —— (1983). An effective strain of *Frankia* from *Casuarina sp. Can. J. Bot.* 61, 2815–21.

Dobereiner, J. (1984). Nodulation and nitrogen fixation in legume trees. *Pesq. agropec. bras. Brasilia*, 19, 83–90.

—— and Campelo, A. B. (1977). Importance of legumes and their contribution to tropical agriculture. In *A treatise on dinitrogen fixation. IV. Agronomy and ecology* (eds R. W. F. Hardy and A. H. Gibson) pp. 191–220. John Wiley, New York.

Domingo, I. (1983). 11. Nitrogen fixation in Southeast Asian forestry research and practice. In *Biological nitrogen fixation in forest ecosystems: foundations and applications* (eds J. C. Gordon and C. T. Wheeler) pp. 295–315. Nijhoff/Junk, The Hague.

Dreyfus, B. L., Alazard, D., and Dommergues, Y. R. (1984). Stem-nodulating rhizobia. In *Current perspectives in microbial ecology* (eds M. J. Klug and C. A. Reddy) pp. 161–9. American Society for Microbiology, Washington DC.

—— and Dommergues, Y. R. (1980). Non-inhibition de la fixation d'azote atmosphérique par l'azote combiné chez une légumineuse à nodules caulinaires, *Sesbania rostrata. C. R. Acad. Sci. Paris* D 291, 767–70.

—— and —— (1981). Nodulation of *Acacia* species by fast- and slow-growing tropical strains. *Appl. Environ. Microbiol.* 41, 97–9.

—— Elmerich, C. and Dommergues, Y. R. (1983). Free-living *Rhizobium* strain able to grow on N_2 as the sole nitrogen source. *Appl. Environ. Microbial.* 45, 711–3.

——, Rinaudo, G. and Dommergues, Y. (1985). Observations on the use of *Sesbania rostrata* as green manure in paddy fields. *MIRCEN J.* 1, 111–21.

Duhoux, E. (1984). Ontogenèse des nodules caulinaires du *Sesbania rostrata* (légumineuses). *Can. J. Bot.* 62, 982–94.

——, Sougoufara, B., and Dommergues, Y. R. (1986). Propagation of *Casuarina equisetifolia* through axillary buds of immature female inflorescences cultured *in vitro*. *Plant Cell Reports* 3, 161–4.

Eaglesham, A. R. J., Ayanaba, A., Ranga Rao, V., and Eskew, D. L. (1982). Mineral N effects on cowpea and soybean crops in a Nigerian soil. II. Amounts of N fixed and accrual to the soil. *Plant Soil* 68, 183–92.

Eardly, B. D. and Eaglesham, A. R. J. (1985). fixation of nitrogen and carbon by legume stem nodules. In *Nitrogen fixation research progress* (eds H. J. Evans, P. J. Bottomley and W. E. Newton) p. 324. Nijhoff/Junk, The Hague.

Elkan, G. H. (1984). Taxonomy and metabolism of *Rhizobium* and its genetic relationships. In *Biological nitrogen fixation, ecology, technology, and physiology* (ed. M. Alexander) pp. 1–38. Plenum Press, New York.

Evans. H. J., Bottomley, P. J., and Newton, W. E. (ed.) (1985). *Nitrogen fixation research progress*. Nijhoff/Junk, The Hague.

FAO (1983). *Technical handbook on symbiotic nitrogen fixation*. Food and Agriculture Organization of the United Nations, Rome.

Felker, P. (1984). Legumes trees in semi-arid and arid areas. *Pesq. Agropec. Bras. Brasilia* 19, 47–59.

——, Clark, P. R., Llag, A. E. and Pratt, P. F. (1981). Salinity tolerance of the tree legumes: Mesquite (*Prosopis glandulosa var, torreyana, P. velutina and P. articulata*), Algarrobo (*P. chiliensis*), Kiawe (*P. pallida*) and Tamarugo (*P. tamarugo*) grown in sand culture on nitrogen-free media. *Plant Soil* 61, 311–17.

Franco, A. (1977). Nutritional restraints for tropical grain legume symbiosis. In *Exploiting the legume-Rhizobium symbiosis in tropical agriculture* (eds J. M.

Vincent, A. S. Whitney and J. Bose) pp. 237–52. College of Tropical Agriculture, University of Hawaii.

Freire, J. R. J. (1964). Producion de inoculantes en Brasil. In *Anais da la Reunion Latinoamericana sobre Rhizobium*. Montevideo.

—— (1977). Inoculation of soybeans. In *Exploiting the legume-Rhizobium symbiosis in tropical agriculture* (eds J. M. Vincent, A. S. Whitney, and J. Rose) pp. 335–79. College of Tropical Agriculture, University of Hawaii.

—— (1984). Important limiting factors in soil for the *Rhizobium* legume symbiosis. In *Biological nitrogen fixation, ecology, technology, and physiology* (ed. M. Alexander) pp. 51–74. Plenum Press, New York.

Ganry, F., Diem, H. G., Wey, J., and Dommergues, Y. R. (1985). Inoculation with *Glomus mosseae* improves N_2 fixation by field-grown soybeans. *Biol. Fert. Soils* 1, 15–23.

Gardner, I. C. (1976). Ultrastructural studies of non-leguminous root nodules; In *Symbiotic nitrogen fixation in plants* (ed. P. S. Nutman) pp. 485–96. Cambridge University Press, London.

Gauthier, D. L., Diem, H. G. and Dommergues, Y. R. (1984). Tropical and subtropical actinorhizal plants. *Pesq. agropec, bras. Brasilia*, 19, 119–36.

——, ——, ——, and Ganry, F. (1985). Assessment of N_2 fixation by *Casuarina equisetifolia* inoculated with *Frankia* ORS021001 using ^{15}N methods. *Soil Biol. Biochem.* 17, 375–9.

Gebhardt, C., Turner, G. L., Gibson, A. H., Dreyfus, B., and Bergersen, F. J. (1984). Nitrogen-fixing growth in continuous culture of a strain of *Rhizobium sp.* isolated from stem nodules on *Sesbania rostrata*. *J. Gen. Microbiol.* 130, 843–8.

Germani, G., Diem, H. G., and Dommergues, Y. R. (1980). Influence of dibromo-3-chloropropane fumigation on nematode population, mycorrhizal infection, N_2 fixation and yield of field-grown ground. *Revue Nématol.* 3, 75–8.

Gibson, A. H. (1977). The influence of the environment and managerial practices on the legume Rhizobium- symbiosis. In *A treatise on dinitrogen fixation. IV. Agronomy and ecology* (eds R. W. F. Hardy and A. H. Gibson) pp. 393–450. John Wiley, New York.

——, Dreyfus, B. L., and Dommergues, Y. R. (1982). Nitrogen fixation by legumes in the tropics. In *Microbiology of tropical soils and plant productivity* (eds Y. R. Dommergues and H. G. Diem) pp. 36–73. Nijhoff/Junk, The Hague.

Graham, P. H. (1981). Some problems of nodulation and symbiotic nitrogen fixation in *Phaseolus vulgaris* L.: a review. *Field Crops Res.* 4, 93–112.

—— and Chatel, D. L. (1983). 3. Agronomy. In *Nitrogen fixation. Vol. 3: Legumes* (ed. W. J. Broughton) pp. 59–98. Clarendon Press, Oxford.

Granhall, U. (1986). Leguminous trees, potential and utilization. In *Manuscript reports. Proc. Seminar on nitrogen-fixing trees, Dakar, March 1986*. IDRC, Ottawa. (in press)

Gresshoff, P. M., Day, D. A., Delves, A. C., Mathews, A. P., Olsson, J. E., Price, G. D., Schuller, K. A., and Carroll, B. J. (1985). Plant host genetics of nodulation and symbiotic nitrogen fixation in pea and soybean. In *Nitrogen fixation research progress* (eds H. J. Evans, P. J. Bottomley and W. E. Newton) pp. 19–25. Nijhoff/Junk, The Hague.

Guevarra, A. B., Whitney, A. S., and Thompson, J. R. (1978). Influence of intra-row spacing and cutting regimes on the growth and yield of Leucaena. *Agronomy J.* 70, 1033–7.

Halliday, J. (1984a). Principles of Rhizobium strain selection. In *Biological nitrogen fixation, ecology, technology, and physiology* (ed. M. Alexander) pp. 155–71. Plenum Press, New York.

—— (1984*b*). Integrated approach to nitrogen-fixing tree germplasms development. *Pesq. Agropec. Bras. Brasilia* 19, 91–117.

—— (1985). Biological nitrogen fixation in tropical agriculture. In *Nitrogen fixation research progress* (eds H. J. Evans, P. J. Bottomley, and W. E. Newton) pp. 675–81. Nijhoff/Junk, The Hague.

Herridge, D. F. and Betts, J. H. (1985). Nitrate tolerance in soybean: variation between genotypes. In *Nitrogen fixation research progress* (eds H. J. Evans, P. J. Bottomley, and W. E. Newton) p. 32. Nijhoff/Junk, The Hague.

Högberg, P. and Kvarnström, M. (1982). Nitrogen fixation by the woody legume *Leucaena leucocephala*. *Plant Soil* 66, 21–8.

Hutton, E. M. (1984). Breeding and selecting Leucaena for acid tropical soils. *Pesq. agropec. brasil. Brasilia* 19, 263–74.

IDRC (1983). *Leucaena research in the Asian-Pacific region*. IDRC, Ottawa.

Jarvis, B. D. W., Gillis, M., and De Ley, J. (1986). Intra- and intergeneric similarities between ribosomal ribonucleic acid cistrons of *Rhizobium*, *Bradyrhizobium* and some related bacteria. *Int. J. System. Bacteriol*. 36 (in press).

Jordan, D. C. (1982). Transfer of *Rhizobium japonicum* Buchanan 1980 to *Bradyrhizobium* gen. nov., a genus of slow-growing root nodule bacteria from leguminous plants. *Int. J. Syst. Bacteriol*. 32, 136–9.

—— (1984). International committee on systematic bacteriology. Subcommittee on the taxonomy of *Agrobacterium* and *Rhizobium*. *Int. J. Syst. Bacteriol*. 34, 248.

Kang, B. T. Spikens, L., Wilson, G. F., and Nangju, D. (1981*a*). Leucaena [*Leucaena leucocephala* (Lam) de Wit] prunings as nitrogen source for maize (*Zea mays* L.) *Fert. Res*. 2, 279–87.

——, Wilson, G. F., and Spikens, L. (1981*b*). Alley cropping maize (*Zea mays*) and Leucaena (*Leucaena leucocephala* Lam) in Southern Nigeria. *Plant Soil* 63, 165–79.

Kirkbride, J. H. (1984). Legumes of the Cerrado. *Pasq. agropec, bras. Brasilia* 19, 23–46.

Kolling, J., Scholles, D., Pereira, J., and Mendes, N. (1985). Eficiência e competitividade de estirpes de *Rhizobium leguminosarum* em ervilha. XIX Reunião Anual de Leguminosas. Porto Allegre 1985.

Koslowski, T. T. and Huxley, P. A. (1983). The role of controlled environments in agroforestry research. In *Plant research and agroforestry* (ed. P. A. Huxley) pp. 551–67. ICRAF, Nairobi.

Kyle, N. E. and Righetti, T. L. (1985). *In vitro* micrografting of actinorhizal desert shrubs. In *Nitrogen fixation research progress* (eds H. J. Evans, P. J. Bottomley and W. E. Newton). p. 364. Nijhoff/Junk, The Hague.

Lalonde, M. and Calvert, H. E. (1979). Production of *Frankia* hyphae and spores as an infective inoculant for *Alnus* species. In *Symbiotic nitrogen fixation in the management of temperate forests* (eds J. C. Gordon, C. T. Wheeler and D. A. Perry) pp. 95–110. Forest Research Laboratory, Corvallis.

—— and Simon, L. (1985). Current research strategies for use of actinorhizal symbioses in forestry. In *Nitrogen fixation research progress* (eds H. J. Evans, P. J. Bottomley and W. E. Newton). p. 667–74. Nijhoff/Junk, The Hague.

LaRue, T. A. and Patterson, G. (1981). How much nitrogen do legumes fix? *Adv. Agron*. 34, 15–38.

Meisner, C. A. and Gross, H. D. (1980). *Some guidelines for the evaluation of the need for and response to inoculation of tropical legumes*. Tech. Bull. No 265. North

Carolina Agricultural Research Service, Raleigh NC.

Munns, D. N. (1977). Mineral nutrition and the legume symbiosis. In *A treatise on dinitrogen fixation. IV. Agronomy and ecology* (ed. R. W. F. Hardy and A. H. Gibson) pp. 353–91. John Wiley, New York.

Nabors, M. W., Gibbs, S. E., Bernstein, C. S., and Meis M. E. (1980). NaCl-tolerant tobacco plants from cultured cells. *Z. Pflanzenphysiol.* 97, 13–17.

National Research Council (1977). *Leucaena: promising forage and tree crop for the tropics.* National Academy of Sciences, Washington DC.

—— (1979). *Tropical legumes.* National Academy of Sciences, Washington DC.

—— (1980). *Firewood crops.* National Academy of Sciences, Washington DC.

—— (1982). *Priorities in biotechnology research for international development.* National Academy of Sciences, Washington DC.

—— (1983). *Mangium and other fast-growing acacias for the humids tropics.* National Academy of Sciences, Washington DC.

Normand, Ph. and Lalonde, M. (1982). Evaluation of *Frankia* strains isolated from provenances of two *Alnus* species. *Can. J. Microbiol.* 28, 1133–42.

Newcomb, W. (1981). Fine structure of the root nodules of *Dryas drummondii* Richards (Rosaceae). *Can. J. Bot.* 59, 2500–14.

Okigbo, B. N. (1984). Nitrogen-fixing trees in Africa: priorities and research agenda in multiuse exploitation of plant resources. *Pesq. agropec. bras. Brasilia* 19, 325–30.

Périnet, P., Brouillette, J. G., Fortin J. A., and Lalonde, M. (1985). Large scale inoculation of actinorhizal plants with *Frankia. Plant Soil* 87, 175–83.

Phillips, D. A. and Teuber, L. R. (1985). Genetic improvement of symbiotic nitrogen fixation in legumes. In *Nitrogen fixation research progress* (eds H. J. Evans, P. J. Bottomley and W. E. Newton) pp. 11–18. Nijhoff/Junk, The Hague.

Puppo, A., Dimitrijevic, L., Diem, H. G., and Dommergues, Y. R. (1985). Homogeneity of superoxide dismutase patterns in *Frankia* strains from Casuarinaceae. *FEMS Microbiol. Lett.* 30, 43–6.

Righetti, T. L. and Hannaway, D. B. (1985). Agriculture and forestry poster discussion. In *Nitrogen fixation research progress* (eds H. J. Evans, P. J. Bottomley and W. E. Newton) pp. 711–12. Nijhoff/Junk, The Hague.

Rinaudo, G., Dreyfus, B., and Dommergues, Y. (1983). *Sesbania rostrata* green manure and nitrogen content of rice crop and soil. *Soil Biol. Biochem.* 15, 111–13.

Roskoski, J., Montano, J., Van Kessel, C., and Castilleja, G. (1982). Nitrogen fixation by tropical woody legumes: potential source of soil enrichment. In *Biological nitrogen fixation technology for tropical agriculture* (ed. P. H. Graham) pp. 447–54. CIAT, Calif.

Roughley, R. J., Bromfield, E. S. .P., Pulver, E. L. and Day, J. M. (1980). Competition between species of *Rhizobium* for nodulation of *Glycine max. Soil Biol. Biochem.* 12, 467–70.

Rundel, P. W., Nilsen, E. T., Sharifi, M. R., Virgina, R. A., Jarrell, W. M., Kohl, D. H., and Shearer, G. B. (1982). Seasonal dynamics of nitrogen cycling for a Prosopis woodland in the Sonoran desert. *Plant Soil* 67, 343–53.

Sanginga, N., Mulongoy, K., and Ayanaba, A. (1986). Evaluation of indigenous strains of *Rhizobium* for *Leucaena leucocephala* (Lam.) De Wit in Nigerian conditions. In *Manuscript reports. Proc. Seminar on Nitrogen-Fixing Trees Dakar March 1986.* IDRC, Ottawa (in press).

Scholla, M. H. and Elkan, G. H. (1984). *Rhizobium fredii* sp. nov., a fast growing species that effectively nodulates soybean. *Int. J. Syst. Bacteriol.* 34, 283–6.

Selbach, P. A., Freire, J. R. J., Scholles, D., and Kolling, J. (1978). Efeito da incorporação de cobertura vegetal nativa sobre a nodulação rendimento da soja em terra de primeiro cultivo. *10 Seminário Nacional de Pesquisa da Soja*, pp. 269–74. Londrina, PR.

Sellstedt, A. and Huss-Danell, K. (1985). Biomass production and nitrogen utilization by *Alnus incana*. In *Nitrogen fixation research progress* (eds H. J. Evans, P. J. Bottomley and W. E. Newton). p. 369. Nijhoff/Junk, The Hague.

Shearer, G. B., Kohl, D. H., Virginia, R. A., Bryan, B. A., Skeeters, J. L., Nilsen, E. T., Sharifi, M. R., and Rundel, P. W. (1983). Estimates of N_2 fixation from variation in the natural abundance of ^{15}N in Sonoran desert ecosystems. *Oecologia Berlin* 56, 365–73.

Simon, A. (1986). Possibilités d'introduction des jachères améliorées en milieu traditionnel rural africain: cas de l'Ouest Cameroun. In *Manuscript reports. Proc. seminar on nitrogen-fixing trees, Dakar, March 1986*. IDRC, Ottawa (in press).

Somasegaran, P. and Hoben, H. J. (1985). *Methods in legume-Rhizobium technology*. University of Hawaii NifTAL Project and MIRCEN, Hawaii.

Sougoufara, B., Duhoux, E., Corbasson, M., and Dommergues, Y. (1987). Improvement of nitrogen fixation by *Casuarine equisetifolia* through clonal selection. *Arid Soil Research and Rehabilitation*, 1, 129–32.

Sprent, J. (1983). Agricultural and horticultural systems: implications for forestry. In *Symbiotic nitrogen fixation in the management of temperate forests* (eds J. C. Gordon, C. T. Wheeler and D. A. Perry) pp. 213–32. Forest Research Laboratory, Corvallis.

Stowers, M. D. (1985). Further studies on the nodulating potential of *Rubus ellipticus* by the actinomycete *Frankia*. In *Nitrogen fixation research progress* (eds H. J. Evans, P. J. Bottomley, and W. E. Newton) p. 702. Nijhoff/Junk, The Hague.

——, Elkan, G. H. (1980). *Criteria for selecting infective and efficient strains of Rhizobium for use in tropical agriculture*. Tech. Bull. No 264. North Carolina Agricultural Research Service, Raleigh NC.

Thompson, J. A. (1980). Production and quality control of legume inoculants. In *Methods of evaluating biological nitrogen fixation* (ed. F. J. Bergersen) pp. 489–533. John Wiley, Chichester.

Tjepkema, J. D. (1983). Hemoglobins in the nitrogen-fixing root nodules of actinorhizal plants. *Can J. Bot*. 61, 2924–9.

—— (1984). Oxygen, hemoglobins, and energy usage in actinorhizal nodules. In *Advances in nitrogen fixation research* (eds C. Veeger and W. E. Newton) pp. 467–73. Nijhoff/Junk, The Hague.

Torres, M. E. (1985). Biological nitrogen fixation in *Prosopis tamarugo* and *Prosopis alba*. Nitrogen fixation by different inoculants in *Prosopis chiliensis*. In *The current state of knowledge on Prosopis tamarugo* (ed. M. A. Habit) pp. 413–25. FAO and Tarapaca University, Chile.

Torrey, J. G. (1985). The site of nitrogenase in *Frankia* in free-living culture and in symbiosis. In *Nitrogen fixation research progress* (eds H. J. Evans, P. J. Bottomley and W. E. Newton) p. 293–9. Nijhoff/Junk, The Hague.

—— and Tjepkema, J. D. (1979). Preface and program. *Bot. Gaz*. 140 (suppl), i–ii.

Trinick, M. J. (1976). *Rhizobium* symbiosis with a non-legume. In *Proc of the first international symposium on nitrogen fixation* (eds W. E. Newton and C. J. Nyman) pp. 507–17. Washington State University Press.

—— (1981). The effective *Rhizobium* symbiosis with the non-legume *Parasponia andersonii*. In *Current perspectives in nitrogen fixation* (eds A. H. Gibson and W. E.

Newton) p. 480. Australian Academy of Science, Canberra.

—— (1982). 3. Biology. In *Nitrogen fixation Vol. 2: Rhizobium* (ed. W. J. Broughton) pp. 77–146. Clarendon Press, Oxford.

—— and Galbraith, J. (1980). The *Rhizobium* requirements of the non-legume *Parasponia* in relationship to the cross-incoulation group concept of legumes. *New Phytol.* 86, 17–26.

Tsien, H. C., Dreyfus, B. L., and Schmidt, E. L. (1983). Initial stages in the morphogenesis of the nitrogen-fixing stem nodules of *Sesbania rostrata. J. Bacteriol.* 156, 888–97.

Turvey, N. D. and Smethurst, P. S. (1983). 8. Nitrogen-fixing plants in forest plantation management. In *Biological nitrogen fixation in forest ecosystems: foundations and applications* (eds J. C. Gordon and C. T. Wheeler) pp. 233–53. Nijhoff/Junk, The Hague.

Vargas, M. A. T., Peres, J. R., and Suhet, A. R. (1982). *Adubação nitrogenada e inoculacão de soja em solos de cerrados.* Circular Técnica No. 13, EMBRAPA.

Williams, P. M. (1984). Current use of legume inoculant technology. In *Biological nitrogen fixation, ecology, technology, and physiology* (ed. M. Alexander) pp. 173–200. Plenum Press, New York.

Wittenberg, J. B., Wittenberg, B. A., Trinick, M., Gibson, Q. H., Fleming, A. I., Bogusz, D., and Appleby, C. (1985). Hemoglobins which supply oxygen to intracellular procaryotic symbionts. In *Nitrogen fixation research progress* (eds H. J. Evans, P. J. Bottomley and W. E. Newton) p. 354. Nijhoff/Junk, The Hague.

Zhang, Z., Lopez, M. F., and Torrey, J. G. (1984). A comparison of cultural characteristics and infectivity of *Frankia* isolates from root nodules of *Casuarina* species. *Plant Soil* 78, 79–90.

2

Arid land biotechnology

A. S. El Nawawy

Introduction

Arid and semi-arid areas face difficult problems in their plans for food safety and efforts to produce food and feed from different resources. The main challenges that these areas have to meet are depletion of the groundwater and insufficiency of water resources, soil erosion, and soil fertility. These areas are faced by desertification and droughts.

However, there are many different approaches for dealing these problems. The potential for developing suitable innovative technologies for water and soil conversion is promising to meet the need of these regions. These should aim to increase the supply of usable water, optimize recycling of used water, and reducing the demand for water as much as possible. At the same time, management of soil fertility and ensuring organic matter in the soil, optimizing biological activity in the soil, will respond in more food production by both plant and animal resources.

In this chapter, emphasis will be placed on the biotechnological approaches for facing the stress of water supply, biological activities affecting soil facility, i.e. organic matter, mineral recycling, rhizosphere, and nitrogen fixation. However, some of these aspects which are given in details in other chapters of this book will be only summarized. The reader is requested to consult the relevant chapters for details (Chapters 1, 3, 8, 11, 13, and 15).

Water supply and limitation in arid climates

Irrigation with saline water

The salt resistance of crops largely determines the suitability of saline irrigation waters. The salt tolerance of crops has now been studied intensively, and adequate information for the selection of crops with suitable resistance to a saline environment is becoming available. Although only a few crops, such as cotton, barley, wheat, sugar beets, rye grass, Bermuda grass, and the wheat-grasses *Agropyron elongatum* and *Agropyron desertorum*, are known to be salt tolerant, they are important in developing countries because they form the basis of much agricultural production. Salt-tolerant trees

Table 2.1

Salt tolerance of agricultural crops

Crop	TR	Crop	TR	Crop	TR	Crop	TR
Alfalfa *Medicago sativa*	MS	Corn (forage) *Zea mays*	MS	Orange *Citrus sinensis*	MS	Peanut *Arachis hypogaea*	S
Almond *Prunus dulcis*	S	Corn (grain, sweet) *Zea mays*	S	Orchardgrass *Dactylis glomerata*	MS	Pepper, bell *Capsicum annuum*	MS
Apple *Malus sylvestris*	S	Cotton *Gossypium hirsutum*	S	Peach *Prunus persica*	T	Plum, common *Prunus domestica*	S
Apricot* *Prunus armaniae*	S	Cowpea *Vigna unguiculata*	S	Fescue, tall *Festuca elatior*	MS	Potato *Solanum tuberosum*	MT
Avacado *Persea americana*	S	Cucumber *Cucumis sativus*	S	Flax *Linum usitatissimum*	MS	Trefoil, birdsfoot narrowleaf *L. corniculatus tenuifolium*	MT
Barley (forage)** *Hordeum vulgare*	MT	Date palm *Phoenix dactylifera*	MT	Grape *Vitis* spp.	T	Vetch *Vicia sativa*	MS
Barley (grain)** *Hordeum vulgare*	T	Boysenberry *Rubus urisinus*	T	Radish *Raphanus sativus*	S	Wheat *Triticum aestivum*	MS
Bean *Phaseolus vulgaris*	S	Broadbean *Vicia faba*	S	Raspberry *Rubus idaeus*	MS	Wheatgrass crested *Agropyron desertorum*	S
Beet, garden *Beta vulgaris*	MT	Broccoli *Brassica oleracea botrytis*	MS	Rhodegrass *Chloris gayana*	MS	Wheatgrass, fairway *Agropyron critatum*	T
Bentgrass *Agrostis palustris*	MS	Hardingrass *Phalaris tuberosa*	MT	Rice *Oryza sativa*	MT	Sugarbeet *Beta vulgaris*	T
Bermudagrass *Cynodon dactylon*	T	Lemon *Citrus lemon*	T	Ryegrass, perennial *Lolium perenne*	S	Sugarcane *Saccharum officinarum*	MT

Common name	Scientific name	Tolerance	TR
Blackberry	Rubus spp.	S	
Bromegrass	Bromus inermis.	MT	
Cabbage	Brassica oleracea capitata	MS	
Canary grass, reed	Phalaris arundinacea	MT	
Carrot	Daucus carota	S	
Clover, alsike, ladino, red, strawberry	Trifolium spp.	MS	
Clover, berseem	Trifolium alexandrinum	MS	
Lettuce	Lactuca sativa	S	
Lovegrass	Eragrostis	MT	
Meadow foxtail	Alopecurus pratensis	MS	
Millet, foxtail	Setaria italica	MT	
Okra	Abelmoschus esculentus	S	
Olive	Olea europaea	MT	
Onion	Allium cepa	MS	
Safflower	Carthamus tinctorius	MS	
Sesbania	Sesbania exaltata	MS	
Sorghum	Sorghum bicolor	MS	
Soybean	Glycine max	MS	
Spinach	Spinacia oleracea	S	
Strawberry	Fragaria spp.	MT	
Sudangrass	Sorghum sudanese	S	
Sweet potato	Ipomea batatas	MT	MT
Tomato	Lycopersicon esculentum	MS	T
Wheatgrass, slender	Agropyron trachycaulum	MT	T
Wheatgrass, tall	Agropyron elongatum	MT	MT
Wildrye, Altai	Elymus angustus	MS	T
Wildrye, beardless	Elymus triticoides	S	
Wildrye, russain	Elymus junceus	MT	

* Tolerance is based on growth rather than yield.

** Less tolerant during emergence and seedling stages.

TR = tolerance rate; MS = moderately sensitive; MT = moderately tolerant; S = sensitive; T = tolerant.

include the date palm, olive, pomegranate, and pistachio (National Academy of Sciences 1974).

In general, irrigation waters whose total dissolved solids are below 600 mg/l may be used on almost any crop. If leaching and drainage are adequate, water with a total dissolved solids content of between 500 and 1500 mg/l is widely used on all but the most salt-sensitive crops. Water with a content of 1000–2000 mg/l can be used for crops of moderate tolerance, especially if frequent irrigation is employed. Water of 3000–5000 mg/l will produce high yields only from highly tolerant crops, as those listed earlier. Despite claims to the contrary, irrigation with undiluted sea water has not proved practical for producing crops. Sea water has a total salt content of about 35 000 mg/l, greatly exceeding the tolerance of even the most salt-tolerant crop studied to date, such as Suwanee Bermuda grass cynodon *dactylon*, which can tolerate about 12 000 mg/l. Table 2.1 lists some crops and their salt tolerance (Mass and Hoffman 1977; Rawling 1980).

The type, as well as concentration, of salt in the water is important. For example, the relative concentration of sodium to calcium and magnesium affect a water's suitability because high sodium ratios affect soil structure and plant nutrition. The anion of the salt (chloride or sulphate) may also be important (National Academy of Science 1974).

In the practice of saline irrigation, the basic premise is that proper irrigation and drainage management prevent salts from building up in the soil. It is essential to avoid this build-up by leaching, applying more irrigation water than the plants requires so that the extra water carries the salt down below the plant's roots. The suitability of saline water for irrigation is thus also governed by the leaching characteristics of the environment, i.e. whether they facilitate or retard the removal of salts from the root zone. If the leaching characteristics of the soil particles or the overall drainage of the area are insufficient, soil salinity will increase, and the final result may be a barren wasteland. Light-to-medium-textured soils that are not subject to structural changes that restrict water flow are more likely to be successfully irrigated with saline waters.

According to recent findings, new irrigation methods can increase a crop's tolerance to salinity. Compared to furrow irrigation, trickle irrigation, for example, has been shown to improve yields of crops irrigated with saline water (2000–2500 mg/l). The salt stresses on a plant are aggravated as the soil dries out and the salt concentration increases. Frequent irrigation (as in the trickle method) minimizes such stresses.

Saline ground-, surface, and estuarine waters are widely available, but not often used for irrigation; new research findings make it possible to use them more widely for agriculture, landscaping, etc.

When irrigation water contains 5000 mg/l or more of salts, the leaching requirements for even highly tolerant crops may be substantial. For example, more than 25 per cent extra water may be needed just to move the salts below the root zone; if less is used, salinization of the soil occurs because salts are not removed as fast as they are added.

Even highly tolerant crops may go through salt-sensitive stages when they need low-salt water. For example, seedlings of small grains and sugar beets are sensitive, and though the adult plants are not, and saline water may impair their development (National Academy of Science 1974).

Re-use of water

Water may have great impact on the future usable water supply in arid areas, either for agriculture, industrial, and/or municipal uses. For agriculture, spread waste water on marginal land to create new farmland may prove particularly important in arid countries. In such areas reclaimed water will probably be used first for irrigation. Filtering waste water through soil removes all particulate matter; most cations and some anions (including phosphates) are strongly adsorbed, and organic matter is decomposed by soil bacteria. These actions can contribute plant nutrients to soil.

Using municipal waste water for irrigation is especially attractive where agricultural lands are located close to cities, because the plant nutrients in sewage would otherwise be wasted. Some biological treatment of sewage should precede land application, but for many crops the degree of treatment required is so low that little technology and capital investment are required. When irrigation systems are already in use, connecting them to municipal systems is fairly simple, though institutional arrangements may prove difficult (National Academy of Sciences 1974).

Recycling irrigation water runoff by pumping it back to the head of the system is another way to reuse water for agriculture, but salinization of the soil is a serious hazard. Industrial waste water may also be fit for irrigation, but it may require treatment when the industrial process adds chemicals detrimental to plant growth or public health.

For industrial use, municipal waste water from secondary treatment plants can be used for cooling, ore separation, and other purposes that do not have severe water quality requirements. Use as boiler feed water requires advanced treatment. The degree and kind of treatment depend on demand and economics of the application. For pulp and paper production the use of waste water after only limited advanced treatment has been found to be economically feasible.

Municipal use places the highest demands on water quality. Waste water must usually undergo secondary and tertiary treatment to make it potable. Processes for removing ammonia, nitrates, and phosphorus are available; residual, potentially toxic compounds and dissolved organic substances can be reduced to very low levels by adsorption on activated carbon. Dissolved mineral matter can, if necessary, be reduced to acceptable levels by ion exchange, electrodialysis, or reverse osmorsis. However, adding these processes can double or triple the capital and operating costs of a conventional treatment plant.

Producing water of the necessary quality requires a large investment in capital equipment, power, and chemicals. The cost of such water is relatively high, but may be lower than the cost of developing alternative supplies of water, though not if treatment is needed to remove dissolved mineral salts, where the costs might be similar to that of desalinated sea water.

Windhoek, South Africa, a metropolitan area of 84 000 people, meets its water needs by treating and recycling into the potable water supply 4 million tonnes of its sewage, which represents one-third of the total daily supply (National Academy of Sciences 1974).

The important advantage of water reuse is that it can, if properly managed, reduced the demand on water from natural sources several-fold. Continuously recycling 50 per cent of the waste water in effect doubles the water supply.

In some arid locations re-using waste water in industry may provide the additional water needed to permit a level of industrialization that would not otherwise be feasible.

However, in any re-use scheme these major constituents of waste water have to be considered: pathogenic bacteria and viruses, parasite eggs, heavy metals, salts, and nitrates. To re-use water without causing environmental disaster calls for good management and a good understanding of the use's requirements. Systems can easily be mishandled and may cause serious disease or harm to the environment. If more than 50 per cent of the water supply is waste water, salt accumulation can cause serious problems whether the water is for agriculture, industry, or municipal use.

The cost and difficulty of reusing water depend on the treatment processes needed. Some secondary and most tertiary treatments require large capital investment and trained, capable personnel. Operating costs are high, in many cases too high for water re-use to be feasible. In arid regions the cost structure is more favourable. Direct reuse for potable supplies may have to overcome ethical objections, even if the water is demonstrably pure. Furthermore, people may object to eating food grown with human waste. Re-using water will often require that all sectors—agriculture, industry, and urban administrations—be integrated in management and policy.

Because water re-use will undoubtedly be highly competitive with its alternatives, it deserves a high research priority. The considerable amount of research under way has not emphasized the revise of water to increase a nation's water supply as much as at least one of its competitors, desalination. Research should centre on ways to reduce the cost, to combine secondary and advanced treatment processes, and to answer concerns about virological hazards.

The importance of continued research to develop treatments to reduce virological hazards, and to determine the residual hazard after treatment, cannot be overstressed.

Research is needed to reduce the cost of tertiary treatments and to develop alternative, less expensive treatment processes. Electrodialysis and reverse osmosis show promise, removing many kinds of dissolved impurities, but better antifouling techniques and membranes that require less pretreatment are needed. Improved biological processes are needed for removing ammonia and nitrates in secondary effluents, as are new low-cost specific ion exchangers for removing mineral salts.

In the agricultural use of waste water our biggest lack of knowledge is on the effects of long-term application. We do not fully understand to what degree continuous application of waste water to land alters the nature of soil. We do not know the capacity of soils to absorb different metals (boron for instance) without permanent damage. Nor have we learned how to prevent or restore altered soils.

Research is also needed to develop improved management techniques and institutional arrangements whereby effluent can be substituted for potable water now used in agriculture and industry.

Other sources of water

There are also other sources of water in arid lands. Three examples are given below.

Groundwater running

Aquifers in arid regions store large quantities of water that can be tapped by deep-well extraction techniques. Such acquifers are present beneath deserts in northern Mexico, southern United States, North Africa, eastern Saudi Arabia, the Sinai, and South-West Asia. However, the water in such aquifers is not naturally replaced once it is withdrawn.

Desalination

The low cost of desalination of sea water would be a main source to arid lands bordering seas or salt lakes. Although new and improved desalination methods using membranes and ion exchange have been developed, but the costs are still high (Fried and Edlund 1971). However, efforts to optimize this technology and reduce the costs are in progress (Rawlins 1981). Problems in the disposal of hot brine, pumping and conveying desalinated water to the point of use, as well as large amounts of energy required for constructing and operating desalination plants should be considered when adopting this technology.

Solar distillation

In this case, the sun's radiation passes through a transparent cover on to a source of brine, water evaporates from the brine, and the vapour condenses on the cover which is arranged to collect and store it. The process is generally in the pilot stages where small community-scale have been developed in the USA, France, Spain, and Australia (National Academy of Science 1974; Proctor 1973).

Water conservation

Arid regions are facing difficult problems with regards to desertification and droughts. Arid lands have under-exploited agricultural potential, which can be developed by concepts and methods specifically suited to dry regions. Thus, fresh and innovative approaches to water technologies, particularly those designed to meet the needs of arid regions, should be emphasized. Besides increasing the supply of usable water, efforts should be made to conserve and reduce the demand for water. There are different approaches too water conservation some of which are mentioned below.

Reducing evaporation from water surfaces

This includes methods to cover the water surface with a barrier that inhibits evaporation, such as liquid chemicals (e.g. aliphatic alcohols which are non-toxic to fish and humans; Frazier and Myers 1968), floating blocks of wax, or solid blocks of light-weight concrete (Cooley and Meyers 1973).

Reducing seepage losses

There are many ways of reducing seepage in waterways, such as soil compaction, chemical treatment of the soil, and soil covers such as butyl-rubber, sheet plastic, asphalt reinforced with plastic or fibre-glass, and ferrocement (Board on Sciences and Technology for International Development 1973; Kraatz 1971; Myers 1963).

Reducing evaporation from soil surfaces

Evaporation from soil surfaces wastes large amounts of water, an important consideration in arid areas where low humidity greatly encourages evaporation. From one-quarter to one-half of the water lost from a crop is due to evaporation from the soil surface (Viets 1966). This loss can be reduced and irrigation water save by placing a water-tight moisture barriers or water retardant mulches on the soil surface (National Academy of Sciences 1974).

In many cases, these barriers will also stabilize loose soils, stop desert enroachment, allow runoff agriculture, aid in landscaping, or reduce salinity build-up. In arid regions, small water savings may be more important to a crop's survival than large improvements at earlier points in the water supply.

Trickle (drip) irrigation

In this method, water is carried in the pipes and drips onto the soil through outlets arranged near each plant; thus, a small amount of soil is watered and evaporation losses are minimized (National Academy of Sciences 1974): Bernstein and Francois 1973). Besides drip irrigation, sprinkle and sub-surface irrigation can be applied as a means of minimizing the amounts of water needed for plants.

Increasing organic matter in soil and improving its water holding capacity

This can be achieved by adding compost as soil conditioner and fertilizer. In some countries, using oil sludge for land treatment is an attractive treatment for modifying physical properties of sandy soils, and increasing its water-holding capacity, thus decreasing rate of evaporation as well as rates of leaching (Arora *et al.* 1982).

Methods of water conservation including reducing cropland percolation losses, especially in sandy soils (Erickson *et al.* 1968; Saxena *et al.* 1967), and reducing transpiration of the plant, are now in practice in some arid lands (Abou-Khaled *et al.* 1970; Davenport *et al.* 1969; Gale and Hagan 1966).

Apart from the above mentioned, it is very important to continue research for the selection and breeding of food, forage, and industrial crops which can use less water per unit of product. Wise selection of cultural practices can do much to improve the efficiency of water use (Arnon 1972; Evenari *et al.* 1971). The investigations on water use patterns by the plant during its life cycle, should include the genetic and physiological basis of water use efficiency (Turner 1979), i.e. the regulation of stomatal behaviour in gas exchange between plant and atmosphere (Raschke 1975).

Water conservation by various technologies mentioned before will help on-farm water demands, and also have a number of incidental effects both on and off the farm. Davenport and Hagan (1981) summarized on- and off-farm impacts of various on-farm water conservation measures (Table 2.2). In general, they stated that reducing recoverable water losses saves water on-farm, but not for the basin, while reducing irrecoverable evapotranspiration losses, conserves water both on-farm and for the basin.

Table 2.2

Various effects of water conservation measures applied on irrigated fields*

Loss	Aim	How	Water conservation effect On-farm	Off-farm (other users)	Basin or state
Recoverable losses					
Runoff	Reduce it	Improve application efficiency by land levelling, determining root zone water deficit, measuring water and changing irrigation system.	Reduce relivery requirements and thus reduced $ paid for water and/or pumping energy.	More water left in sources for others to use.	Does not conserve water unless there is loss to saline sinks or loss of ponded water to evapotranspiration
	Re-use it	Tailwater system	Reduced delivery requirement and thus reduced $ paid for water and/or pumping energy, but may need energy for tailwater pumpback	Less tailwater for other users and for in-stream flow maintenance.	Does not conserve water unless there is loss to saline sinks or loss of ponded water to evapotranspiration
Deep percolation	Reduce it	Improve application efficiency by land levelling, determining root zone water deficit, measuring water and changing irrigation system.	Reduced delivery requirement and thus reduced $ paid for water and/or pumping energy.	Could reduce groundwater recharge and result in greater pumping costs.	Does not conserve water unless there is loss to saline sinks or non-recoverable formations.

Re-use it	Collect in drains, or by well, and pump as needed to place of re-use.	Reduced delivery requirement and thus reduced $ paid for water and/or pumping energy to cover it.	If intercepted by tile drains, could reduce groundwater recharge	Does not conserve water unless there is loss to saline sinks or non-recoverable formations.
Irrecoverable losses				
Evaporation (from soil) Reduce it	Less surface wetting by drip irrigation or reduced irrigation frequency. Mulches, etc.	Will save some water (and thus reduce delivery requirement somewhat), usually without depressing crop yield. Reduces irrecoverable loss to atmosphere but not as much as transpiration reduction.	Little, if any, effect.	Reduces irrecoverable loss to atmosphere, but savings relatively small
Transpiration (from crops and weeds) Reduce it	Reducing area rate, time duration of transpiring surfaces, e.g. by crop selection and management, deficit irrigation, antitranspirants, and weed control.	Has potential for substantial water saving (and thus reduce water demands and/or $ paid for water and/or pumping energy), but generally reduces crop yield proportionately.	More water left in source for others to use.	Potential for large reduction of water that is irrecoverably lost from the state to the atmosphere, but will likely reduce crop production.

*After Davenport and Hagan (1981).

Effects of water stress and salinity on plant physiology

The capacity of a plant to withstand stress conditions is mainly dependent upon the efficiency of its association with root micro-organisms. Accordingly, the plant's resistance to water stress and excessive acidity or alkalinity can be increased by the availability of nutrients and/or by protecting the root against pathogenic micro-organisms. In this case root nodule bacteria, rhizosphere micro-organisms, play an important role in helping the plant to withstand stress conditions (Dommergues 1978).

As soil is the environment for root micro-organisms then soil characteristics will affect their activity, especially those related to form, content, and distribution of organic and inorganic compounds or to physical characteristics, namely soil structure, water potential, gas phase composition, texture, pH, and temperature (Dommergues 1978).

Thus, in arid land, where there is water stress, and hence a need to use saline water for irrigation, and possible lack of organic matter, it is expected that plant physiology will be modified to allow for this. For example the root exudates will increase in dry soil and, accordingly, more loss of solutions from the root will occur. The root exudates contain amino acids, sugars, organic acids, proteins, polysaccharides, growth substances, growth inhibitors, attractants, and repellents (Hale *et al.* 1978).

Rains (1981), described some of the physiological and biochemical mechanisms, associated with salt tolerance in plants as well as in micro-organisms. He divided these mechanisms into osmoregulation by the organism through the uptake and/or formation of organic molecules; osmoregulation by accumulation of inorganic ions and intracellular distribution, and compartmentation of both essential ions; and biochemical and structural properties of cells and their constituents, including acid/base ratio of amino acids, specialized proteins, and alterations in membrane structure and composition.

Organic matter in arid lands

Organic matter in the soil

In arid zones, natural vegetation may be almost nil in some areas, yet in other parts very dense. The high temperature enhances chemical and biochemical reactions in soil. Therefore, humus content is mostly relatively low (0.1–1.0 per cent). In arid regions, soil transport of salts and sodium ions (chloride, hydrogen bicarbonate, etc.) occurs by capillary movement as a precursor of evapotranspiration. These process retard plant growth and therefore humus dynamics (Flaig *et al.* 1977).

Humus content differs widely between 0.5 and 10 per cent of the dry matter of soil (Kononova 1975). The addition of organic material to the soil and its transformation in soils, which is known as 'humus husbandry', plays an important role in soil productivity. Most of the reactions for the transformation of organic materials such as plant residues after harvest, stable manure, excreta of animals and man, municipal waste products, etc., are caused by microbial activity.

Synthesis of humic substances can be summarized in a scheme (Flaig *et al.* 1975). The main organic constituents of the dead organisms in soils are cellulose, lignins, and proteins. By biochemical degradation different products are formed which serve the micro-organisms as carbon and nitrogen sources, mainly cellulose and proteins. Different phenolic compounds are formed from lignin. Microbial synthesis is another way of forming phenolic compounds. These types of phenolic substances are mainly transformed by oxidation, hydroxylation, and decarboxylation. During these transformations, ring cleavage also occurs and aliphatic compounds are formed, which are used by the micro-organisms as carbon sources. Finally, the humic substances are degraded to CO_2, ammonia, and water.

The formation and degradation follow a dynamic course, which is influenced by climatic factors, vegetation, and composition of the mineral part of soil. Organic matter plays an important role in the life processes of the soil and has significant role in plant nutrition. The plant is composed mainly of organic compounds, containing carbon, hydrogen, oxygen, and nitrogen. The first three elements are taken up by plants as CO_2 and water which are the final products of the decomposition or organic matter. Most of the CO_2 is released into the atmosphere and enters into the common cycle, and only traces of CO_2 enter the plant through the root system (Sauerbeck and Fuhr 1966). Organic matter in the soil is the main natural source of nitrogen. Nitrogenous constituents of soil organic matter serve as a slow release nitrogen source for plant nutrition.

Variable amounts of nitrogen can be fixed by micro-organisms from the air and small amounts can enter the soil as a result of rainfall. In addition to nitrogen, organic materials and soil organic matter provide other plant nutrients—P, K, Ca, Mg, and trace elements. Furthermore, soil organic matter provides the energy source for the growth of the microbial population which is responsible for a number of significant biochemical processes. Soil organic matter also provides plant nutrients from sources other than itself (e.g. release of P or K from soil inorganic materials). The availability of phosphorus is increased by the addition of organic materials, partly by a possible increase of microbial activity.

Organic matter also checks soil erosion and help soil conservation. Soil erosion is a serious problem in tropical and subtropical climatic conditions. Decreasing water runoff is an important element for arid lands where water convervation is an essential element for agricultural development.

Thus, there should be national and regional strategies for improving soil productivity with organic wastes. These strategies will help the process of halting the decline in soil productivity and to restore a long-term, stable, sustainable, and productive agricultural system (Parr and Colacicco 1982). These strategies should consider four items: (i) survey of organic materials, types, amounts, and availability; (ii) re-introduction of organic fertilizers in soil, including animal manures, crop residues, green manures, and compost; (iii) re-introduction of better management practices, and thus better soil and water conservation practices, decreasing the rate of soil erosion, and preventing irreparable damage to crop land, sedimentation, and nutrient run-off; (iv) development of a more sustainable organic farming system.

There are certain areas of research that should be continuously explored, such as (i) rates of decomposition for each type of waste under different soil regimes and cropping systems; (ii) rates at which the plant nutrients are mineralized, recycled, and utilized by both current and subsequent crops; (iii) the potential toxic effects of certain wastes on plants and micro-organisms; and (iv) the impact of organic waste management on the control of plant insects and diseases. Each type of organic wastes has unique properties that should be thoroughly investigated in the soil/water/plant ecosystem (Abou El Fadl *et al.* 1958, 1968; Parr and colacicco 1982; Riad 1982).

Returning organic matter to land is one method of maintaining soil productivity and reducing the dependence on chemical fertilizers. Some well known examples are composting of solid wastes (Flaig *et al.* 1977), co-composting of solid wastes and sewage sludges (Kuchenrither *et al.* 1984), and biogas technology, which besides producing fuel provides the remaining solids, considered to be good fertilizers, and residual water, suitable for irrigation (Alaa El-Din 1982).

Other uses of organic matter

Besides returning organic matter to the land, there are other several biotechnological approaches which add significantly to better conservation of natural resources. A wide variety of biological, chemical, and physical process can be followed. Organic wastes from agriculture, agro-industrial processing, animals, and humans can fulfil the requirements for food, feed, and fuel.

Unproductive means of waste disposal can be replaced by methods that boost crop yield, save energy, improve the environment and strengthen the independence and well being of individual farmers and villages. Of course, the reader is aware of the many publications which are continuously published in this area. In this chapter, the main approaches for solid waste recycling are summarized.

Table 2.3

Mushroom cultivation conditions

Species	Temperature (°C)		Level of environmental control required	Waste substrate
	Spawn running	Fruiting		
Agaricus bisporus (common mushroom)	20–27	10–20	++++	Composed horse manure or rice straw
Agaricus bitorquis	25–30	20–25	++++	Composed horse manure or rice straw
Auricularia spp. (ear mushrooms)	20–35	20–30	+++	Sawdust-rice bran
Coprinus fimetarius	20–40	20–40	+	Straw
Flammulina velutipes (winter mushroom)	18–25	3–8	+++	Sawdust-rice bran
Lentinus edodes (shiitake mushroom)	20–30	12–20	++	Logs or sawdust-rice bran
Pholiota nameko (nameko mushroom)	24–26	5–15	+++	Logs or sawdust-rice bran
Pleurotus ostreatus (oyster mushroom)	20–27	10–20	+	Straw, paper, sawdust-straw
Stropharia rugosoannulata	25–28	10–20	++	Straw, paper, sawdust-straw
Tremella fucitormis (white jelly mushroom)	20–25	20–27	+++	Logs or sawdust-rice bran
Volvariella volvacea (straw mushroom)	35–40	30–35	+	Straw, cotton wastes

Food production

Agro-industrial wastes and manures can all be bioconverted to food.

1. For mushroom production on sawdust, cotton wastes, bagasse, shredded paper, and wood (Atal *et al.* 1978; Chang 1980; Hatch and Finger 1979; Kurtzman 1979; Zakhary *et al.* 1984), examples of which are given in Table 2.3.

2. For the production of food-grade yeasts on whey, molasses, potato, and cassava wastes (Reed and Peppler 1973; Wiken 1972; El-Nawawy 1982; Litchfield 1979).

Production of feed

Many agro-industrial wastes, particularly those derived from food processing, are potentially useful as animal feed. Through microbial conversion, they can be upgraded for producing protein fodder or protein-enriched fodder, e.g. production of single cell protein from starch wastes (Rolz 1982), lignocellulosic wastes (Bellamy 1976; Dyer *et al.* 1975; Han *et al.* 1971, El-Nawawy 1978; El-Nawawy *et al.* 1974), and production of algae on waste water (Oswald *et al.* 1978).

Production of fuel

Gaseous, liquid, and solid fuels can all be produced from wastes, for example, biological production of biogas from agriculture, animal, and human wastes (Alaa El-Din 1982; Bryant 1979; National Academy of Sciences 1977), and biological production of ethanol from sugar, starch, and cellulose (Phillips and Humphrey 1985).

The fermentation of wastes to generate methane, rather than their direct use as fuel or fertilizer, yields a number of benefits, including:

(i) producing an energy resource that can be stored and used more efficiently in many applications;

(ii) creating a stabilized residue that retains the fertilizer value of the original material;

(iii) reducing faecal pathogens and improving public health; and

(iv) reducing transfer of plant pathogens from one year's crop to the next.

In conclusion to this section, it is clear that organic matter is very important for the development of agriculture and food in many countries, and of special interest in arid zones, where resource conservation by all means are needed. There should be an integrated system for the optimal use of resources. According to the status and need in each country, the approach for integration may be different. However, in any integrated system, the available organic wastes can be used to produce food, feed, fuel, and fertilizers. The conversion processes are combined and balanced to maximize self-sufficien~·

Mineral cycling by micro-organisms in soil

Plant and animal residues are decomposed in soil by soil micro-organisms, which accordingly release carbon, nitrogen, sulphur, phosphorus, and trace elements from organic materials in forms that can be absorbed by plants. This is the process of mineralization which is the primary source of atmospheric carbon dioxide and consequently maintains the carbon cycle, the most important biological process on earth.

Microbial activities also transform soil nitrogen, sulphur, phosphorus, and other elements. The transformtion of nitrogen and sulphur are similar in the sense that both elements can be oxidized and reduced. Sulphur reduction is necessary for the synthesis of sulphur-containing amino acids. Under anaerobic conditions, however, sulphur reduction may produce hydrogen sulphide, which can be harmful to plants. It accumulates in wet soils, such as the rice paddies cultivated in certain semi-arid zones, and may cause straight-thread disease of rice and other physiological plant disorders (National Academy of Sciences 1979). Joshi and Hollis (1977) noticed that hydrogen sulphide might be oxidized by a bacterium belonging to genus Beggiatoa, thus detoxifying flooded rice soils, and its capacity to influence plant growth favourably deserves further study. Sulphur may also be oxidized in soils by *Thiobacillus* sp. producing sulphuric acid, which can dissolve minerals when otherwise would not be available for plant growth.

Micro-organisms are also able to promote phosphorus solubilization by the production of chelators, which form complexes with metal ions and increase their solubility. Solvent action by micro-organisms is also characteristic of many members of the rhizosphere population and can be accomplished in part by plant roots as well.

Investigators in the USSR tried to make use of phosphate-solubilizing bacteria in the growth of plants. They selected an efficient strain *Bacillus megatherium* var. *phosphaticum*, which is very active in solubilization of insoluble phosphate. They made a preparation of this organism adsorbed on kaolinite and named it phosphobacterin. This preparation was used on a wide scale in agriculture. According to Novikova (1960), phosphobacterin gave a steady yield increases in cultivated plates. A 13–14 per cent increase in yield was found (Mishustin and Naumova 1962). The increase in yield by phosphobacterin inoculation cannot be due only to the increase in phosphorus uptake, but may also be due to the inhibition of plant pathogens and synthesis of plant growth regulators (Mishustin and Naumova 1962).

Studies under Egyptian soil conditions indicated that these soils harbour a high density of phosphate-dissolving bacteria, and are preferentially stimulated in the rhizosphere (Abd El Hafez 1966; Taha 1969; Mahmoud 1973). These studies indicated that the numbers of phosphate-dissolving bacteria in the soils of Egypt are affected by the kind of soil, the fertility status, the kind of growing plants, and the stage of its development (Mahmoud 1973). It is interesting to note that the density of phosphate-dissolving bacteria in the rhizosphere increased with the increase in plant development and that the densities of these micro-organisms in the rhizosphere were affected by manuring and fertilizer treatments, the highest counts were recorded for organic manure. The presence of high densities of phosphate-dissolving bacteria in Egyptian soils may explain the ability of the growing plants to take up their phosphorus requirements in Egyptian alkaline soils where the level of available phosphorus is low.

Studies also indicated that the most dominant phosphate-dissolving bacteria were members of the genera Bacillus and Streptomyces with low percentages of Gram-positive and Gram-negative rods, *Micrococcus* and *Sarcina*. This result is completely different from those recorded for temperate regions where the most abundant phosphate-dissolving bacteria were pleomorphic organisms and *Pseudomonas* (Mahmoud 1973).

However, micro-organisms, especially in the rhizosphere, will compete with roots for nourishment in case of short supply of nitrogen, phosphorus, or sulphur. Because of their abundance, small size, and relatively large surface area, and because they surround the absorbing part of the root microbes will absorb nutrients at the expense of the plant. In this case, the plant will display signs of nutrients deficiency and crop yields may decrease.

The autotrophic nitrifying bacteria are responsible for the formation of nitrate in soils, sewage, and aquatic environments. Nitrate serves as the principal source of nitrogen to higher plants. The ammonium nitrogen complexes form parallel to decomposition of the organic nitrogen complexes of the soil organic matter. When fresh organic residues are added to soil, ammonia is evolved in much large amounts than in untreated soil. Under acid or anaerobic soil conditions, the ammonia formed will constitute the end product of nitrogen mineralization, while under fertile soil, ammonia will be oxidized rapidly, due to the activity of nitrifying organisms to nitrite then to nitrate.

Nitrate in soil is subject reduction either through nitrogen assimilation by plants and soil micro-organisms, and/or through nitrate reduction to nitrite, then ammonia by anaerobic and facultative bacteria, which use nitrate as a hydrogen acceptor. If nitrate is reduced to nitrogen gas or gaseous oxides of nitrogen, a process called denitrification, it leads to a serious loss of available nitrogen. This happens mainly under anaerobic conditions and alkaline reaction. More information is needed on the mechanism of denitrification. About 50 per cent of the fixed nitrogen applied as fertilizers to a plant system is unaccounted for and does not pass through the plant system. If denitrification could be controlled, it would have important economic effects in agriculture, since as much as 25 per cent of the energy expenditure in cereal crop production is associated with fertilizer application (Humphrey 1982). In arid areas, the fate of nitrogen cycle in the soil should be more emphasized.

The sum of the various inter-relationships of rhizosphere micro-organisms and roots can benefit plant growth by influencing the availability of essential nutrients, by producing plant growth regulators and by suppressing root pathogens. Gray and Williams (1975) had compared the number of various micro-organisms in the rhizosphere of wheat and control soil, and had the result shown in Table 2.4.

In the rhizosphere oxygen consumption occurs more rapidly than diffusion, so that anaerobic sites may form in the root. Such reduced conditions could be important, for instance, in making ferrous ions from ferric, which increases iron solubility. Wheat roots have high populations of denitrifying bacteria, so oxygen-free conditions must exist in their presence.

Micro-organisms are prolific producers of vitamins, amino acids, hormones, and other growth-regulating substances. Many bacteria and fungi isolated

Table 2.4

Comparison of the numbers of various groups of organisms in the rhizosphere of spring wheat and in control soil*

Organisms	Numbers per g in rhizosphere soil $(\times 10^6)$	Numbers per g in rhizosphere soil $(\times 10^6)$	Approximate rhizosphere soil ratio
Bacteria	1200	53	23:1
Actinomycetes	46	7	7:1
Fungi	12	0.1	120:1
Protozoa	0.0024	0.001	2:1
Algae	0.005	0.027	0.2:1
Bacterial groups			
Ammonifiers	500	0.04	12,500:1
Gas-producing anaerobes	0.39	0.03	13:1
Anaerobes	12	6	2:1
Denitrifiers	126	0.1	1,260:1
Aerobic cellulose decomposers	0.7	0.1	7:1
Anaerobic cellulose decomposers	0.009	0.003	3:1
Spore formers	0.930	0.575	2:1
'Radiobacter' types	17	0.01	1,700:1
Azotobacter	0.001	0.001	?

* Gray and Williams (1975).

from soil are able to synthesize compounds that provoke a growth response in plant tissue. Some produce indoleacetic acid or gibberellins, the hormones controlling plant growth, while other produce vitamins. Many may also produce unidentified growth factors. Rhizosphere micro-organisms are variously credited with promoting increased rates of seed germination, root elongation, root-hair development, nutrient uptake, and plant growth (National Academy of Sciences 1979).

Role of Mycorrhizae in plant growth

Most plants, both wild and cultivated, have roots infected with fungi that increase nutrient and water uptake, and may also protect the roots from certain diseases. These infected roots are called mycorrhizae. Although the mycorrhizal fungi probably increase the uptake of all the essential elements, they are usually most important in improving phosphorus nutrition. There are a number of types of mycorrhizae—endomycorrhizae and the ectomycorrhizae are the most important ones.

Endomycorrhizae of the vesicular-arbuscular (VA) type occur on most crop plants. VA mycorhizal fungi are present in almost all soils and they are not host-specific. The mycorrhizal condition is normal for most plants, and absence or scarcity of mycorrhizal fungi can greatly limit plant growth. VA mycorhizal fungi survive in soil as resting spores. They obtain their food from the plant roots and they are unable to grow independently in soil. It is unlikely that they obtain much, if any, organic nutrient from soil. VA fungi have not been grown in pure culture, which presents an obstacle to artificial inoculation. However, these fungi produce the largest spores of any known fungi, some being 0.5 mm or more in diameter. The spores can be easily extracted from the soil with sieves and then propagated on the roots of living plants. The greatest opportunity for the use of VA fungi is in soils low in available phosphorus, which includes many untreated soils in tropical regions.

Ectomycorrhiza is the second most common type of mycorrhiza. It occurs on roots of pine, spruce, fir, larch, hemlock, willow, poplar, hickory, pecan, oak, birch, beech, and eucalyptus. The fungi that form ectomycrorrhizae produce mushrooms and puffballs as their reproductive structures (fruit bodies). Ectomycorrhizae benefit trees by increasing nutrient and water absorption from soil; increasing the tolerance of the tree to drought and extremes of soil conditions (acid levels, toxins, etc.); increasing the length of the feeder root system; and protecting the fine feeder roots from certain harmful soil fungi.

The mycorrhizal mycelia thus serve as highly efficient extension of the root systems. Mycorrhizae are an important consideration in maximizing range productivity, because mycorrhiza-dependent plants cannot succeed without their fungal associate when growing in nutrient deficient soil. The ecological requirements of mycorrhizal fungi are particularly relevant to programmes for improving rangelands degraded by erosion, compaction, overgrazing, or contamination with salt, oil, heavy metals, etc. The native mycorrhizal fungi may not be well adapted to such conditions so new fungi may have to be introduced to the soil to establish desired vegetation. However, little is known about mycorhizal ecology of semi-arid and arid rangelands. Baseline data are urgently needed on the strains, distribution, and adaptability of the mycorrhizal fungi, as well as on the relative mycorrhizal dependence of both desirable and undesirable rangeland hosts (Trappe 1981).

The importance of mycorrhizae to the nutrition of most vescular plants and to the health of ecosystems has been overwhelmingly demonstrated in recent decades (Marks and Kozlowski 1973; Sanders *et al*, 1975; Trappe and Fogel 1977). Research on mycorrhizae and their application to enhancing plant productivity has been mostly concerned with forest cultivated and pasture crops, rather than arid and semi-arid rangelands.

Mycorrhizal fungi grew between or into cortical cells of host rootlets and out into the surrounding soil (Ames *et al.* 1982). Within the root cortex, nutrients absorbed by the fungus from the soil are translocated to the host, and photosynthates and their derivatives are extracted from host tissues by the fungus (Ho and Trappe 1974). The fungal hyphae extending into the soil serve as extensions of the root systems, extensions that are both physiologically and geometrically more effective for nutrient adsorption than the roots themselves.

Isotope and plant nutritional experiments confirm that N, P, K, Ca, S, Zn, Cu, and Sr absorbed from soil by mycorrhizal fungi are translocated to the host plant (Bowen 1973; Cooper and Tinker 1978; Gray and Gerdemann 1973; Jackson *et al.* 1973; Rhodes and Gerdemann 1978; Sanders *et al.* 1975). Minerals more than 4 cm distant from the nearest host root can be absorbed by the hyphae and translocated to the root. Mycorrhizal fungi are not only structurally efficient for extraction of nutrients from exchange sites in soil, they also produce exogenous enzymes such as phosphatases, phytases, and nitrate reductase, which are important in the uptake and metabolism of nutrients (Gianinazzi-Pearson and Gianinazii 1978; Ho and Trappe 1975: Theodorou 1968). Recent work on organic acid secretion by mycorrhizal fungi (Cromack *et al.* 1979; Graustein 1977) has opened new possibilities for understanding their particular effectiveness in extraction of nutrients from soil.

A distinctive type of mycorrhizae is formed in the Arabian Peninsula by *Helianthemum* spp. with the desert fungi, *Terfezia* and *Tirmania* spp. (Awameh *et al.* 1981). The same fungi also occur in North Africa, but are apparently associated with other host genera. Other taxa of desert truffles occur in deserts of southern Africa, central Australia, and southwestern United States, but their mycorrhizal association are unknown.

Indigenous mycorrhizal fungi are not necessarily the best for optimum growth of desired forage species in a given soil (Abbot and Robsonn 1978; Powell 1976). The introduction of more efficient fungi to a site to selectively promote desire forage species in arid lands deserves research attention.

In P-deficient soils it is tempting to fertilize. This should be done circumspectively: in some situations it may not disrupt mycorrhizal populations (Porter et al. 1979); in others, spore production or mycorrhizal formation may be drastically reduced (Menge et al. 1978). In the latter case, continuing and even increasing P fertilization may be required to sustain growth of hosts deprived of the mycorrhizal fungi they normally require for P uptake. Endomycorrhizae do not only substantially increase P uptake by the plant, but also lower the host root's resistance to water transport, probably by improving the nutrient uptake (Safir et al. 1971).

Besides, attention has been given to the effect of endomycorrhizae on the legume-Rhizobium symbiosis. Endomycorrhizae improve both nodulation and N_2 fixation (Mosse et al. 1976) Dart and Day 1975). Suitable mycorrhizal inoculation, together with application of some form of organic matter and phosphorus fertilizer, should allow the extension of legume crops in poor tropical soils. However, more research is needed in order to produce endomycorrhizal inoculants on a proper scale.

When mycorrhiza-dependent forage species are desired for prompt revegetation of such sites, they must be supplied with mycorrhizal fungi (Aldon 1975; Hall and Armstrong 1979; Lindsey et al. 1977; Williams et al. 1974). This may be done by inoculation of the soil or by planting desired hosts with mycorrhizae already established. In either event, the mycorrhizal fungi must be adapted to the site. Preliminary evaluation of inocula in the greenhouse followed by field studies will be required initially. The accumulation of data over time will ultimately enable choice of an appropriate inoculum without extensive preliminary trials.

Inoculum can be in the form of soil from analogous sites already supporting mycorrhizal hosts (Black and Tinker 1977). Other forms of inocula are in various stages of development (Gaunt 1978; Hall 1979; Hattingh and Gerdeman 1975; Menge et al. 1977; Ganry et al. 1982, 1985).

The mycorrhizal status of lands irrigated with saline water should be carefully monitored. A build-up of salts may be detrimental to indigenous fungi. When that is the case, productivity of the crop plants may decline or require repeated phosphorus fertilization unless mycorrhizal fungi more adapted to high salt concentrations are introduced.

There is need for more investigations of the effects of vesicular-arbuscular mycorrhiza (VAM) fungi on root exudation and rhizosphere microflora. In theory, a reduction in root exudation from mycorrhizal plants should cause a reduction in the micro-organisms that utilize exudates as their primary food source. However, specific organisms may actually increase, as was the case for an *Azotobacter* sp. in the study of Bagyaraj and Menge (1978), and for the *Pseudomonas* sp., as reported by Barea et al. (1975) and Azcon et al. (1976).

Many kinds of plant-microbial interactions occur in the plant rhizosphere that are under the direct influence of root exudation. It is important therefore, to learn more about the function of mycorrhizas as they relate to plant health and microbial activities (Ames *et al.* 1984; Mahmoud *et al.* 1985; Ishae *et al.* 1986).

Biological nitrogen fixation

Fixation of nitrogen in soils for the nutrition of plants occurs by micro-organisms mainly through:

(i) free-living organisms, such as blue-green algae and *Azotobacter* bacteria (asymbiotic N_2 fixation);

(ii) micro-organisms living in symbiosis with special species of plants such as leguminosae (symbiotic N_2 fixation); and

(iii) nitrogen-fixing micro-organisms living mostly in the rhizosphere of tropical gramineae (rhizospheric N_2 fixation).

The effect of biological nitrogen fixation in soil depends on environmental conditions, e.g. composition of the soil, soil structure, and climatic factors.

Asymbiotic fixation

The free-living nitrogen-fixing micro-organisms have their optimal growth under different soil conditions. The C/N ratio of soil organic matter as the source of energy must be high to allow N_2 fixation. Among the bacteria *Azotobacter* has its best living conditions in well-aerated soils with neutral reaction (Brown *et al.* 1964). *bacillus amylobacter* and *Clostridium* (Ridge and Rovira 1968) occur in many soils with a wide pH range, *Beijerinckia* (Becking 1977) is adapted to acid soils because of the low calcium requirement and is distributed in lateritic soils in the tropics.

Other nitrogen-fixing organisms have a special importance for food production in semi-arid countries, such as rice in Egypt, where blue-green algae such as *Nostoc linckia, Anabaena variabilis, Aulosira fertilissima, Tolypothrix tenuis* (El-Nawawy *et al.* 1958, 1968, 1972, 1973, El-Nawawy and Hamdi 1975; El-Borollosy 1972; Shaalan 1980; El-Sayed 1978) proved to have positive effect in providing rice with considerable part of its nitrogen requirements.

Symbiotic fixation

Each cross-inoculation group of leguminosae needs its own species of *Rhizobium*, an organism that acts through nodules on the roots. The nitrogen-fixing capacity of the bacteria depends on the plant's growth conditions. In some cases, about 50 per cent of the relatively large amount of nitrogen fixed remains in the roots and stubble on the fields. During humification of these materials, the next crop (e.g. cereals) can make use of the nitrogen fixed for its own nutrition. (Dommergues 1980)

About 13 000 leguminosae are known, but only a very few are used in agricultural production. The selection of new varities for food production is an important task for plant breeders. The species of *Rhizobium* must be adapted to the particular leguminous plant. The inoculation of seeds increases nitrogen fixation if the specific bacteria are not present in the soil to be cultivated.

Rhizospheric nitrogen fixation

It was found for many years that bacteria living in the rhizosphere of tropical grasses fix a remarkable amount of nitrogen. According to first calculations, the fixation reaches 60 kg N/ha in rice fields (Yoshida and Ancajas Rosabel 1973). *Azotobacter, Beijerinckia, Pseudomonas*, and *Arthrobacter* have been identified (Balandreau *et al.* 1975) as fixing bacteria.

Furthermore, it has been demonstrated that several nitrogen-fixing bacteria occur on the roots of sugarcane, Bahia-grass, and other grasses, which are used to feed cattle (Dobereiner 1968, 1977). The same author found that temperatures below 27°C during the day and below 18°C at night inhibit nitrogen fixation. By this 'Rhizospheric' nitrogen-fixing system, up to 1 kg N/ha/day was supposed to be fixed by tropical grasses for feedstuffs. In one case it was calculated, by using ^{15}N, that about 75 per cent of the fixed nitrogen is utilized by the plant.

Azospirilla grasses association (Neyra and Dobereiner, 1977) was assumed to be very promising systems. It was reported (Dobereiner and De-Polli 1980) that cereal grains showed significant amounts of nitrogen through such associative symbiosis with *Azospirilla*. Furthermore, inoculation of such plants with these particular micro-organisms has been tried in an effort to increase their population density and to intensify their activities on roots (Hegazi *et al.* 1979*a*, *b*). Promising results were obtained, particularly in the tropics. Results of the inoculation of wheat (Khawas 1981; Monib *et al.* 1981) and zea maize (Hegazi *et al.* 1979*a*, *b*) were very encouraging as it was found that microbial inoculation might partially, if not completely, replace nitrogen fertilization. However, it is now widely recognized that inoculation with Azospirillum or other nitrogen-fixing micro-organisms increases the yield of grass crops (an increase of 10–20 per cent), but that nitrogen fixation is not the mechanism producing the increase (Diem and Dommergues 1979; Smith *et al.* 1984).

To summarize the potential of biological nitrogen fixation system, Hamdi

(1982) had stated that Rhizobia-grain legume fix between 41 and 552 kg N/ha. Forage legumes fix between 62 and 897 kg N/ha. Frankia non-legume associations fix between 2 and 300 kg N/ha/year. Response to inoculation with Azotobacter showed about a 10 per cent increase in yield. Inoculation with Azospirillum showed variable results according to location and crop variety.

Nitrogen fixation by blue-green algae in culture media ranges between 3.6 and 330 mg N/100 ml within 2 months according to species and time of inoculation. Response of the rice crop to algae inoculation ranged between 4.2 and 368 per cent increase in grain yield in pot experiments. Under field conditions a range of yield increases up to 32 per cent were reported.

Azolla is a symbiotic algal association that grows on the surface of water. Under field conditions rates of nitrogen fixation were between 103 to 1216 kg/ N/ha/year according to location, time, and variety of azolla. Green manuring with 10 t/ha azolla is efficient as the basal application of 30 kg. Rice grain yield increased as much as 54 per cent with manuring with azolla.

Generally, if nitrogen-fixing bacteria can function in subarctic temperate and tropical regions, various strains differ significantly in their tolerance to elevated temperature. In many cases, semi-arid soils reach high temperature and desicated condition due to exposure to radiant heating; thus, survival and die-off studies concerned with the behaviour of temperature-tolerant strains in such soils, are very important. The same criteria should be investigated also with regard to range of moisture and salinity conditions; especially when considering the free-living nitrogen-fixing organisms, either in soil or in rhizosphere.

Genetic manipulation of plants to increase their resistance to water stress and salinity

Besides the selection and breeding of plants for salt tolerance, i.e. barley and wheat (Epstein et al. 1979; Norlyn 1980), there are new approaches to genetic manipulation for salt tolerance. This approach depends on the presence of genetic variability in plant genotypes. Natural variability has provided an enormous assay of genetic variants which have been used for selection and breeding of the desired agronomics traits (Rains 1981). This lies in one of the following techniques.

Tissue culture technique

The primary goal of the cell selection programme is to obtain lines of cells with the desired traits, to regenerate these cells to whole plants, and to evaluate the potential of genetic transfer of this characteristic to other genotypes by standard breeding techniques. This technique provides a number of advantages for the development of salt tolerant plants.

1. The culture of plant cells under controlled environments on rigidly defined media permits a uniform and precise treatment with salt.

2. Culture cells, being relatively undifferentiated, reduce the complications inherent in studying complicated organisms of varying morphology and stages of development. Cellular physiology can be observed directly and genetic markers identified.

3. Cell culture techniques permit the manipulation of literally billions of cells using relatively small space and numbers of person hours. The variability of these cells can be increased by treating the cells with mutagenic agents (chemical or radiation). The cell population enriched in mutants provides the experimental plant material with an increased potential for selection of salt tolerant variants.

4. Cell suspensions provide the ideal material for the application of molecular genetic techniques, such as recombinant DNA and plasmid transfer, to higher plants. In recent years, the potential for the manipulation of plant germplasm for crop improvement has been widely discussed. The numerous rapid advances to recombinant DNA technology in microbial and other systems has added to this growing enthusiasm (Abelson and Butz 1980). The methods for generating a new plant varieties are summarized in Fig. 2.1 (Kamel 1986).

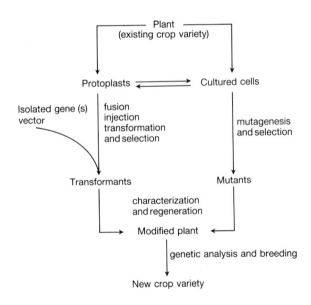

Fig. 2.1 Advanced biotechnology methods for generating a new plant variety.

5. The use of cell selection and tissue culture techniques have advantages; in particular, the ability to do physiological studies on the mechanism responsible for salt tolerance. Comparative studies using the unselected and selected cell lines provide a unique opportunity to conduct theses investigation.

It is now possible to regenerate cells cultured for over 3 years back to whole plants (Stavarek *et al.* 1980). The cell culture approach has proven feasible for selection of salt-tolerant cell lines of a number of plant species. A number of examples are presented in Table 2.5.

Table 2.5

Examples of salt-tolerant cell lines selected through tissue culture*

Species	Callus or suspension	NaCl level tolerated (w/v.%)
Nicotiana tabacum	Suspension	0.52
Nicotiana sylvestris	Callus	1.0
	suspension	2.0
Capsicum annuum	Callus	1.0
	suspension	
Medicago sativa	Callus	1.0
Oryza sativa	Callus	1.5

* After Rains (1981).

Exposing cultured alfalfa cells to an agar-solidified nutrient medium containing a typically lethal concentration of NaCl permitted identification of a cell line with an elevated tolerance to salt. The tissue to be screened for salt tolerance was increased in quantity through suspension culture. Inoculating salt-containing media in petri dishes with a slurry of these suspension-grown cells allowed this extremely large number of individuals to be screened for salt tolerance. The screening consisted of visual identification of the cells that maintained both growth and a healthy appearance despite exposure to high salinity. Consistent subculture to saline nutrient medium permitted enrichment of the most tolerant cell lines, ultimately resulting in isolation of the few salt-tolerant types present in the very large original population. (Levit 1972)

Salt-tolerant lines of rice cells have been selected using haploid cells. The selected cells were capable of surviving and growing in 1.5 per cent NaCl. These cells demonstrated similar growth responses as alfalfa cells at low NaCl concentration. The salt-selected rice cells did not grow as well as the unselected cells at low NaCl concentration and apparently required the presence of at least 0.5 per cent NaCl for optimal growth (Rains 1979, 1981).

Gene transfer technology

Genetic engineering can play a significant role in developing new varieties of agricultural crops suitable for arid land stresses. Gene transfer technology combined with the techniques of tissue culture, protoplast fusion and regeneration of plants from a single isolated protoplast are now important tools in the development and growth of a complete fertile and tolerant variety of crop suitable for use in arid and desert conditions (Kamel 1986).

There are a number of plants that grow well in desert conditions, such as cactus. It is possible that through the transfer of certain genes from these varities to crop plants, or perhaps entire chromosomes or section of chromosomes, transfered through protoplast fusion (Fig. 2.1). Thus, the multifaceted nature of advanced biotechnology including recombinant DNA techniques, and its application in agriculture is one of the advances in science which might help improvement in plants for nutritional contents and other crop improvement, i.e. nitrogen fixation, increased inherent photosynthetic efficiency, plant disease control, etc., which are examples of its possible implication on agricultural growth in the world. This may open new avenues of hope for those countries which have suffered environmental stresses to become more and more vulnerable towards dependency in food production.

Conclusion

Modern agricultural productivity amplifies the capture of solar energy in plants by the application of fertilizer, irrigation water, and other protective materials, when needed (Rabson and Rogers 1981). Research bearing on environmental stress relating to biomass productivity includes the following studies, especially for arid land.

1. Investigations on water use patterns by the plant during its life cycle, including the genetic and physiological basis of water use efficiency (Turner 1979); i.e. the regulation of stomatal behaviour in gas exchange between plant and atmosphere (Raschke 1975).

2. Much more needs to be known about mineral cycling in representative ecosystems. Studies are needed on the relationships among plants, soils and the inclusive microflora, the enclosing atmosphere, precipitation and runoff with respect to key nutrients to avoid serious problems of soil depletion and degradation (Jackson *et al.* 1975).

3. Determination of the underlying mechanisms of how plants adapt to highly saline conditions. Model systems for investigating salinity damage should be devised in which lesions are defined biochemically, (Rains 1979).

4. Investigations on the several roles of soil micro-organisms with respect to plant nutrition and growth. The interactions between plant and microbes in soils, either in the case of nitrogen fixation and also in the mycorrhizal activity

which affects phosphorous availability to roots are good indicators of the importance of such studies. Considering the importance of roots in the acquisition of water and nutrients, these relationships require further attention.

5. Considerable attention should be directed towards membrane studies since changes in plant composition and structure seem to occur during temperature stress (Saxon *et al*. 1980).

6. Application of advanced genetic engineering for selecting and improving plant yields under environmental stress in arid and semi-arid lands. Gene transfer technology combined with the techniques of tissue cultue, protoplast fusion, and regeneration of plants from a single isolated protoplast are efficient tools to develop, and grow fertile and tolerant varieties of crops suitable for arid and semi-arid lands.

Thus, genetic engineering can play a significant role in developing new varities of agriculture crops with stress tolerance.

References

Abbot, L. K. and Robson, A. D. (1978). Growth of subterranean clover in relation to the formation of endomycorrhizas by introduced and indigenous fungi in a field soil. *New Phytol.* 81, 575–85.

Abd El Hafez, A. M. (1966). Some studies on acid producing micro-organisms in soil and rhizosphere with special reference to phosphate dissolvers. PhD thesis, Faculty of Agriculture, Ain Shams University, Cairo.

Abelson, J. and Butz, E. (eds) (1980). Recombinant DNA issue. *Science* 209, 1317–434.

Abou El Fadl, M., Fahmy, M. and Hafez, M. (1958). A study on the control of the pink bollworm (*Pectinophora gossypiella* (Saurd) through composting the infested bolls into organic manure. *Agric. Res. Rev. Min. Agric., Giza, Egypt*, 36, 331–6.

——, Rizk, S. G., Abdel-Ghani, A. F., El-Mofty, M. K., Kadr, M. F., Shehata, S. M., and Farag, F. A. (1968). Utilization of water-hyacinth as an organic manure with special reference to water-borne helminths. *Egypt. J. Microbiol.* 3, 27–34.

Abou Khaled, A., Hagan, R. M., and Davenport, D. C. (1970). Effects of kaolinite as a reflective antitranspirant on leaf temperture, transpiration, photosynthesis, and water use efficiency. *Water Resources Res.* 6, 280–9.

Alaa El-Din, M. N. (1982). Biogas technology with respect to Chinese experience. *FAO Soils Bull.* 45, 175–86. FAO Rome 1982.

Aldon, E. E. (1975). *Endomycorrhizas enhance survival and growth of fourwing saltush on coal mine spoils*. U.S. Forestry Services Research Note RM–194.

Ames, R. N., Ingham, E. R., and Reid, C. P. P. (1982). Ultraviolet-induced autofluorescence of arbuscular mycorrhizal root infection: an alternative to clearing and staining methods for assessing infections. *J. Microbiol.* 28, 351–5.

——, Reid, C. P. P., and Ingham, E. R. (1984). Rhizosphere Bacterial population responses to root colonization by a vesicular arbuscular mycorrhizal fungus. *New Phytol.* 96, 555–63.

Arnon, I. (1972). *Crop production in dry regions, Vol. I: Background and principles*. Barnes & Noble, New York.

Arora, H. S., Cantor, R. R., and Nemeth, J. C. (1982). Land treatment: A viable and successful method of treating petroleum industry wastes. *Environ. Int.* 7, 285–91.

Atal, C. K., Bhat, B. K., and Kaul, T. N. (1978). *Indian mushroom science–1 (Agaricus, Pleurotus, Volvariella, Native Indian Species)*, Indo-American Literature House, Globe, Arizona.

Awameh, M., El-Kholy, H., and Al-Slamain, F. (1981). Ecological studies of desert truffles. *Annual Research Report, 1981*, pp. 40–3. Kuwait Institute for Scientific Research, Kuwait.

Azcon, R., Barea, J. M., and Hayman, D. S. (1976). Utilization of rock phosphate in alkaline soils by plants inoculated with mycorrhizal fungi and phosphate-solubilizing bacteria. *Soil Biol. Biochem.* 3, 135–8.

Bagyaraj, D. J. and Menge, J. A. (1978). Interaction between a VA mycorrhiza and Azotobacter and their effects on rhizosphere microflora and plant growth. *New Phytol.* 80, 567–73.

Balandreau, J, Puinaudo, G., Fares-Hamad, I., and Dommergues, Y. R. (1975). Nitrogen fixation in the Rhizosphere of rice plants. In *Nitrogen fixation by free living microorganisms*, (eds W. D. P. Stewart) IBP Vol. 6, pp. 57–70 Cambridge University Press.

Barea, J. M., Azcon, R., and Hayman, D. S. (1975). Possible synergistic interaction between Endogene and phosphate-solubilizing Bacteria in low-phosphate soils. In: *Endomycorrhizas* (eds F. E. Sanders, B. Mosse, and P. B. Tinker), pp. 409–17. Academic Press, London.

Becking, J. H. (1977). Dinitrogen fixing associations in higher plants other than legumes. In *A treatise on dinitrogen-fixation section. III. Biology* (eds R. W. F. Hardy and W. S. Silver), pp. 185–275. Wiley, New York.

Bellamy, W. D. (1976). Production of single-cell protein for animal feed from lignocellulose wastes. *WI Anim. Rev.* 18, 39–42.

Bernstein, L. and Francois, L. E. (1973). Comparisons of drip, furrow, and sprinkler irrigation. *Soil Sci.* 115, 73–86.

Black, R. L. B. and Tinker, P. B. (1977). Interaction between effects of vesicular-arbuscular mycorrhiza and fertilizer phosphorous on yields of potatoes in the field. *Nature* 267, 510–1.

Board on Sciences and Technology for International Development (BOSTID) (1973). *Ferro-cement, application in developing countries*. National Academy of Sciences, Washington, D.C.

Bowen, G. D. (1973). Mineral nutrition of ectomycorrhizae. In *Ectomycorrhizae. Their ecology and physiology* (eds G. C. Marks and T. T. Kozlowski), pp. 151–205. Academic Press, New York.

Brown, M. E., Burlingham, S. K., and Jackson, R. M. (1964). Studies on *Azotobacter* species in soil. III. Effects of artificial inoculation on crop yields. *Plant Soil* 20, 194–214.

Bryant, M. P. (1979). Microbial methane production-theoretical aspects. *J. Anim. Sci.* 48, 1–10.

Chang, S. T. (1980). Cultivation of Volvariella mushrooms in Southeast Asia. *Mushroom Newslett. Tropics* 1, 5–10.

Cooley, K. R. and Meyers, L. E. (1973). Evaporation reduction with reflective covers. *J. Irrigation Drainage Div. Am. Soc. Civ. Eng.* 99, 353–60.

Cooper, K. M. and Tinker, P. B. (1978). Translocation and transfer of nutrients in vesicular-arbuscular mycorrhizas. II. Uptake and translocation of phosphorus, zinc, and sulphur. *New Phytol.* 81, 43–52.

Cromack, K., Sollins, P., Craustein, W. C., Speidel, K., Todd, A. W., Spycher, G., Li, C. Y., and Todd, R. I. (1979). Calcium oxalate accumulation and soil weathering in mats of the hypogeous fungus *Hysterangium crassum*. *Soil Biol. Biochem.* 11, 463–8.

Dart, P. J. and Day, J. M. (1975) Nitrogen fixation in the field other than by nodules. In *Soil microbiology: a critical review*. (ed. N. Walker) pp. 225–52. Butterworths Scientific Publications, London.

Davenport, D. C. and Hagan, R. M. (1981). Concepts for conserving agricultural water. In *Advances in food producing systems for arid and semi-arid lands. Part A* (eds J. J. Manassah and E. J. Briskey) pp. 329–90. Academic Press, London.

——, ——, and Martin, P. E. (1969). Antitranspirants research and its possible application in hydrology. *Water Resources Res.* 5, 735–43.

Diem, H. J. and Dommerges, Y. (1979). Significance and improvement of rhizospheric N_2 fixation. In *Recent advances in biological nitrogen fixation* (ed. N. S. Subba Rao) pp. 190–226. Oxford and IBH, New Delhi.

Dobereiner J. (1968). Non-symbiotic nitrogen fixation in tropical soils. *Pesq. agrogec. bras. Brasilia.* 3, 1–6.

—— (1977). Nitrogen-fixation association with non-leguminous plants. In: *Genetic engineering for nitrogen-fixation.* (ed. A. Hollaender) pp. 451–61. Plenum Press, New York and London.

——, and De-Polli, H. (1980). Diazotrophic rhizocoenoses. In *Nitrogen fixation* (eds W. D. P. Stewart and J. R. Gallon), pp. 301–15. Academic Press, London.

Dommergues, Y. R. (1978). The plant-microorganisms system. In *Interaction between non-pathogenic soil microorganisms and plants* (eds Y. R. Dommergues and S. V. Krupa) pp. 1–37. Elsevier, Amsterdam.

—— (1980). The effect of edaphic factors on nitrogen-fixation with special emphasis on organic matter in soils. *FAO Soil Bull.* 43, 145–57, FAO, Rome.

Dyer, I. A., Riquelme, E., Baribo, L., and Couch, B. Y. (1975). Waste cellulose as an energy source for animal protein production, *Wld Anim. Rev.* 15, 39–43.

El-Borollosy, M. A. (1972). Studies on N_2-fixing blue-green algae in A. R. E. Msc thesis. Faculty on Agriculture, Ain Shams University, Cairo, Egypt.

El-Nawawy, A. S. (1978). Comparative studies on some yeast and their efficiency in utilizing Rub el Tamer wastes for production of food yeasts. *XII International Congress of Microbiology*, Munich, September, 1978.

—— (1982). The promise of microbial technology. *Impact Sci. Soc.* 32, 157–66.

—— and Hamdi, Y. A. (1975). Research on blue-green algae in Egypt 1958–72. In *Nitrogen-fixation by free living micro-organisms* (ed. W. D. P. Stewart), IBP, Vol. 6, pp. 219–28. Cambridge University Press.

——, ——, El-Sayed, M., and Shaalan, S. N. (1973). Growth of the blue-green algae, *T. tenuis* as affected by cobalt and molybdenum. *Zenbl. Bakt. Abt. II, Bd*, 182, S. 452–6.

——, Ibrahim, A. M., and Abou El Fadl, M. (1968). Nitrogen fixation by *Calothrix* sp. as influenced by certain sodium salts and nitrogenous compounds. *Acta Argon. Hung.* 17, 323–7.

——, Kamal, R. M., and Abou El Fadl, M. (1972). Growth and N-fixation by blue-green algae, *T. tenuis* as affected by phosphorus content of media. *Agric. Res. Rev.* 50, 111–15.

——, Lofti, M., and Fahmy, M. (1958). Studies on the ability of some blue-green algae to fix atmospheric nitrogen and their effect on growth and yield of paddy. *Agric. Res. Rev.* ARE 36, 308–49.

——, Mahmoud, S. A. Z., Mashoor, W. M., and Ibrahim, E. M. (1974). Utilization of rice bulls for the production of microbial protein. *2nd Rice Conf. Proc.* pp. 100–10. Cairo, December 1974.

El-Sayed, M. (1978). Studies on N_2-fixing blue-green algae in Egyptian soils. Ph.D. thesis, Ain Shams University, Egypt.

Epstein, E., Kingsbury, R. W., Norlyn, J. D. and Rush, D. W. (1979). Production of food crops and other biomass by seawater culture. In *The Biosaline concept: an approach to the utilization of under-exploited resources*, (ed. A. Hollaender), pp. 77–85. Plenum Press, New York.

Erickson, A. E., Hansen, C. M., and Smucker, A. J. M. (1968). The influence of sub-surface asphalt barriers on the water properties and the productivity of sand soils. In *Transaction of the 9th International Congress of Soil Science*, Adelaide, Australia, 1, 331–7.

Evenari, M., Shanan, L., and Tadmor, N. (1971). *The Negev: The challenge of a desert.* Harvard University Press, Cambridge, Massachusetts, pp. 229–300.

Flaig, W., Beutelspacher, H., and Rietz, E. (1975). Chemical Composition and physical properties of humic substances. In *Soil components. I. Organic components* (ed. J. E. Gieseking) pp. 1–211. Springer-Verlag, Berlin.

——, Nagar, B., Sochtig, H., and Tietjen, C. (1977). Organic materials and soil productivity. *FAO Soils Bulletin* 35, FAO/Rome 1977.

Frazier, G. W. and Myers, L. E. (1968). Stable alkanal dispersion to reduce evaporation. *J. Irrigation Drainage Div. Am. Soc. Civ. Eng.* 94, 79–89.

Fried, J. J. and Edlund, M. C. (1971). Desalting technology for Middle Eastern agriculture; an economic case. *Special studies in international economic and development*, Praeger, New York. p. 113.

Gale, J. and Hagan, R. M. (1966). Plant antitranspirants. *Ann. Rev. Plant Physiol.* 17, 269–82.

Ganry, F., Diem, H. G., and Dommergues, Y. R. (1982). Effect of inoculation with *Glomus mosseae* on nitrogen fixation by field grown beans. *Plant Soil* 68, 321–9.

Gaunt, R. E. (1978). Inoculation of vesicular-arbuscular mycorrhizal fungi on onion and tomato seeds. *N Z J. Bot.* 16, 69–71.

Gianianazzi-Persson, V. and Gianianazzi, S. (1978). Enzymatic studies on the metabolism of versicular-arbuscular mycorrhiza. II. Soluble alkaline phosphatase specific to mycorrhizal infection in onion roots. *Physiol. Plant Pathol.* 12, 45–53.

Graustein, W. C., Cromack, K., and Sollins, P. (1977). Calcium oxalate: occurrence in soils and effect on nutrient and geochemical cycles. *Science* 198, 1252–4.

Gray, L. E. and Gerdemann, J. W. (1973). Uptake of sulphur-35 by versicular-arbuscular mycorrhizae. *Plant Soil* 39, 687–9.

Gray, T. R. C. and Williams, S. T. (1975). *Soil microorganisms.* Longman, New York.

Hale, M. G., Moore, L. D. and Griffin, G. J. (1978). Root exudates and exudation. In *Interactions between non-pathogenic soil microorganisms and plants* (eds Y. R. Dommergues and S. V. Krupa) pp. 163–203. Elsevier, Amsterdam.

Hall, I. R. (1979). Soil pellets to introduce vesicular-arbuscular mycorrhizal fungi into soil. *Soil Biol. Biochem.* 11, 85–86.

—— and Armstrong, P. (1979). Effect of versicular-arbuscular mycorrhizas on growth of white clover, lotus, and ryegrass in some eroded soils. *N Z J. Agric. Res.* 22, 479–84.

Hamdi, Y. A. (1982). Application of nitrogen fixing systems in soil improvement and management. *FAO Soil Bull.* 49, 1–188, FAO Rome 1982.

Han, Y. W., Dunlap, C. E., and Callihan, C. D. (1971). Single-cell protein from cellulosic wastes. *Food Technol.* 25, 130–3, 154.

Hatch, R. T. and Finger, S. M. (1979). Mushroom fermentation, In *Microbial technology*, 2nd ed, (eds H. J. Peppler and D. Perlman) pp. 179–99. Academic Press, New York.

Hattingh, M. J. and Gerdeman, J. W. (1975). Inoculation of Brazilian sour orange seed with an endomycorrhizal fungus. *Phytopathol.* 65, 1013–16.

Hegazi, N. A., Eid, M., Farag, R. S., and Monib, M. (1979a). Asymbiotic N_2-fixation in the rhizosphere of sugarcane planted under semi-arid conditions of Egypt. *Rev. Ecol. Bull. Soc.* 16, 23–37.

——, Monib, M., and Vlassack, K. (1979b). Effect of inoculation with N_2-fixing *Spirilla* and *Azotobacter* on N_2-ase activity on roots of maize grown under subtropical conditions. *Appl. Environ. Microbiol.* 38, 621–5.

Ho, I. and Trappe, J. M. (1974). Translocation of ^{14}C from *Festuca* plants to their endomycorrhizal fungi. *Nature* 244, 30–1.

——, and ——, (1975). Nitrate reducing capacity of two vesicular-arbuscular mycorrhizal fungi. *Mycologia* 67, 886–8.

Humphrey, A. E. (1982). Biotechnology: the way ahead. *J. Chem. Tech. Biotechnol.* 32, 25–33.

Ishac, Y. Z., El-Haddad, M. E., Daft, M. J., Ramadan, E. M., and El-Demerdash, M. E. (1986). Effect of seed inoculation, mycorrhial infection and organic amendment on wheatgrowth. *Plant and Soil* 90, 373–82.

Jackson, N. E., Miller, R. H., and Franklin, R. E. (1973). The influence of vesicular-arbuscular mycorrhizae on uptake of ^{90}Sr from soil by soybeans. *Soil Biol. Biochem.* 5, 205–12.

Jackson, W. A., Knezek, B. D., and van Schilfgaarde, J. (1975). Water, soil, and mineral input. In *Crop productivity—research imperatives* (eds A. W. A. Brown, T. C. Byerly, M. Gibbs, A. San Pietro, E. Lansing). Michigan Agric. Expt. Sta., pp. 201–74.

Joshi, M. M. and Hollis, J. P. (1977). Interactions of *Beggiatoa* and rice plants: detoxification of hydrogen sulfide in the rice rhizosphere. *Science* 197, 179–80.

Kamel, W. (1985). The potential of biotechnology for the Gulf region and the role of the international centre for genetic engineering and biotechnology. In *Perspectives in biotechnology and applied microbiology* (eds D. I. Alani and M. Moo-Young). Elsevier, London and New York. pp. 369–77.

Khawas, H. M. (1981). Ecological Studies on asymbiotic N_2 fixing bacteria in soil and Rhizosphere of certain plants. MSc Thesis, Faculty of Agriculture, Cairo University, Egypt.

Kononova, M. M. (1975). Humus of virgin and cultivated soils. In *Soil components. 1. Organic components* (ed. J. E. Gieseking) pp. 475–526. Springer-Verlag, Berlin.

Kraatz, D. D. (1971). *Irrigation canal lining.* Irrigation and drainage Paper No. 2, Water Resources and Development, FAO, Rome, 170p.

Kuchenrither, R. D., Martin, W. J., Smith, D. G., and Psaris, P. J. (1984). An economic comparison of composting and dual utilization. *Biocycle*, July-August 1984, 33–7.

Kurtzman, R. H., Jr (1979). Mushrooms: single-cell protein from cellulose. In *Annual report on fermentation processes*, (ed. D. Perlman), vol. 3. Academic Press, New York.

Levit, J. (1972). *Responses of plants to environmental stresses.* Academic Press, New York.

Lindsey, D. L., Cress, W. A., and Alson, E. F. (1977). The effects of endomycorrhizae on growth of rabbitbrush, fourwing saltbush, and corn in coal mine spoil material. *USDA For. Serv. Res. Note* RM–343, pp. 6–7.

Litchfield, J. H. (1979). Production of single-cell protein for use in food or feed. In *Microbial Technology*, 2nd ed (eds H. J. Peppler and D. Perlman) pp. 93–156. Academic Press, New York.

Mahmoud, S. A. Z. (1973). Inorganic insoluble phosphate dissolving bacteria in soils of Egypt and the rhizosphere of broad bean and wheat. *Agrokemia Es Talajtan Tom*, 22, 351–6.

——, Ishac, Y. Z., Ramadan, E. H. and Daft, M. J. (1985). Occurrence and infectivity of Endomycorrhizas in Egyptian Soils. *Egypt. J. Microbiol.*, special issue, pp. 47–56.

Marks, C. G. and Kozlowski, T. T. (eds) (1973). *Ectomycorrhizae. Their ecology and physiology*. Academic Press, New York and London.

Mass, E. V. and Hoffman, G. J. (1977). Crop Salt tolerance current assessment. *J. Irrigation Drainage Div. Am. Soc. Civ. Eng.* 103 (IR2), 115–34.

Menge, J. A., Labanauskas, C. K., Johnson, E. L. V., and Platt, R. G. (1978). Partial substitution of mycorrhizal fungi for phosphorus fertilization in the culture of citrus. *Soil Sci. Soc. Am. J.* 42, 926–30.

——, Lembright, H. and Johnson, E. L. V. (1977). Utilization of mycorrhizal fungi in citrus nurseries. *Proc. Int. Soc. Citricult.* 1, 129–32.

Mishustin, E. N. and Naumova, A. N. (1962). Bacterial fertilizers, their effectiveness and mode of action. Translated Mikrobiologya (USSR). *Nat. Sci. Found. NY* 31, 442–52.

Monib, M., Hegazi, N. A., Shokr, S., and Khawas, H. M. (1981). Response of field grown wheat to inoculation with *Azospirilla (A. brasilense)*. *Res. Bull. Fac. Agric. Ain Shams Univ.* 1535, 1–23.

Mosse, B., Powell, G. L., and Hayna, D. S. (1976). Plant growth response to versicular-arbuscular mycorrhiza IX. Interactions between VA mycorrhiza, rock phosphate and symbiotic nitrogen fixation. *New Phytol.* 76, 331–42.

Myers, L. E. (ed). (1963). *Seepage Symposium, Phoenix, Arizona*, Proc. Agric. Res. Serv. Resport, A.R.S. 41–90. U.S. Department of Agriculture, Washington, D.C.

National Academy of Sciences, U.S.A. (1974). *More water for arid lands, promising technologies and research opportunities*. N.A.S., Washington D.C.

—— (1977). *Methane generation from human, animal, and agricultural wastes.* Report of an *Ad Hoc* panel of the Advisory Committee on Technology Innovation, Board on Science and Technology for Intern. Development, Commission on International Relations, Washington, D.C., N.A.S.

—— (1979). *Microbial processes: promising technologies for developing countries.* N.A.S., Washington D.C.

Neyra, C. A. and Dobereiner, J. (1977). Nitrogen fixation in grasses. *Adv. Agron.* 29, 1–38.

Norlyn, J. D. (1980). Breeding salt-tolerant crop plants. In *Genetic engineering of osmoregulation: impact on plant productivity for food, chemicals, and energy* (eds D. W. Rains, R. C. Valentine, and A. Hollaender), pp. 293–8. Plenum Press, New York.

Novikova, A. T. (1960). The effectiveness of phosphobacterin on soil of the Kustanaiv region in relation to their method of cultivation. *Agrobiologiya*, 4, 604–14.

Oswald, W. J., Lee, E. W., Adnan, B., and Yao, K. H. (1978). New wastewater treatment method yield of harvest of saleable algae. *WHO Chron.* 32, 348–50.

Parr, J. F. and Colacicco, D. (1982). Organic materials as fertilizers and soil conditioners. *Ind. Environ.* 5, 23–6. UNEP Office, Paris.

Phillips, J. A. and Humphrey, A. E. (1985). Microbial production of energy: I. Liquid fuels. In *Biotechnology and Bio process engineering, Proceedings of the VII International Biotechnology Symposium, New Delhi* Feb. 1984 (ed. T. K. Ghose) pp. 157–86.

Porter, W. M., Abott, L. K., and Robson, A. D. (1979). Effect of rate of application of superphosphate on populations of vesicular-arbuscular endophytes. *Aust. J. Exp. Agric. Anim. Husb.* 18, 573–7.

Powell, C. (1976). Mycorrhizal fungi stimulate clover growth in New-Zealand hill country soils. *Nature* 264, 436–8.

Proctor, D. (1973). The use of waste heat in a solar still. *Solar Energy* 14, 433–49.

Rabson, R. and Rogers, P. (1981). The role of fundamental biological research in developing future biomass technologies. *Biomass.* 1, 17–37.

Rains, D. W. (1979). Salt tolerance of plants: strategies of biological systems. In *The biosaline concept* (ed. A. Hollaender), pp. 47–67. Plenum Press, New York.

——, (1981). Salt tolerance—new development. In *Advances in food producing systems for arid and semi-arid lands*, Parts A (eds J. J. Manassah and E. J. Briskey), pp. 431–53. Academic Press, New York.

Raschke, K. (1975). Stomatal action. *Ann. Rev. Plant Physiol.* 26, 309–40.

Rawlins, S. L. (1981). Principles of salinity control in irrigated agri-culture. In *Advances in food producing systems for arid and semi-arid lands*, Part A, (eds J. J. Manassah and E. J. Briskey) pp. 391–418, Academic Press, New York.

Reed, G. and Peppler, H. J. (1973). *Yeast technology*, AVI Publishing Company, Westport, Connecticut.

Rhodes, L. H. and Gerdemann, J. W. (1978). Translocation of calcium and phosphate by external hyphae of vesicular-arbuscular mycorrhizae. *Soil Sci.* 216, 125–6.

Riad, A. (1982). Potential sources of organic matter in Egypt. In *FAO Soil Bulletin 45. Organic materials and soil productivity in the near East.* FAO, Rome 1982.

Ridge, E. H. and Rovira, A. D. (1968). Microbiol inoculation of wheat: *9th Int. Congr. Soil Sci. Trans.* Vol. II, pp. 473–81.

Rolz, C. (1982). Microbial biomass for renewables. *Adv. Biochem. Eng.* 21, 2–53.

Safir, G. R., Boyer, J. S., and Gerdemann, J. W. (1971). Mycorrhizal enhancement of water transport in soybean. *Science* 172, 581–3.

Sanders, F. E., Mosse, B., and Tinker, P. B. (eds). (1975). *Endomycorhizas.* Academic Press, New York.

Sauerbeck, D. and Fuhr, F. (1966). The interference of carbon–14 labelled carbon dioxide in studies on the uptake of organic substances by plant roots. In *The use of isotopes in soil organic matter studies*, Report of the FAO/IAEA Tech. Meeting, pp. 61–72. Pergamon, Oxford.

Saxena, G. K., Hammoud, L. C., and Lundy, H. W. (1967). Response of several vegetable crops to underground asphalt moisture barrier in Lakeland and fine sand. *Florida State Horticult. Soc. Proc.* 80, 1211–7.

Saxon, M. J., Breidenbach, R. W., and Lyons, J. M. (1980). Membrane dynamics: effect of environmental stress. In *Genetic engineering of osmoregulation* (eds D. W. Rains, R. C. Valentine and A. Hollaender), pp. 203–33. Plenum Press, New York.

Shaalan, N. S. (1980). Some studies of metabolites of nitrogen fixing blue-green algae. PhD thesis, Faculty of Agriculture, Ain Shams University, Egypt.

Smith, R. L., Schank, S. C., Milam, J. R., and Baltensperger, A. A. (1984).

Responses of *Sorghum* and *Pennisetum* species to the N_2 fixing bacterium *Azospirillum brasilense*. *Appl. Env. Microbiol.* 47, 1331–6.

Stavarek, S. J., Croughan, T. P., and Rains, D. W. (1980). Regeneratation of plants from long-term cultures of alfalfa cells. *Plant Sci. Lett.* 19 (3), 253–61.

Taha, S. M. (1969). Activity of phosphate dissolving bacteria in Egypt. *Plant Soil* 31, 149–60.

Theodorou, C. (1968). Inositol phosphates in needles of *Pinus radiata* D. Don and the phytase activity of mycorrhizal fungi. *9th Int. Congr. Soil Sci. Trans.* 3, 483–90.

Trappe, J. H. (1981). Mycorrhizae and productivity of arid and semi-arid rangelands. In *Advances in food producing systems for arid and semi-arid lands* (eds J. T. Manassah and E. J. Briskey) pp. 581–99. Academic Press, New York.

Trappe, J. M. and Fogel, R. D. (1977). Ecosystematic functions of mycorhizae. *Col. State Univ. Range Sci. Dept. Sci. Serv.* 26, 205–14.

Turner, N. C. (1979). Drought resistance and adaptation to water deficits in crop plants. In *Stress physiology in crop plants* (eds H. Mussell and R. C. Staples) pp. 343–72. John Wiley, New York.

Viets, F. G., Jr. (1966). Increasing water use efficiency by soil management. In *Plant environment and efficient water use*, pp. 270–5. American Society of Agronomy and Soil Science, Madison.

Wiken, T. O. (1972). Utilization of agricultural and industrial wastes by cultivation of yeasts. In *Fermentation technology today*, Proc. 4th int. Fermentation Symp. pp. 569–96.

Williams, S. E., Wollum, A.G., II, and Aldon, E. F. (1974). Growth of *Atriplex canescens* (Pursh) Nutt. improved by formation of vesicular-arbuscular mycorrhizae. *Soil Sci. Soc. Am. Proc.* 38, 962–4.

Yoshida, T. and Ancajas Rosabel, R. (1973). Nitrogen fixing activity in upland and flooded rice fields. *Soil Sci. Soc. Am. Proc.* 37, 42–6.

Zakhary, J. W., ElMahdy, A. R., Abo-Baker, T. M., and El Tabaei Shehata, A. M. (1984). Cultivation and Chemical composition of the paddy straw mushroom 'Volvariella volvaceae'. *Food Chem.* 12, 2–12.

3

Biotechnological practices in integrated rural development

C. W. Lewis

Introduction—the need for an integrated rural development

In their push for high economic growth rates, many of the world's developing countries have, over the past 20 or 30 years, opted for programmes of rapid industrialization and have largely neglected to invest adequately in their rural sectors. Most have lived to regret this course of action. In the words of Mao Tse-Tung, 'Take grain as the key link and assure all-round development . . . Industry must develop with agriculture'. Mao has since been proved correct in this assessment time after time, by country after country where these words have not been heeded.

An example of a comparatively successful, newly-industrializing nation is the Republic of (South) Korea which has achieved its success via a carefully balanced approach to growth. Originally, its development priority had been industry, ranging from iron and steel manufacture and shipbuilding to electronics and petrochemicals, but as world grain prices increased during the early 1970s more attention was paid to agriculture, so that self-sufficiency (if only temporarily) in grain was achieved by 1976. It had previously been falsely assumed that industrial growth would in turn provide an impetus to agricultural growth, but when this turned out not to be the case then prompt action was taken and the nation's improving prosperity continued unabated (Cole *et al.* 1980).

Other countries, however, have given scant attention to rural development and found themselves paying out huge food import bills as a result. Whereas food production per capita in South Korea was 26 per cent higher at the beginning of the 1980s than it had been a decade earlier, it was lower in most other developing countries, e.g. by 27 per cent in Mozambique, 26 per cent in Ghana, 19 per cent in Algeria and Morocco, 16 per cent in Peru, 15 per cent in Kenya, etc. (World Bank 1983). This is a disturbing trend exacerbated by high average population growth rates throughout the decade, at 2.5 per cent per annum outside China (1.5 per cent) and India (2.1 per cent), adverse climatic conditions and deteriorating soil fertility. It is clear that the low income economies (gross national product—GNP—per capita below US $ 410 in 1981) especially are unable to raise their food output to keep pace with population growth. It is also clear that inequalities in income and resource distribution between urban and rural sectors means that the centralized approach to development now almost universally in practice will not improve matters.

Hence, there is a need in these countries for a co-ordinated decentralized policy of integrated rural development to maximize the use of local resources, many of which are in the form of biomass of one type or another and therefore amenable to microbial technology utilization. Biotechnological practices at the village level could certainly help to satisfy some domestic and agricultural basic needs such as the provision of energy, food, animal feed, fertilizers, pesticides, clean water, and medicines; but these would still need to be augmented by better educational and health facilities, housing, transportation and communication networks, and the various organizational and institutional bodies to effect an integrated development which benefits all classes and strata of society. A more prosperous rural sector with improved material quality of life standards should also go some way towards stemming the mass migration to already overcrowded cities, which are growing at over 4 per cent each year and which cannot cater for the massive influx in terms of housing, employment, and so forth. Development in the villages will also aid the development of the cities, and hence of the nation as a whole.

Energy—the key constraint

Seventy-nine per cent of the people living in the world's low income economies are rural dwellers with generally more limited access to commercial (mainly fossil) fuels than have the 21 per cent urban residents. Yet agriculture is the mainstay of these economies, employing 70 per cent of their labour force and furnishing 37 per cent of their GDP. [The respective values for industrial market economies (the West) are only 6 and 3 per cent (World Bank 1983).] However, the average Third World farmer can no longer afford the amounts of diesel or electricity for irrigation pumping and the energy-intensive fertilizers and pesticides required to maximize crop production. Therefore, the localized use of available biomass energy resources through biotechnology seems the logical solution.

Around 43 per cent of energy consumption in the developing countries accrues from biomass: mostly wood, dung, and crop residues. On a heat-supplied basis, this works out at the equivalent of over eight million barrels of oil each day. However, traditional methods of bioenergy production and utilization are both inefficient and environmentally degrading. Thus, the application of modern biotechnology has great scope to increase both food and fuel production in Third World rural communities via the generation of methane gas and nitrogenous fertilizer from animal and plant residues; the cultivation of nitrogen-fixing trees, legumes, and algae; the production of bioinsecticides; the production of gas from wood and agricultural residues; and, potentially, the fermentation of low-cost substrates to power alcohol and single cell protein.

Table 3.1, derived from the World Bank's Development Report 1983, shows just how much the developing countries depend on their rural sectors, and more specifically on agriculture, for both national income and employment, as compared to the industrialized West. This is particularly true of the poorer, chiefly agrarian economies such as Bangladesh, China, Ghana, India, Pakistan, and Uganda, but even the newly industrializing nations like Brazil and Mexico, which are more urbanized than the others, still have around one-third of their labour force employed in agriculture.

It is clear from Table 3.1 that there is some correlation between degree of urbanization and *per capita* commercial energy consumption—though China with its extensive system of rural communes provides an obvious anomaly. 'Commercial' here refers to certain primary energy sources only: coal and lignite, petroleum, natural gas and natural gas liquids, and hydroelectric and nuclear power. Traditional fuels like firewood and charcoal, though themselves often qualifying as commercial too, are omitted, according to the World Bank 'because reliable and comprehensive data are not available.' Nevertheless, such statistics are now coming to light in the case of more and more developing countries, for example in Hall *et al.* (1982). Essentially, however, the 'official' commercial fuels plainly do not find their way in any significant quantities to the mass of the people, particularly those living in the rural areas, and this palpably reduces agricultural and quality of life potentials. Even in the petroleum-exporting nations of Nigeria, Indonesia, Egypt, and to a lesser extent Mexico, precious little of this oil percolates out of the refineries into the villages—and the same can be said of the oil revenues.

A plausible solution to this energy scarcity in the rural Third World lies in the decentralized, indigenous production of fuels from locally available resources, and the most prevalent and convenient resource is generally the chemical energy stored in biomass—that is, energy derived from the sun via the photosynthesis of plants. However, the energy transformation efficiency of photosynthesis is low, at only 5–6 per cent in theory, while in practice typical annual conversion efficiencies for subtropical and tropical plants are only in the range of 0.5–2.5 per cent (Hall 1980). Therefore, land area is a major constraint to large-scale bioenergy production schemes and, as Table 3.1 infers, those countries with a high population density, such as Bangladesh, have limited biomass energy potential compared to Brazil, Mexico, and Egypt, for instance, where population pressure is much less. However, the type of land available must also be considered, as must the requirements for growing food and cash crops, while the population growth rate is obviously ,crucial to what can be achieved on a future per capita basis. Although the overall population growth rate of the Third World is lessening, it is estimated

Table 3.1

Population, wealth, agriculture, and energy indicators of 15 selected developing countries for 1981 (from World Bank, 1983)

Country	Population (millions)	Population density (no per km²)	Projected annual population growth rate, 1980–2000 (%)	GNP per capita (US$)	% rural dwellers	% labour force in agriculture	% contribution of agriculture to GDP	Commercial energy* consumption per capita (GJ)
China	991.3	104	1.0	300	79	69	35	17
India	690.2	210	2.0	260	76	69	37	6
Indonesia	149.5	78	2.0	530	79	55	24	7
Brazil	120.5	14	2.1	2220	32	30	13	30
Bangladesh	90.7	630	2.9	140	88	74	54	1
Nigeria	87.6	95	3.5	870	79	54	23	5
Pakistan	84.5	105	3.0	350	71	57	30	6
Mexico	71. 2	36	2.6	2250	33	36	8	45
Philippines	49.6	165	2.3	790	63	46	23	10
Thailand	48.0	93	2.0	770	85	76	24	10
Turkey	45.5	58	2.1	1540	53	54	23	21
Egypt	43.3	43	2.1	650	56	57	21	16
South (Republic of) Korea	38.9	397	1.6	1700	44	34	17	42
Uganda	13.0	55	3.5	220	91	83	75	1
Ghana	11.8	49	3.9	400	63	53	60	7
Industrial market economies	719.5	23	0.7	11120	22	6	3	202

* 1980 figures

that 245 000 km^2 (approximately the size of Great Britain) of tropical forests are still lost each year, and it is no coincidence that in the country with the highest annual growth rate from 1970 to 80, the Ivory Coast at 5 per cent, there will be no fuel wood left at the end of the century at the present rate of deforestation (Sedjo and Clawson 1982). It therefore follows that efficiencies of biomass energy consumption must be accounted as high a priority as efficiencies of production if localized bioenergy operations are to be effective. The application of biotechnology, allied to improved agricultural practice and other appropriate energy technology utilization, can make real progress towards the attainment of these dual and complementary aims.

Energy supply is the key to rural development because with energy, water availability and crop productivity can be increased, machinery can be powered for cottage industries, good quality lighting made possible in the hours of darkness, and a multitude of tasks formerly difficult or impossible to perform become that much easier. Yet even where a high grade form of energy is, in theory, freely available as exemplified by the substantial electricity networks of developing countries like India, the process of reaching remote areas is often a slow and tortuous one. Today only 44 per cent of nearly 600 000 Indian villages receive electricity, and within the villages themselves only 14 per cent of the households can afford to pay for connection and supply. This means that only 16 per cent of India's electricity reaches the countryside, where 76 per cent of the people live (Reddy, 1981*a*). Moreover, overall transmission losses are 19.81 per cent of total generation, and anyway, electricity is a most uneconomic and wasteful method of supplying energy to villages, where 75–80 per cent of the inanimate energy demand is for medium-grade heat for cooking. So far as oil products such as kerosene (for lighting) and diesel (for irrigation pumps) are concerned, these are becoming less affordable in adequate quantities as prices rise and so 'home-grown' substitute fuels are needed with increasing urgency. Such fuels are best produced within what Reddy and Subramanian (1979) have termed 'rural energy centres' or integrated energy systems.

Integrated energy systems

What then is an integrated energy system (IES) and what should be its goals? An obvious function of such a system is to provide a community with the energy it needs, initially to meet current requirements and subsequently to enable that community to develop in both a technological and economic sense via the establishment of agro- or other industries. In other words, it should supply reliable and renewable heat and work energy sources such that the villagers are able to satisfy their basic domestic and agricultural needs, and additionally perform tasks currently out of their reach owing to today's energy constraints. The mix of fuels supplied should cater for, and be appropriately used for, the various end-use demands according to the grade of energy required. Thus, the combustion of wood or use of a simple solar collector should provide low temperature heat rather than burning biogas or locally-fermented ethanol. In thermodynamic terms, 1 GJ of heat from the combustion of ethanol can do much more work than 1 GJ from the combustion of wood, and so the ethanol should be used for a more appropriate, more demanding task, such as powering machinery for example. The matching of fuel sources with tasks should increase the overall efficiency of the energy system and so minimize waste.

With respect to the technologies used, these vary with the local climate, available resources, and other relevant factors including sociological considerations. Clearly, windmills, direct solar energy capturing devices, hydropower installations, and so on are all possibilities in addition to biomass-based technologies, though biogas plants particularly are often well-suited for incorporation within an IES.

A simple example of an IES might in fact be the production of biogas and an excellent nitrogen fertilizer sludge from the digestion of cow dung in a biogas plant. This plant could have a shallow water receptacle on top of its gas holder covered with a transparent glass or polythene material to reduce convection and radiation heat loss, thereby increasing the rate of biogas output and at the same time solar-heating the water contained in the receptacle. By converting the transparent cover into a tent-shape, the solar heater can serve the function of a solar still to yield distilled water which could have numerous applications, notably in health care (Reddy *et al.* 1979). The biogas could be used for cooking, heating, lighting, generating electricity, and powering a variety of mechanical devices, such as irrigation pumps and so on, while the by-product, N-rich sludge, might be used to grow high-protein algae for feeding to livestock or fish within a truly integrated system! In this way the system becomes not only an integrated energy system, but more of an integrated biosolar system, supplying much more than energy alone to the community.

Integrated biotechnology systems

Integrated biotechnology systems may be considered as extensions of integrated energy systems and, like them, their component parts will vary from location to location. Therefore, the assertion made by Reddy and Subramanian (1979) that 'Rural energy systems must be society-specific and culture-specific. There cannot be standardized designs and packages for universal application. Rural energy centres cannot be mass produced' holds true for integrated biotechnology systems also. At the village level too, simplicity and low cost are prime requirements for technological (including biotechnological) success. This is well summed up by Seshadri (1979) when he says 'Technology should aim only at modest targets. To do this best the technology must be local, adaptable, and evolutionary. These three qualities do not preclude sophistication of analysis or thought'.

A great advantage of integrated systems is that they invariably act synergistically and increase overall productivity over and above that which would be obtained from the sum of the component subsystems operating on their own. At the village level, the whole concept is based upon the

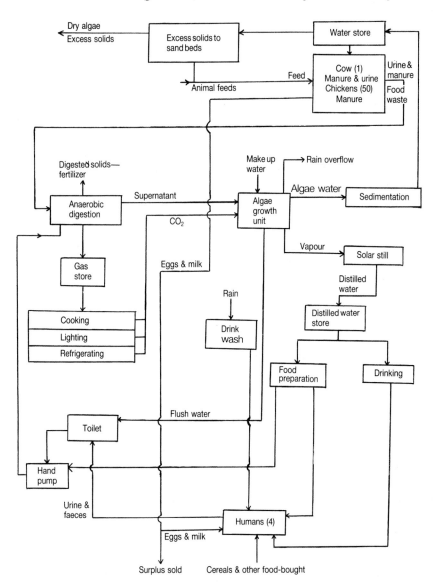

Fig. 3.1. Integrated biotechnology system for rural Third World family.

maximization of energy and materials production and the minimization of waste, with an emphasis on recycling. Figure 3.1 shows an integrated (mainly) biotechnology system aimed at virtual self-reliance in food and energy for a family living in the rural Third World. In the figure, a biogas generator (represented by 'anaerobic digestion') and algal growth system act as the fulcra around which the various constituent subsystems revolve. The algae, in symbiotic association with degradative bacteria, utilize waste products from various sources within the system and upgrade them to algal cellular matter which may be used as food, fuel, or fertilizer. Simultaneously, the incoming waste water is purified and can then be recycled. Water hyacinths, in addition to algae, and other under-utilized biomass resources can also be introduced. Further subsystems which may be added could include an aquaculture component for rearing algae-fed fish and a biofertilizer component for increasing the nitrogen budget of the soil by growing free-living nitrogen-fixing algae and bacterial/legume symbiotic associations. The permutations are almost endless, and some of the above biotechnological practices will be discussed subsequently in more detail, as will biotechnological practices at the community or village level, where economies of scale come into play and where they are an effective adjunct to integrated rural development in the true meaning of the term.

These innovative technologies cannot and should not, however, be divorced from the more conventional agronomic considerations where, for example, crop yields can be maximized through selecting genetically-improved strains, using optimum levels of fertilizer and pesticide application, and operating efficient irrigation schemes. Biotechnology and sound agricultural practice should complement each other, but the shrewd adoption of biotechnologies appropriate to a given situation should also offer the prospect of increasing output, and hence prosperity, over and above that from good agricultural practice alone.

A good example here is afforded by paddy rice cultivation in India where, in 1975–76, the national fertilizer industry was able to furnish at most 43 kg/ha of nitrogen fertilizer. Since the newer hybrid rice varieties can gainfully utilize much higher nitrogen levels than this, their full potential was not being exploited, and actual land productivity fell well short of the attainable. However, organic nitrogen fertilizer obtained through the anaerobic digestion of cattle dung in biogas plants throughout India could be more than double the nitrogen available from the chemical fertilizer industry. Therefore, through what might be termed biotechnological intervention, that is the strategic installation of biogas plants, there is the potential to fertilize the rice fields at an average level of well over 100 kg N/ha if desired, thereby realizing the latent benefits of the high yielding varieties (Seshadri 1977).

Biotechnological progress in the Third World

Since individual microbial technologies are discussed in some detail in other chapters of this book only brief descriptive passages will be included here. Applications are confined to the rural environment and only where they benefit the local population, so that the Brazilian National Alcohol Programme and other such large-scale enterprises (for example in Zimbabwe), which are not specifically concerned with integrated rural development, are excluded.

Biotechnological progress in the developing countries has been much slower than in developed economies because of lack of expertise in all but a few Third World institutions. Yet the value of such technology actually promises much more by way of increased food and fuel productivity clean water, medicines, and other basic commodities which are largely taken for granted in the West. This situation was fully recognized by the United Nations Educational, Scientific, and Cultural Organization (UNESCO) which, in 1975 along with the United Nations Environment Programme (UNEP), convened a meeting of many of the world's foremost microbiologists. Their mandate was to formulate a plan of action aimed at establishing a world-wide network of Microbiological Resources Centres (MIRCENs), and thereby to preserve microbial gene pools and make them accessible to the developing countries. The ever-widening gap between the microbiological capability of the industrialized and developing worlds had for some time been a cause of great concern to UNESCO and thus the MIRCEN network concept was born.

The declared aims of MIRCEN are as follows:

1. Provide the infrastructure for a network to facilitate the management, utilization, and distribution of the microbial gene pool among collaborating organizations on a worldwide scale, and at the same time promote regional and inter-regional co-operation on specific research projects.

2. Assist in the conservation of micro-organisms, especially *Rhizobium* gene pools, in the agriculturally-based economies of the Third World so as to increase the nitrogen budget of the soil for enhanced crop productivity.

3. Encourage the development of novel inexpensive technologies native to particular localities.

4. Realize the potential of applied microbiology for improving rural economies.

5. Provide focal centres for manpower training and the widespread dissemination of microbial technology knowhow.

Every facet of how microbiological knowledge can be applied to the needs of developing countries is theoretically covered here, although emphasis is particularly placed on realizing the potential gains of biological nitrogen fixation through optimizing the *Rhizobium*-legume symbiotic system. This is a feature of several MIRCENs, and indeed that established at the University of Hawaii is commonly referred to as the Niftal MIRCEN, with the acronym denoting Nitrogen Fixation by Tropical Agricultural Legumes. Six other MIRCENs have been set up in developing countries and are located at Bangkok, Cairo, Guatemala City, Nairobi, Porto Alegre (Brazil), and Dakar. The remainder are found within existing institutions in the industrialized world at Brisbane, Stockholm, Hawaii, Beltsville (Maryland), and Aston in Birmingham, England. These 11 MIRCENs can be subdivided into their main areas of activity as follows:

  (i) World Data Centre—Brisbane (now relocated at Tokyo);
  (ii) *Rhizobium*—Hawaii, Nairobi, Porto Alegre, Beltsville, Dakar;
  (iii) fermentation, food, and waste recycling—Bangkok;
  (iv) biotechnology—Cairo, Guatemala City, Stockholm;
  (v) biodeterioration—Aston.

It can be seen from the geographical distribution of the MIRCENs that UNESCO has indeed succeeded in its objective of establishing a global network with the aim of realizing the potential of applied microbiology for improving rural economies being particularly relevant to this chapter (Bull and Da Silva 1983; Lewis 1983). This is not of course to suggest that before UNESCO's initiative in setting up the MIRCENs network, biotechnological practices were unheard of in the developing countries. For example, India had developed a programme of research and development into biogas production from livestock manure as far back as the 1930s, and this pioneering work eventually gave rise to the family-sized Gobar gas plants introduced by the Khadi and Village Industries Commission (KVIC) in 1954 (Barnett *et al.* 1978) and later to the Gobar Gas Research Station in 1961 (National Academy of Sciences, 1977). The impetus for the production of biogas plants came from concern over the loss of fertilizer inherent in the direct burning of dung as an energy source. This problem is overcome in a biogas plant where the post-digested sludge retains the original nitrogen, phosphorus, and potassium content of the dung. In a family-sized Indian plant delivering around 30 m^3 of gas from 50 kg of manure per day, the residual sludge would contain the equivalent of 100 kg of urea, 250 kg of superphosphate and 50 kg of potassium phosphate over a period of one year (Da Silva *et al.* 1978).

Over 75 000 family-sized units existed in India by 1979 and the Government hoped (rather optimistically) to establish a further half-million by 1983 along with a few larger community digesters to cater for the energy needs of whole villages (Agarwal 1979). There are many times more than this number in mainland China where approximately one million units were installed each year over a 7-year period up to 1978, a large proportion of which were built for Sichuan Province. Taiwan, too, has been active in this area and from 1955 to 1978 had constructed about 7500 plants adjoining pig farms. The Republic of Korea now has nearly 30 000 plants, having started a programme in 1969, while lesser numbers are to be found in countries as diverse as Bangladesh,

Indonesia, Iran, Israel, Malaysia, Nepal, Pakistan, Singapore, Sri Lanka, and Thailand in Asia, Cameroon, Ethiopia, Kenya, Lesotho, Rwanda, Senegal, Tanzania, Uganda, Upper Volta, Zaire, and Zambia on the African continent, and on islands including Cook, Fiji, Papua-New Guinea, and the Philippines. Research, and development and demonstrations are also taking place in Latin America, especially in the Caribbean states of Barbados, Guyana, Jamaica, and Trinidad (Da Silva *et al.* 1978: Da Silva and Doelle 1980).

Community biogas plants would seem to be the answer for those families who own no or few cattle, but they pose their own problems as witnessed by cost–benefit analyses of a system operating at Fatehsingh-ka-Purwa, India. Two quite large prototype plants of daily biogas output 35 and 45 m^3 have been constructed in this small village of only 177 inhabitants. All cooking and lighting requirements are now completely met by gas, and there is sufficient extra power available for pumping water, street lighting, and operating some farm machinery (Hayes and Drucker 1980). (One kWh of mechanical power can almost be supplied by the daily dung output of a single cow.) However, the conclusions of the economic analyses indicate that the overall costs outweigh the benefits of increased fertilizer availability and the release of traditional fuels for alternative uses. The whole argument is put into perspective by Reddy (1981*b*; see also Hayes and Drucker, 1980) however, who stresses that one cannot generalize such analyses, and analyses have been made on other systems which contradict the findings at Fatehsingh-ka-Purwa.

For a village of 500 people, plants totalling 200 m^3/day in gas output would be necessary to meet domestic energy demands. A plant of capacity 140 m^3/day has been constructed for 123 households in Kubadthal village, Gujarat, but with limited success. In Sri Lanka at Pattiyapola a large biogas unit exists within a scheme also including wind and solar energy converters for electricity generation. The resulting total of about 60 000 kWh/year is more than enough to light 85 homes and pump drinking water into many more (Anon. 1984).

Energy tree plantations containing fast-growing species, preferably with the ability to fix atmospheric nitrogen, can both prevent deforestation and reduce the drudgery of fuel wood collection. *Leucaena leucocephala*, *Sesbania grandiflora*, and *Casuarina equisetifolia* are just three species which have proved successful in the tropics for fuel wood, fodder production, and soil fertilization (National Academy of Sciences, 1980). It has been estimated that the latter can supply 700 GJ of wood energy and 80 kg of fixed nitrogen per hectare year (Seshadri *et al.* 1978). Should the wood be used for gas production by heating to just over 1000°C in the presence of limited quantities of air, then the resulting producer gas (containing around 25 per cent CO, 10 per cent H_2, with the remainder being mostly N_2, and lesser amounts of other gases) can be used for supplying mechanical power within an internal combustion engine. The Philippines and Brazil are particularly interested in such an approach to energy supply since they are quite well-endowed with forests, though this is not typical of most of the Third World, where countries are more concerned with reafforestation programmes, China, India, Nepal, and Thailand just being four examples. Additionally, the energy content of producer gas is rarely more than 7 MJ/m^3 compared to the 20–25 MJ/m^3 of biogas, although it is claimed that the overall efficiency of energy conversion from biomass feed to mechanical power is twice as great using the producer gas rather than the biogas route (Datta and Dutt 1981).

The cultivation of algae for protein supplements to livestock and fish feed and for nitrogenous biofertilizers is a valuable biotechnology in reducing the indirect energy-intensive (and hence financially expensive) imports of high-grade animal feeds and chemical fertilizers. The blue-green *Spirulina* is easily grown and can contain up to 70 per cent protein on a dry weight basis, while N-fixing algal packages containing mixed cultures of free-living blue-greens such as *Nostoc*, *Anabaena*, *Plectonema*, and others, are cheap and can substitute for 30 per cent of chemical nitrogen fertilizers in rice paddy fields with no decrease in yields. Over 15 000 km^2 of cultivated land in India and Burma have been successfully inoculated with such cultures (Indian Agricultural Research Institute, 1978), while the Chinese and Vietnamese have been utilizing the N-fixing properties of the algal symbiont *Anabaena azolla* in association with *Azolla* water fern on their rice fields.

Well over 70 per cent of the nitrogen supply to plants accrues from nitrogen-fixing micro-organisms. These are mainly in symbiotic systems with leguminous plants as hosts, but a substantial contribution is also made from free-living bacteria and blue-green algae. It is only relatively recently that research at the International Rice Research Institute in the Philippines has begun to show the real extent of the natural indicence of nitrogen-fixing organisms, and where values of up to 80 kg N/ha/year have been measured in rice paddies (Venkataraman 1977). This is on a par with the amounts fixed within the root nodules of well-known legumes such as the soybean and alfalfa by rhizobial bacteria. However, *Rhizobium* species can effect a more rapid transfer of nitrogen because of their intimate association with the host plant from which sugars, the products of plant photosynthesis, are acquired. There is a precise relationship existing between legumes such as the cowpea, ground nut, and soybean and their bacterial symbiotic partners so that higher levels of nitrogen are fixed when a specific *Rhizobium* strain infests a plant than if another strain does so. Hence, the recently developed technology of seed inoculation of legumes with rhizobia has much potential in developing countries' agricultural systems, although so far this has been largely a neglected technique in the tropics (Cherry 1981). However, the *Rhizobium* MIRCENs mentioned earlier have already begun to make an impact here in a very short space of time.

Indigenous bioinsecticide production through growth of the neem tree, *Azadirachta indica*, the seed kernel and leaf extracts of which possess prodigious pest control properties, can substitute for energy-intensive pesticide imports. Though the neem is a native of Burma, it is also widely planted in India and West Africa, and is increasingly being used by the small farmer in many locations.

When some of the above measures have been enacted, a favourably-placed rural community should be in a position to expand its biotechnological expertise into the more complex areas of fermentation, perhaps to single cell protein and power alcohol. The latter, of course, would be a great boon towards the further mechanization of local agriculture and considerably aid the problems of heavy expenditure on petrol for transport. Already, villagers throughout the Third World are familiar with the fermentation of carbohydrate substrates to ethanol and the subsequent distillation to give rise to a whole range of portable spirits; hence village ethanol plants might not be so distant a possibility as some people would have us believe.

Biotechnology for rural industries

Biotechnological practices can, and must, extend beyond their roles as food and energy suppliers at the village level and be employed as instigators of rural, mainly agro-, industries. Certainly those small-scale industries which can be developed should be geared to cater for a real demand, and be flexible and adaptable enough to both compete with products from the large-scale sector of the economy and to modify output where necessary. Such industries furnish employment and hence income for the small landowners, as well as for the landless for whom casual agricultural labouring is generally the only money earner. Those countries which can successfully develop their rural industry offer themselves the chance to create a new and more satisfying pattern of life (Harrison 1979), and biotechnology can make a significant contribution in this extension of the integrated rural development concept.

Agro-industries based around the production of algal biofertilizers, high-protein algal supplemented feeds or foods, and neem leaf-extract bioinsecticides (all described above) for export are all possibilities. *Spirulina* algae have long been a traditional source of food in Chad, while various seaweeds are eaten as food supplements in many parts of the world, as are fermented foods like miso, natto, and tempeh, particularly in south-east Asia. In Asia, too, mushroom cultivation on various composted substrates such as rice straw, manure, vegetable residues, and organic soil components can give rise to an edible fungal product containing 16–25 per cent protein (Alicbusan 1979). *Volvariella volvaceae* and *Lentinus elodes* are the favoured species. Such cultivating units can be operated at the family and village levels, as can the intensive biodynamic gardening practices carried out in the area south of Madras by the Murugappa Chettiar Research Centre (MCRC) for the production of marketable vegetables and fruits such as tomatoes, chillies, radishes, sweet potatoes, water melons, cowpeas, beans, and so on (Seshagiri and Chitra 1983). It might be argued that these practices are agricultural rather than biotechnological or industrial, but they do generally employ some aspects of microbial technology in their operation, and while mushroom cultivation in compost is an agro-industry, biodynamic vegetable growing is clearly not mainstream agriculture as practised in the tropics.

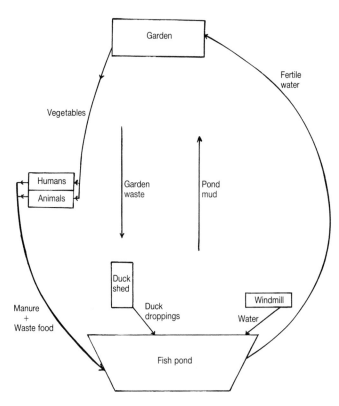

Fig. 3.2. Integrated duck–fish–garden system (after MCRC 1982).

The MCRC, Madras, has taken the biodynamic garden concept further by integrating it with duck rearing and fish farming (Raja 1982). It is thus analogous to the integration of paddy cultivation with duck and fish ponds practised in many countries of South-East Asia. The MCRC system, depicted in Figure 3.2, is based on the fertilization of the fish pond by duck droppings to increase the nitrogen and phosphorus content of the water. Algae, and other pond flora and fauna then grow profusely, and form the feed for the fish. The nutrient-rich water is passed to a vegetable garden, as is the pond mud which also makes a good fertilizer. In turn, waste plant matter from the garden is made available to the fish pond. Such a scheme could be labelled as an integrated agro-industrial system producing a rich variety of (mainly) food or feed products for home consumption, and the surplus can additionally be exported.

Another feasible agro-industry is afforded by the production of livestock feeds from crop by-products once the local people can satisfy the nutritional needs of their own livestock population. In 1978, for instance, on an all-India basis, there was estimated to be a shortfall of 44 per cent in feed concentrates, 44 per cent in dry fodders, and 38 per cent in green fodder. While efforts are being made to make good these shortages, for many years to come crop residues and by-products will contribute the principal components of ruminant animal diets in the subcontinent. For every tonne of rice paddy grown, for example, there is approximately 2.35 tonnes of accompanying straw. Adult males often each receive around 1 kg of straw per day when available, but it is of poor nutritive value with a low nitrogen and a high lignin content. Straw can nevertheless be made more palatable by soaking for 1–2 hours in water (1 kg straw/1 litre water) before consumption, and feeding trials on zebu and buffalo calves have shown that, while the digestibilities of crude protein and fibre were depressed, the digestibility of the nitrogen-free extract increased significantly on soaking. The availability of digestible and metabolizable energy and the production rates of total volatile fatty acids were all higher after the soaking treatment, which also removed soluble oxalate from the rice straw. Such a simple feed processing treatment could easily be within the capabilities of Third World rural communities, which might then go on to more involved treatments including alkali spraying (3.3 per cent sodium hydroxide) for delignification and subsequent impregnation with urea and molasses, the former to increase the feed's nitrogen content and the molasses to provide soluble carboyhdrates for effective urea utilization. Rice husks and rice bran, too, might also be incorporated into the feeds, although the husks' use is limited owing to its silicon content of 14.5–17.5 per cent which has to be reduced by treatment with alkali and steam under pressure.

The efficacy of an agro-industry based on nutritionally-improved rice straw is given credence by work in the United Kingdom carried out by Unilever Ltd, who have developed an alkali treatment of straw to produce nutritionally improved straw aimed as a conserver of forage. The straw's energy value can be increased by 59 per cent on a metabolizable energy basis; the product is easily stored and transported and is efficiently utilized by a wide range of farm livestock. It is also competitively priced.

Sugar-cane, by-products such as molasses, bagasse and sugar-cane tops, tree fodder, grasses, neem seed cake, water weeds, seaweeds and other crop residues may also be considered as livestock feed constituents, often after processing, depending on the particular locality. However, the effective utilization of crop by-products at the small farm level is really still in its infancy throughout the world and there would seem to be much scope here provided sufficient quantities are locally available (Ranjhan 1978). Again, while not strictly within the realms of biotechnology, such a technology would fit in well within an integrated biotechnological and agro-industrial system for rural development. Extraction of medicinal plant products might also fit into this category.

The major debilitating diseases of developing countries are the infectious, parasitic kinds already eradicated in the West. Over the years, traditional medicines have been developed and used with apparent success to combat these diseases. In India and other countries local herbalists and their plant-derived medications have long been valued as contributors to the primary health care of the people. Many of these traditional treatments are now coming under more scientific scrutiny and have been found to indeed have beneficial effects. Examples are extracts of the Indian rauwolfia, which reduces blood pressure, and the flesh of the pawpaw or medicine fruit, which seems to heal recalcitrant wounds (Harrison 1980). A concentrated effort at growing reasonably large quantities of plants such as these could be another avenue to further development in the rural environment—the growth, literally, of a pharmaceutical industry. There is in fact a whole range of plants, some exploited more than others, able to produce valuable products such as oil, cork, industrial fibres, wax, gum, rubber, and so forth, which could form the basis of a local industry within an integrated rural development strategy; and while the microbiological connotations of such industries may be small or even non-existent, they do come within the broader definition of biotechnology as being 'the application of biological organisms, systems, or processes to manufacturing and service industries,' (Smith 1981).

With respect to small-scale fermentation processes as the basis of a rural agro-industry, experience of both biogas and soil conditioner production through the anaerobic digestion of livestock manure, plus the fermentation and distillation of potable alcoholic liquor (arak in India, rakshi in Nepal, grog in Latin America) has certainly been acquired in numerous villages throughout the developing world. The fermentation of crop residues to produce single cell protein, industrial ethanol and/or power alcohol, and value-added chemicals would seem to be a natural extension of these activities. In selecting strategies for deciding whether one or more of these processes is appropriate for a particular locality every raw material locally available must be detailed in an inventory and undergo a cost–benefit analysis involving each conversion process for which it is a candidate. All possible products, their costs and demand, should be established, a task requiring research and hard data, allied to a knowledge of local conditions. Indeed,a review should be made of the socio-economic status of the whole village and finally, of course, it is imperative that local personnel receive sufficient training to both run the process selected and, as far as possible, be able to solve the problems which will inevitably arise (Rolz et al. 1979).

Xinbu and Kubadthal—two contrasting case studies

The biogas plant is the one application of biotechnology which is at the centre of most integrated rural development systems, particularly in Asia where the largest such establishments is located in the Philippines at Maya Farms south of Manila. With 15 000 pigs supplying the input feed manure to 48 batch plants, the gas generated is sufficient to supply a canteen, meat processing plant, soup cannery, and worker dormitories as well as running diesel pumps and electricity generators to provide water and power for the farm itself (DaSilva 1981). However, not all biotechnological practices in integrated rural development are so effective, though others certainly are. The experience of Kubadthal village in Gujarat State, India comes into the former category, while that of Xinbu village in Guangdong Province, China belongs in the latter.

Xinbu, mostly through biotechnological practices, claims to be the world's first truly integrated energy village. In fact, though, biotechnology at Xinbu has achieved much more than the satisfying of energy demand alone, in that fertilizers, food, and a host of agricultural products are also produced within a quite complex system as shown in Fig. 3.3 (United Nations University 1982). The system is based around a complex of seven community anaerobic digesters of total gas capacity 200 m^3, plus 86 individual 6 m^3 family plants, with the input feed consisting mainly of night soil, pig waste, food waste, napier grass (planted as an energy crop), and banana waste. Gas from the community digester powers a 12 kW generator at the rate of 1.5 kWh of electricity per m^3 of biogas input to yield 55 kWh each day for electric lighting and operating a pulverizer used in processing food. Waste heat from the electricity generation equipment is utilized in silkworm cocoon drying while live silkworms are kept warm by biogas heating during cold weather. The biogas from the family units is used exclusively for domestic cooking and lighting.

A waste recycling scheme, illustrated by Fig. 3.3, operates whereby mulberry leaves are fed to silkworms whose droppings in turn are passed on to the digester (along with banana waste) and to fish ponds. Each year the village produces 135 tonnes of sugar cane, the leaves of which are also fed to the fish (carp) ponds, while the pond sludge, plus that from the biogas plants act as fertilizer for bananas, mulberries, sugar cane, and napier grass. Water hyacinths and napier grass are fed to the pigs. The result of all this organized activity is that, whereas previously around 50 per cent of the villagers' energy and fertilizer needs were imported at ever-increasing prices, today the people are on their way to becoming self-sufficient in both. Furthermore, with the community biogas plant complex paying for itself (including the wages of the main operator and a technician) within 4 years, and the family plants doing so in under 18 months, the whole project is financially sound (Madeley 1982).

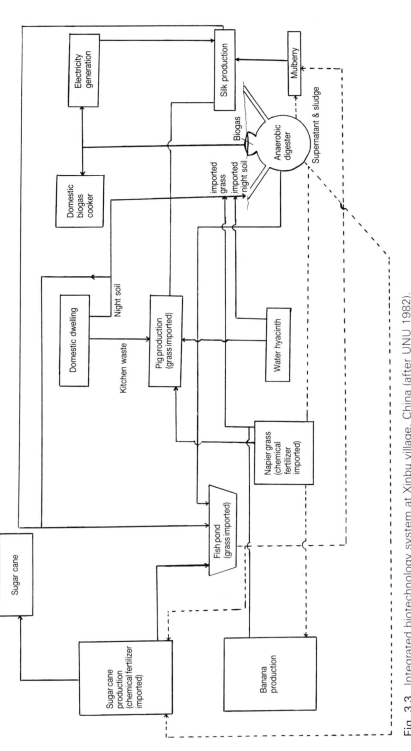

Fig. 3.3. Integrated biotechnology system at Xinbu village, China (after UNU 1982). ----- materials used as plant fertilizer; —— other materials.

Of course, the degree of organization, co-operation, and participation required from the people themselves for the successful implementation of a scheme such as this is vital. The fact that 86 out of a total of 90 households in Xinbu village installed their own biogas plants as part of the system testifies to the willingness of the people to participate. Of course this kind of concept fits in well with the whole philosophy behind the Chinese commune system. With a little help and encouragement from the Guangzhou Institute of Energy Conversion, Chinese Academy of Science, and the United Nations University, the inhabitants of Xinbu transformed their whole village economy from the presence of poverty to the prospect of prosperity—all within a period of 5 years and all through a programme of biotechnological practices within an integrated rural development.

To achieve the same level of co-operation among people living within a hierarchical social structure, such as exemplified by the caste system of India, is a much more difficult proposition. This proved to be the case (though technological and economic difficulties also contributed) in the operation of a community biogas plant installed in the village of Kubadthal, Gujarat State, some 35 km from the city of Ahmedabad (Moulik 1982). The plant was donated in good faith by a voluntary agency, the Vimla Gram Seva Samaj (VGSS) Trust as part of their rural development programme, with the objectives of providing better fuel and richer manure, preventing deforestation and environmental pollution, and improving the general health and economic condition of the villagers. As 175 (33 per cent) of the village's 534 households were of so-called backward castes, with the majority caste being predominantly Hindu, the VGSS Trust hoped for sufficient participation from all classes and a commitment to benefit the poorer people. In May 1980 the villagers met with VGSS Trust officials and 123 households (23 per cent) expressed a desire to participate in the scheme. Construction began in the following month of a 140 m^3 per day gas output plant with a daily dung requirement of 3.46 tonnes, easily met from the cattle owned by the 123 participating households. However, delays in completing the plant construction, finally achieved in December 1980, substantially raised the capital cost by 88 per cent. This invalidated previous financial agreements between the Trust and the villagers who were now faced with a 22 per cent increase in gas prices and a 150 per cent increase in the price of digester manure as fertilizer, plus higher costs for gas connection. These new financial burdens were largely opposed by the villagers and the Trust was forced to back down. At this time, the main would-be consumers agreed to supply gas to the poorer landless labourers at only 50 per cent of the current rate.

Then came a real bombshell. The Trust informed the people that their investment in the project was actually to be a loan which would have to be repaid in instalments at an 11 per cent interest rate. By this time, 83 participating families had paid their initial deposits, almost all of which were promptly withdrawn and the plant, already commissioned with cow dung feed, remained closed for almost 6 months until June 1981 after 51 of the families only had paid back their deposits, none of whom being the poorer landless labourers the project was supposedly to benefit the most. In the intervening period, the villagers of Kubadthal had tried unsuccessfully to raise the price to be paid for their cow dung and the Trust had retaliated by attempting to buy dung from surrounding villages with the idea of supplying gas to the poor landless labourers virtually free of charge. However, the landed Hindu caste households, not wishing to share the technology or the gas with the backward castes, sought to prevent the poorest strata of society from obtaining gas connections. Eventually, the VGSS Trust had to accede to a 150 per cent increase in dung prices in order to avert the impending impasse, though clearly the friction between the village castes, and between the Trust and the villagers as a whole would not be resolved easily. Nevertheless, with the plant operating to the benefit of 51 households it was hoped that others would soon become more interested in joining the scheme.

The Trust did, however, eventually have to raise the price paid by the villagers for the post-digested manure, and again the villagers were reluctant to acquiesce. Another problem lay in a shortage of water for the plant's operation, only overcome by purchasing the required amount from the well of a private landlord and pumping it to the plant by a diesel engine which resulted in the estimated annual surplus revenue generation of the plant being cut by 44 per cent. In turn this meant that the plant has been unable to generate enough surplus to repay the loan instalment plus interest, thus creating a major economic constraint on the successful operation of the entire project. The Trust also failed to involve the village community in the management and operation of the plant, partly because of the difficulties posed by class and caste differences, and partly because no educational and extension services were provided. It is, however, hoped that active community participation will gradually be achieved in the future.

Mistakes were also made in the design and operation of the digester in that it was a novel one, and so had not been adequately tried and tested before installation. The resulting intermittent and inadequate gas supply did nothing to gain the villagers' confidence in such 'new' technology. Thus, the VGSS Trust made the two cardinal sins of installing a virtually experimental design of plant, which was only a partial success and therefore limited in its appeal to the villagers, and, by not providing relevant educational instruction on the maintenance and operation of the plant, they failed to fully integrate the technology into the village community.

Nevertheless, the community biogas plant at Kubadthal village has made some positive impact on both the villagers and the village. Biogas users have saved appreciable amounts of money on their fuel bills. The village has become cleaner, with cow dung assuming the status of a priced commodity, and the collection and sale of dung has helped to generate income for the poorest people, while some employment opportunities associated with the biogas plant became apparent (though more would have arisen if the villagers had wholly been made responsible for plant operation and maintenance). Biogas made a particularly favourable impression on the women (and this has occurred in other villages too), who preferred this clean-combusting fuel to traditionally-used kerosene, dung cakes, and agricultural wastes, and they themselves generated more enthusiasm for the biogas plant throughout the general village community. The caste structure within the village was clearly brought out into the open so that efforts generally had to be made to try to minimize the negative impact this had on the organizational and participating operations of the project. These social factors are more difficult to overcome than are the technological and economic problems, particularly where technologies have to be integrated into the whole village framework, such as with a community biogas plant. It is therefore of paramount importance that the social issues are known, and resolved where practicable as quickly as possible and not shelved. Finally, though no two villages are exactly alike, the experience gained in the Kubadthal village community biogas plant project should be invaluable for other such future initiatives throughout India. Let us hope that some of the lessons learnt at Kubadthal will be well-assimilated, so that similar mistakes need not be made elsewhere. Technological, economic, and social viability are all part of successful biotechnological practices and they must all be taken into account since the failure of one means the failure of the whole. It is as simple, or as difficult, as that!

Pura, Ungra, and Injambakkam—three biotechnological success stories

Ungra and the nearby village of Pura, Karnataka State, are separated by less than 400 km from the Tamil Nadu village of Injambakkam in southern India. These three villages have probably received more exposure to biotechnological practices than any others in the developing world, with the possible exception of Xinbu described earlier. They have been the focus of much attention in the scientific and popular literature, discussed extensively at scientific conferences throughout the world, and the Karnataka villages were the subject of a well-produced and well-received television documentary film entitled 'West of Bangalore'. This film has been acclaimed throughout the world by people interested in Third World rural development, renewable energy technologies, and biotechnological practices in action. The reason for all this attention and excitement lies at the door of two successful and eminent Indian scientists who decided that the time had come to channel some of their time, energy, enthusiasm, and expertise into solving the problems of India's rural poor—the most difficult challenge by far which the subcontinent has to face.

Ungra, initially was chosen as the site for a rural extension centre by Dr Amulya Reddy of the Indian Institute of Science, Bangalore, the convener of ASTRA, the Institute's Cell for the Application of Science and Technology to Rural Areas formed in 1974. ASTRA's overriding priority is to promote the self-reliance of Indian village communities, thereby reducing the inequalities between both different sectors of rural society, and between rural and urban areas. From 1976 to 1980, in-depth surveys were made of the energy and fuel production and consumption patterns of six villages in the area of the extension centre, including Pura and Ungra villages (Reddy and Subramanian 1979; Ravindranath *et al.* 1981; Reddy 1981*b*). Particular attention was paid to biomass production, and its utilization by humans and livestock as part of a study of the whole village ecosystem in the case of Ungra. Data were obtained on land use and cropping patterns, plant biomass productivity, the disaggregation of plant biomass into various components, the utilization of these components, the food consumption by human beings and livestock, the materials and energy flows through the ecosystem, and its imports and exports. Reddy considers that an understanding of the logic of village agricultural ecosystems should be the basis of rural development (Ravindranath *et al.* 1981). The complex linkages between humans, livestock, land, energy, and water and between food, fuel, and fodder are not random within the village system, and so must be appreciated and comprehended as a precondition for intervention into village ecosystems. Otherwise, the results of interventions, almost certainly including the operation of biotechnological practices in agricultural communities, may not have the desired effect.

Reddy and his team have also designed a rural energy centre or integrated energy system (as described earlier) for Pura village based around a community biogas plant delivering 37 m^3/day of biogas. This is sufficient for all village cooking with an 11 m^3/day surplus for water pumping, ball milling rice husk, ash, and lime in cement production, and electric lighting. Thus, the drudgery of firewood collection is eliminated and that of obtaining water alleviated. Additionally, 3 kg/day of nitrogen is obtained from the sludge as fertilizer, and all at around 75 per cent of the capital cost of supplying electricity which, in the case of Pura and most other villages, would be a less versatile and useful energy vector than biogas. In fact, ASTRA firmly believes that community biogas plants are the top priority for villages like Pura, though it does conceivably take time to engender sufficient community spirit for such a technology to be an unqualified success in India, as was shown in the case of Kubadthal.

A methodology has also been developed and adopted for selecting the various technological options for meeting the village energy requirements for all kinds of end-uses involving high temperature heating, stationary power, and motive power. An integrated system is then built up involving a mix of energy sources to match each particular grade of energy required in accordance with maximizing, as far as practical, thermodynamic efficiencies.

In other words, where available, the energy path providing the highest efficiency is the one chosen out of those on offer. Figure 3.4, based on biogas, illustrates Phase I of the rural energy centre being designed for Pura. Phases II and III are now receiving attention. An energy forest is being considered for Phase II, specifically for providing a fuel suitable for low temperature heating (for example, of water for bathing) along with windmills for further water lifting. Phase III will be concerned specifically with motive power requirements, again being met through a biological fuel-driven device, with the contenders being biogas, producer gas, and/or ethanol engines. For stationary power, medium temperature heating, and lighting, biogas (in the case of lighting, biogas is used to generate electricity) has already been chosen in Phase I. Biogas or possibly charcoal are envisaged as the fuels most appropriate for high temperature heating (Reddy and Subramanian 1979).

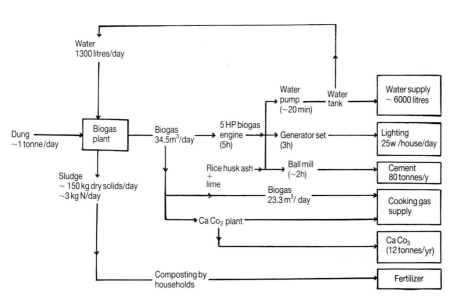

Fig. 3.4. The community biogas plant's contribution to Pura's energy supply (after Reddy and Subramanian 1979).

Ungra, Pura, and neighbouring villages, thanks to the mainly biotechnological practices of Dr Reddy and his colleagues at ASTRA, now seem set fair for a more prosperous future with energy production, distribution, and utilization in particular aimed not only at the satisfaction of the basic needs of the poor, but also at increasing output (and hence wealth) and self-reliance within an environmentally sustainable strategy. Likewise, a similar strategy of integrated rural development has been devised by Dr C. V. Seshadri and his team at the Shri A. M. M. Murugappa Chettiar Research Centre (MCRC), Madras, for Injambakkam village whose population unusually includes over 50 per cent of the Harijan (Untouchable) caste. The MCRC was registered as a non-profitmaking research institute in 1973 funded by companies associated with the late Shri A. M. M. Murugappa Chettiar. Dr Seshadri's Photosynthesis and Energy Division of MCRC was subsequently spawned in 1977 with the goal of applying science and technology to the solving of the problems of the rural poor in Tamil Nadu in particular and India, in general.

Akin to ASTRA's philosophy the MCRC aims 'To enable people to create and acquire scientific tools of thought and action to answer basic human needs and to go beyond and create surpluses to benefit others; to do so collectively, using elegant, ecological and low-cost means which are local, specific and add value to the local economy; to do so in a manner that impacts minimally on society, culture, and lifestyles detrimentally; in a manner that enhances people's ability and sustains their self-respect, confidence and their ability to be creative masters of their own destiny' (MCRC 1982). Thus, self-reliance and quality of life improvement figure highly on MCRC's agenda for development.

Because Injambakkam and the eight other villages in the district in which MCRC has direct influence are basically agricultural, most, though not all, of the installed technologies have had some biotechnological connotations. Windmills for water pumping and solar driers for fish, vegetables, and grain are the two main exceptions. Otherwise, MCRC successes include a low-cost biogas plant, small lot forestry for fuel wood and fodder, aquaculture (pisiculture) in village ponds, the utilization of water hyacinths for biogas production, animal feed, fertilizer, and leaf protein, use of neem leaf extract as a pesticide, the mass cultivation of the high protein alga, *Spirulina fusiformis* and, as mentioned previously, intensive biodynamic horticulture—all within an integrated system partially shown in Fig. 3.2.

At the same time, further innovations are continually being developed and tested in the research centre itself before the villagers are exposed to them—as we have seen and noted a prerequisite for any successful biotechnological practice in the village itself. Emphasis is placed on educating and instructing the villagers to participate in all such schemes and attention to the social aspects of the MCRC programme has been meticulous, with some of the MCRC scientists and social workers living in the villages to experience the local problems at first-hand. All this activity (which has included an in-depth survey of 70 individual households) led by 1984 to improved food and energy production, and indirectly to improved agricultural practice, nutrition, and sanitation, and which in turn has improved the health and income of the villagers. The MCRC strategy of technology implantation allied to social awareness, specifically in Injambakkam, has been further refined by the application of systems analysis as described below.

Systems analysis as an aid to biotechnological practice

A prerequisite to formulating a strategy for satisfying the energy and other needs, initially present and ultimately future, of rural communities via integrated biotechnological systems, is the acquisition of relevant production and consumption data. Detailed village surveys should therefore ideally be carried out, but these are invariably time-consuming and, for various reasons, not always very accurate. However, it is sometimes possible to garner sufficient regional data from a state or district records office which could cover dozens, or perhaps even thousands, of similar villages within a particular area of climatic, soil type, and natural resource similarity. At the village level too there are often 'accountants' who can provide some pertinent village statistics. Clearly, to carry out a detailed survey of every village within a region would be an almost impossible task, but a small number could be performed for comparison with the broadly-based regional data obtained. Crucial energy-related statistics on quantity and quality of fuelwood, tree species, livestock numbers, meteorological data, and so forth could additionally be found within local forestry, agricultural, and meteorological centres.

The technologies most appropriate as components of an integrated biotechnological system for the villages of a particular region can be evaluated on the basis of available information and various technological scenarios explored through the methodology of systems analysis. Systems modelling has already been used to ascertain the best mix of technologies for individual village situations, and a generic model, MERDA (Model for Energy and Resource Development Analysis), developed to provide the results of installing various integrated biotechnological systems in terms of energy, food, fertilizer, etc., supply both quantitatively and qualitatively through time. Such a technique could be easily extended to cover the villages of a whole region with the caveat that no two villages are identical and so the model may need slight modifications to accommodate communities which differ appreciably from the norm.

The systems-modelling technique has both versatility and speed, while at the same time being very flexible. It allows for the examination of a large number of scenarios and their future effects upon a community in a matter of a few seconds of computer time, while to achieve the same spread using solely experimental work in the field would be a huge task, necessitating the setting up and monitoring over decades of a very large number of rural communities each implanted with a different technological mix—clearly an impossible undertaking! It also allows feedback from and to the target village, and this has been the case between the MCRC in Injambakkam village and the systems analysis team at the Energy Studies Unit, University of Strathclyde during a 4-year Self-Reliant Development Programme sponsored by the Swedish-based International Federation of Institutes for Advanced Study (IFIAS).

The coastal village of Injambakkam itself lies south of Madras City on the Bay of Bengal. It has a population of around 950, of whom 41 per cent are children under 15 years of age, and a total land area of 352 ha, of which only 16 per cent is suitable for growing rice, the staple crop. There are 80 tonnes of cereals produced each year. Adult cattle number 200 and calves, goats, and sheep are owned in small numbers together with 2600 poultry. The main difficulties confronting the villagers of Injambakkam are: a 56 per cent overall calorie deficiency; an inability to pay for increasingly expensive diesel fuel (for irrigation), kerosene (for lighting), and chemical fertilizers (for growing high-yielding rice varieties); and the drudgery of collecting fuel wood from farther and farther afield due to deforestation in the immediate vicinity of the village. Nevertheless, the wood and fossil fuel overall consumption figure of 6.4 GJ (5.3 GJ indigenous) per person per year is comparatively high by Indian standards though, as with most other small villages in the country, there is no electricity network to the people's homes. To ensure that a sufficient supply of energy is available to the villagers, a rural energy centre based upon the biogas plant and some of the technologies described previously has been devised. The integrated system can be simulated and, in the case of Injambakkam, has been so for the following selected technological scenarios (among others) within a system dynamics model. Most of the technologies mentioned have already been installed within the village boundary, while some are still undergoing tests at the nearby research centre of the MCRC.

Year	Technological implantation
1	Community biogas plant installed to receive cattle dung; cost Rupees 1000 000; dung collection efficiency increased from 50 to 75 per cent.
2	Energy crop plantation: area 10 ha producing per hectare-year 44 tonnes dry weight *Casuarina* wood, plus leaves with a high nitrogen content (1.8 kg of nitrogen per tonne of wood) subsequently to be used as a fertilizer. Imported insecticides to be faded out over next 10 years due to increased use of locally-produced neem leaf extract.
3	Stoves twice as energy efficient as the traditional kind used gradually throughout the village over a 5-year period. A new rice variety introduced with 57 per cent net protein utilization (up from 50 per cent). Procedures implemented to reduce post-harvest crop losses from 20 to 10 per cent.
4	Nitrogen-fixing algae established in fields over a 4-year period providing 30 kg of nitrogen per crop-hectare.
5	Food legume (rice bean) grown as a third crop on 10 ha at 1.5 tonnes/ha per year having three times the protein content

of rice and also providing 33 tonnes/ha per year of green foliage available for animal feeding.

6   Second community biogas plant installed to receive both human waste directly from latrines plus water hyacinths (50 tonnes dry weight). The first of a hundred windmills to be introduced over 5–6 years is installed to ultimately irrigate 15 per cent of the rice crop area and to release some biogas for domestic use.

7   Two Fiat Totem units (or other appropriate internal combustion engines), for electric lighting and water heating, run on excess biogas.

The above scenario results, at zero population growth, in an increase in food production of 350 per cent, indigenous energy production by 85 per cent, and overall solar energy capture by over 150 per cent (with great potential for further rises) all within 12 years. In this time the community moves from a net importer of energy (as diesel fuel, and kerosene, and wood) to a net exporter (in the form of wood), and virtually all energy needs are met locally through solar energy captured in the storable form of biomass. Additionally, all capital loans plus interest are paid off through the sale of village exports. The nutritional standard of the people's diet improves dramatically, while problems of unemployment and underemployment are greatly eased. Sensitivity testing of the major parameters shows that, even with an annual population growth rate of 2 per cent, the main scenario proposed is sufficiently robust to maintain these improved living standards well into the next century.

The impact of such a scenario on the village's energy supply can be gauged from Tables 3.2 and 3.3.

Table 3.2 illustrates the 1980 import and production levels of energy in Injambakkam village.

In 1980 the community was a net importer of fuel. What little indigenous fuel there was, was gathered from the scattered trees. Within 12 years all could be changed (Table 3.3). The community moves from an annual import of 1070 GJ (Table 3.2) to a situation where it can export wood for cooking and reduce its diesel and kerosene imports to zero (Table 3.3).

Thus, at year 12, 100 per cent of the village's direct energy supply could be solar driven, mostly in the form of bioenergy. The output from the two biogas plants alone reaches 2.72 TJ/y. This corresponds, on a *per capita* basis within the village, to 39 per cent of the *per capita* average Indian fuel consumption— perhaps a surprisingly high proportion. At the same time, the net energy balance of the community has substantially changed: 920 GJ of biogas energy are converted into 260 GJ of electrical energy for lighting by the Fiat Totem modules and enough heat (110 GJ) recovered to eliminate the need for water heat previously provided by wood.

Table 3.2

1980 Energy imports and production in Injambakkam village

Fuel	Quantity	GJ/y
Wood	18 tonnes	285
Kerosene	12 200 litres	450
Diesel oil	6 400 litres	245
Electricity	6.43 MWh*	90
Total imports		1070
Indigenous wood production	311 tonnes	4980
Total energy utilized by population or 6.4 GJ/capita		6050

* 14 MJ/kWh is the primary energy input per kWh (e) produced and delivered.

Table 3.3

Total energy production in year 12 of scenario for Injambakkam village

Fuel	Quantity	GJ/y
Wood export	331 tonnes	5300
Wood consumed	107 tonnes	1700
Biogas production (used as gas)		1800
Electricity converted		
from biogas		260 (e)
+ heat		110
Total energy available or 9.7 GJ/capita		9170

The scenario has also succeeded in enhancing greatly the solar energy capture within the village system boundary. This is probably the key parameter. At the 1980 condition, 0.035 per cent only of the incident solar energy was converted to useful biomass. Within 12 years it will have risen to 0.089 per cent: a 154 per cent increase. The latter is still though only half of the 0.18 per cent obtained in the UK for intensively grown crops achieved with high energy inputs (Blaxter 1975), but with much less solar energy input. Further increases in solar energy capture to about 300 per cent of the present level should therefore be achievable in Injambakkam village to its continued development. Worldwide 2.5 billion people could benefit from this kind of approach, which is more fully reported by Slesser *et al.* (1981, 1982), and Lewis (1984).

The social challenge

As has already been intimated during this chapter, the social acceptability of biotechnological practices is just as important as proving technological feasibility, if not more so. The people themselves must show a willingness to change—to try something new. However, the social question cannot be solved exogenously, but rather must be tackled from within by the local people for the local people. For outsiders to do so would not only be presumptuous and inappropriate, but lead to almost certain failure. Matters to be looked at would include the effect of introducing biotechnologies on income distribution, labour patterns, social structure, inter-caste/religious/ tribal relationships, the role of women, education, health, attitudes towards community organization and co-operation, and so on. The list is virtually endless.

One of the most intractable social problems in India especially, the supposedly rigid caste system, is proving to be not so rigid after all. Co-operation is vital for the successful introduction of biotechnological practices in integrated rural development—and that means inter-caste co-operation, something almost unheard of in the subcontinent until the last few years. However, recently there have sprung up in various parts of the country dozens and dozens of rural, mostly dairy, co-operatives, with emphasis being on the word 'co-operative'. Whereas previously both rich and poor farmers had sold all their cows' and buffaloes' milk to a freelance milk trader who invariably set a price favourable only to himself, they can now be assured of a fair price by joining and selling their milk to the local co-operative. Everyday the people stand in line, landed Brahmins alongside landless labourers of the Harijan Untouchable caste, and their milk ends up in the same vat, a previously unheard of occurrence.

It would seem, therefore, that if they can benefit from a co-operative venture such as this then the people are willing to forget traditional, outmoded caste distinctions and work for the betterment of both themselves and their community as a whole. This, in essence, is the crux of integrated rural development. Seeing is believing and the best incentive to make innovative biotechnological practices acceptable is to demonstrate to the people that it would benefit them individually, their families, and their village community. Much has been written about the social and cultural impediments to fresh ideas, previously unknown technologies, new ways of doing things, and so forth, which doubtless do exist in the minds of the 2.5 billion or so current inhabitants of the rural Third World. Yet more and more instances of changes being accepted are now coming to light, whether it be the acceptance of new crop varieties, biogas plants, improved cooking stoves, new preventive medicines, the need to read and write, plus the many more innovations introduced recently all over the Third World.

Rather than discussing at length, and only in abstract, the theories of changing social and cultural patterns of behaviour which can be found in the textbooks, I shall conclude by dwelling at some length on the experiences of Nalini Keshavaraj and her husband, whom this author had the pleasure of meeting in Injambakkam village. They are essentially social workers who, over a period of 4 years have worked in the villages of the Coromandel Coast, including Injambakkam, and produced in October 1982, a Murugappa Chettiar Research Centre monograph entitled *Social Factors in Technology Transfer* (Keshavaraj and Keshavaraj 1982). The monograph presents the actual experiences and practical difficulties encountered by the social worker in smoothing the way for the transfer of technologies, mostly biotechnologies, at the village level. They write at the outset 'The introduction of science and technology into the village becomes a difficult task for it involves getting many individuals to accept changes in custom or practice for reasons which they do not initially understand. Moreover when people find reasons not to believe in a system, they reject the ideas and motives, which becomes a still more difficult problem.'

Nalini Keshavaraj then goes on to relate how, as a newcomer, she firstly introduced herself to the village primary school teachers as a prelude to visiting the schoolchildren, teaching them stories and songs, and gradually gaining their confidence, at which point the mothers too became interested in her and her reasons for being there. She was invited for meals into people's homes where she could talk informally and also begin to explain the purpose of the MCRC objectives through their introduction of mainly biosolar technologies. Nalini also talked to the people, both doctors and patients, at the local medical clinic and went to the panchayat members for discussions about the village, its problems, and what were priority needs as they saw them.

Having established a good rapport with the villagers, Nalini and her colleagues embarked upon a massive socioeconomic survey of 70 sample households (MCRC 1980) which provided the data base for the IFIAS-SRD programme. Responses varied from friendly and co-operative down to genuinely hostile, and some questions provoked more suspicion than others. However, in the end most people could be persuaded that the survey was aimed to benefit them. Where villagers were reluctant to answer questions about land ownership (for the presumably irrational fear that the Government, if informed, might take away some of their land), the desired information could simply be gleaned from the records of a village officer.

Despite encouraging them with tales of the potential benefits of technology, no one was at the outset at all interested, even though there was a desperate need for some of the hardware which MCRC had to offer. Nalini lists five major constraints for the introduction of technologies at the village level.

1. The people are unaware of the need for change.
2. They will benefit only if they maintain the system properly; otherwise, benefits will take a very long time to accrue.

3. The people will have to undergo many changes in their existing way of life at different stages owing to the introduction of the new technologies.

4. People will have to learn new and possibly rather complex and difficult skills; although, in fact, it may appear to them a simple matter to learn how the technology works.

5. The people may need to organize themselves into a co-operative or an association for achieving new goals.

Since the villagers were still showing a reluctance to become even remotely interested in the introduction of the technologies offered by MCRC, it was decided to increase the rapport between them and the MCRC. Thus, some of the former (14 local youths) were taken on to the MCRC payroll, while a few of the MCRC team including social workers, civil engineers, and an administrative assistant actually went to live in the village, where they could experience the villagers' problems at first-hand and also carry out their work more efficiently. The main advantage of having some of the villagers themselves participating in the MCRC programme was that these people could then begin to explain why certain things were done to their fellow villagers, a task which might sound like preaching if coming from an outsider. One of the local employees actually wrote a book on the construction of the biogas plant and its uses in the village language, for which he received a small remuneration as an incentive for both himself and his colleagues. By such means, inroads were made by the MCRC social workers into the confidences of the Injambakkam people.

But perhaps the biggest coup of the MCRC was in gaining the interest of the village women. They, and the landless agricultural labourers, have most to gain from the improvements which a 'biotechnological package' could make to village life, by reducing the drudgery of activities such as wood gathering and the health risks associated with cooking for long hours in a smoky atmosphere, while at the same time increasing the opportunities for obtaining extra income. Adult education classes for women were organized in collaboration with the district Women's Voluntary Services, initially much against the wishes of the panchayat president. They were taught about biogas plants, windmills, inland pisciculture, energy conservation, and the importance of people's participation and co-operation in community development projects. Opposition came from the men, particularly the landed farmers, who wished to maintain the *status quo*, which was of course to their personal advantage; thus, much work is still to be done before the whole village community accepts the technology transfer operation in which MCRC is engaged. However, enough of the people are co-operating at present to make self-reliant, integrated rural development feasible and workable in Injambakkam—and hopefully others will join in when the benefits are demonstrated to them and seen through their own eyes.

Microbial technology has much to offer the developing world, but only if it is accepted by the people it is intended to help. The process of attempting to ensure the acceptability is a social challenge of the utmost importance and, this author believes, is an integral part of the biotechnological practices themselves.

Acknowledgements

The author would like to thank his colleagues at the Energy Studies Unit, University of Strathclyde, Professor Malcolm Slesser and Ian Hounam, without whom the systems analysis aspects of the work could not have been reported. Thanks are also due to Dr C. V. Seshadri and fellow members of the Murugappa Chettiar Research Centre, Madras with my respect and admiration for their pioneering work in Injambakkam village. I also owe a great debt to the International Federation of Institutes for Advanced Study, Sweden and its Executive Director, Dr Sam Nilsson for financial support throughout the course of the IFIAS SRD study. Travel support from the Royal Society and the Carnegie Trust for the Universities of Scotland is also greatly appreciated.

References

Agarwal, A. (1979). Can biogas provide energy for India's rural poor? *Nature* 281, 9–10.

Alicbusan, R. V. (1979). Mushroom production technology for rural development. In *Bioconversion of Organic Residues for Rural communities*, (ed. C. A. Shacklady) pp. 99–104. United Nations University, Tokyo.

Anon (1984). *The Rural Energy Center in Pattiyapola, Sri Lanka*. Natural Resources, Energy and Science Authority of Sri Lanka, Columbo.

Barnett, A., Pyle, L., and Subramanian, S. K. (1978). *Biogas Technology in the Third World: A Multidisciplinary Review*. International Development Research Centre, Ottawa.

Blaxter, K. L. (1975). The energetics of British agriculture. *Biologist* 22, 14–18.

Bull, A. and Da Silva, E. (1983). World networks for microbial technology. *Soc. Gen. Microbiol. Q.* 10, 6–7.

Cherry, M. (1981). The technology of nitrogen fixation: encouraging its use in the tropics. *Span* 24, 114–16.

Cole, D. C., Lim, Y., and Kuznets, P. W. (1980). *The Korean Economy—Issues of Development*. Institute of East Asian Studies, University of California, Berkeley.

Da Silva, E. J. (1981). Microbial biotechnology: a global pursuit. *Process Biochem.* 16, 38–41, 44.

—— and Doelle, H. W. (1980). Microbial technology and its potential for developing countries. *Process Biochem.* 15, 2–6, 9.

——, Olembo, R., and Burgers, A. (1978). Integrated microbial technology for developing countries: springboard for economic progress. *Impact of Science on Society* 28, 159–82.

Datta, R. and Dutt, G. S. (1981). Producer gas engines in villages of less-developed countries. *Science* 213, 731–6.

Hall, D. O. (1980). Biological and agricultural systems: an overview. In *Biochemical and Photosynthetic Aspects of Energy Production* (ed. A. San Pietro), pp. 1–30. Academic Press, New York.

——, Barnard, G. W., and Moss, P. A. (1982). *Biomass for Energy in the Developing Countries*. Pergamon, Oxford.

Harrison, P. (1979). Small is appropriate. *New Scientist* 82, 14–16.

—— (1980). *The Third World Tomorrow*. Penguin, Harmondsworth, England.

Hayes, P. and Drucker, C. (1980). Community biogas in India. *Soft Energy Notes* 3, 10–13.

Indian Agricultural Research Institute (1978). *Algal Technology for Rice*. Indian Agricultural Research Institute, New Delhi.

Keshavaraj, N. and Keshavaraj, R. (1982). *Social Factors in Technology Transfer*. Murugappa Chettiar Research Centre, Madras.

Lewis, C. W. (1983). Taming of the microbe. *Development Forum* 11, 5–6.

—— (1986). Third World development: the road to self-reliance. *Sci. Am.* (In press).

—— (1984). *Toward Third World Self-Reliance: Development of the People, by the People, for the People*. Report to IFIAS, Solna, Sweden.

Madeley, S. (1982). How the village of Xinbu learned to love the biogas unit. *The Guardian*, London, 5 Aug. 1982, p. 18.

MCRC (1980). *First Annual Report of the IFIAS Self-Reliant Development Project*. Murugappa Chettiar Research Centre, Madras.

—— (1982). *Technical Notes No. 8*, Murugappa Chettiar Research Centre, Madras.

Moulik, T. K. (1982). *Biogas Energy in India*. Academic Book Centre, Ahmedabad.

National Academy of Sciences (1977). *Methane Generation from Human, Animal and Agricultural Wastes*. National Academy of Sciences, Washington, D.C.

National Academy of Sciences (1980). *Firewood Crops*. National Academy of Sciences, Washington, D.C.

Raja, G. (1982). *Studies in Pisciculture*. Murugappa Chettiar Research Centre, Madras.

Ranjhan, S. K. (1978). Use of agro-industrial by-products in feeding ruminants in India. *W. Anim. Rev.* No. 28, 31–7.

Ravindranath, N. H., Nagaraju, S. M., Somashekar, H. I., Channeswarappa, A., Balakrishna, M., Balachandran, B. N., Reddy, A. K. N., Srinath, P. N., Prakash, C. S., Ramaiah, C., and Kothandaramaiah, P. (1981). An Indian village agricultural ecosystem—case study of Ungra village, Part 1: main observations. *Biomass* 1, 61–76.

Reddy, A. K. N. (1981a). India's good life without oil. *New Scientist* 91, 93–5.

—— (1981b). An Indian village agricultural ecosystem—case study of Ungra village, Part II: discussion. *Biomass* 1, 77—88.

——, Prasad, C. R., Rajabapaiah, P., and Sathyanarayan, S. R. C. (1979). Studies in biogas technology. Part IV. A novel biogas plant incorporating a solar water heater and solar still. *Proc. Ind. Acad. Sci.* C2, 387–93.

—— and Subramanian, D. K. (1979). The design of rural energy centres. *Proc. Ind. Acad. Sci.* C2, 395–416.

Rolz, C., Menchu, S., de Cabrera, S., de Leon, R., and Calzada, F. (1979). Strategies for developing small-scale fermentation processes in developing countries. In *Bioconversion of Organic Residues for Rural Communities*, (ed. C. A. Shacklady) pp. 36–40. United Nations University, Tokyo.

Sedjo, R. A. and Clawson, M. (1982). The world's forests. *Resources* No. 71, 15–7.

Seshadri, C. V. (1977). *A Total-Energy and Total-Materials System Using Algal Cultures*. Murugappa Chettiar Research Centre, Madras.

—— (1979). Analysis of bioconversion systems at the village level. In *Bioconversion of Organic Residues for Rural Communities*, (ed. C. A. Shacklady) pp. 144–51. United Nations University, Tokyo.

Seshadri, C. V., Venkataramani, G., and Vasanth, V. (1978). *Energy Plantations—A Case Study for the Coromandel Littoral*. Murugappa Chettiar Research Centre, Madras.

Seshagiri, S. and Chitra, M. (1983). *Biodynamic Horticulture: Improvements and Extension*. Murugappa Chettiar Research Centre, Madras.

Slesser, M., Lewis, C. W., Hounam, I., Seshadri, C. V., Roy, R. N., Geethaguru, V., Jeeji Bai, N., Keshavaraj, N., Manoharan, R., Raja, G., Subramani, V., Thomas, S., and Venkataramani, G. (1981). *Self-Reliant Development*. University of Strathclyde, Glasgow, and Murugappa Chettiar Research Centre, Madras for IFIAS, Stockholm.

Slesser, M., Lewis, C. W., Hounam, I., Seshadri, C. V., Roy, R. N., Geethaguru, V., Jeeji Bai, N., Keshavaraj, N., Manoharan, R., Raja, G., Subramani, V., Thomas, S., and Venkataramani, G. (1982). Biomass assessment in Third World villages via a systems methodology. *Biomass* 2, 57–74.

Smith, J. E. (1981). *Biotechnology*. Edward Arnold, London.

United Nations University (1982). Reproduced in *Impact of Science on Society* 32, 167.

Venkataraman, G. S. (1977). *Algal Fertilizer for Rice*. Indian Agricultural Research Institute, New Delhi.

World Bank (1983). *World Development Report 1983*. Oxford University Press, New York.

4

Microbial bioinsecticides

H. D. Burges and J. S. Pillai

1. Introduction

For present purposes, the following major global areas can be recognized: the developed world which would include North America, Europe, Japan, USSR, Australia and New Zealand; the developing world which would include the African continent (with some exceptions), Central and South America, the Indian sub-continent and bordering countries, and the People's Republic of China and bordering states.

Modern agriculture, worldwide, typically involves sizeable areas of one cultivar of a plant. The plants range from corn, grown as far as possible in a healthy succulent state for maximum yield, to forests of rapidly growing conifers crowded together to provide long, straight lengths of timber. These crops appear ideal breeding and feeding places for a wide range of insects. Why then are these crops not obliterated? Many factors curb the insects. These factors include adverse weather, inability of some insects to breed rapidly, and a vast array of natural enemies that kill them. Prominent among the enemies are diseases caused by many pathogenic viruses and micro-organisms.

In the face of so many adverse factors, one wonders how some insect species succeed in taking advantage of the abundant food to become serious pests. Although diseases and other natural enemies prevent many species reaching pest proportions, the complex of diseases that may assail an insect often strikes too late to avoid economic or crippling damage to plants. To prevent some pests exceeding economically important numbers, man has harnessed, among many other factors, some of the diseases for his own benefit. This chapter describes the present microbial control of pests and discusses the extensive—still largely untapped—potential, which is probably greatest in the developing world.

1.1. Strategy

To use pathogens for pest control the basic requirements for successful microbial control must be met. These requirements are (i) an adequate reservoir of pathogens in the pest population or in the environment; (ii) good transmission of pathogens from reservoir to healthy pests; and (iii) rapid build-up of disease to prevent the pests becoming economically important or—with insect vectors of human disease—to prevent the vector reaching significant numbers at the stage when it infects man. The requirements can be fulfilled in a number of ways: a pathogen may be introduced into a new geographical area; a strain more efficient than the native strain of a pathogen may be introduced; the natural pathogen reservoir may be supplemented or the environment altered to achieve disease earlier than usual; and, finally, if these methods are inadequate, a pathogen may be applied extensively to swamp the pest, i.e. used as a microbial insecticide, which may have to be applied frequently during the pest season.

1.2. Transmission of pathogens

A spore (Figs 4.1 and 4.3) or equivalent stage, such as an inclusion body (Fig. 4.2), is part of the life-cycle of most pathogens. This stage often tolerates a hostile environment. It survives and frequently becomes more widely spread when the pathogen is not growing or replicating in an insect.

Spores may spread to healthy hosts by many routes. They may be eaten on food contaminated by spores from the faeces of other herbivorous or predatory insects, birds, or other animals. Spores, free or in soil particles, may be carried onto insect food by wind, water, or animals. Healthy hosts may acquire infection by eating live or dead infected insects. Spores from an infected male adult may be transmitted to a female at copulation. Females may transmit pathogens in or on eggs to progeny, or progeny may eat dead, diseased adults or other life stages, or their infected faeces. Only some of these modes can be used by infective stages unable to withstand adverse environments.

Infection of healthy insects is commonest by the oral route. Only some fungi and nematodes attack through the insect's cuticle, which enables them to infect insects that suck blood or sap, or bore into wood, stems, leaves, or roots, all foods that would not normally contain insect pathogens (Burges 1981c; Cunningham and Entwistle 1981; Fine 1984; Fuxa and Tanada 1986).

1.3. Stages in the investigation of microbial control of a pest

The first step in the evaluation of potential microbial control is to obtain information, either from literature or by research, about the biology of the pest in question and its associated pathogens. Emphasis should be placed on detecting the infective stages of the pathogens and the weak points in the

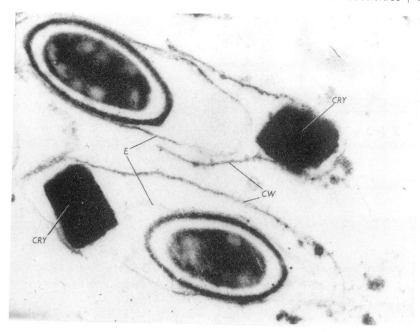

Fig. 4.1. *Bacillus thuringiensis*: *upper,* section through two sporulating rod-shaped bacteria, showing the spore with double-layered tough coat and the crystal of toxic protein; CRY = protein crystal; CW = cell wall. E = endospore membrane; (x 30 000); *lower,* shadowed crystal showing protein molecules (x 162 000). Photographs by J. R. Norris.

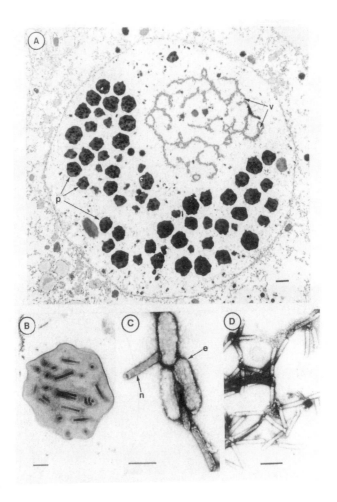

Figure 4.2. Baculovirus: A, inclusion bodies (p) embedded in a fat-body cell of an infected larva of the moth *Spodoptera frugiperda*. v = virogenic stroma. Marker bar = 1 μm. B, sectioned inclusion body of virus from *Spodoptera exempta*; bar = 200 nm. C, purified virus particles (e) after release by treatment of inclusion bodies with alkali (n, capsid of DNA and protein released from its lipoprotein envelope); bar = 200 nm. D, purified capsules after treatment with a detergent to remove the envelope; bar = 200 nm. Photographs from C. C. Payne.

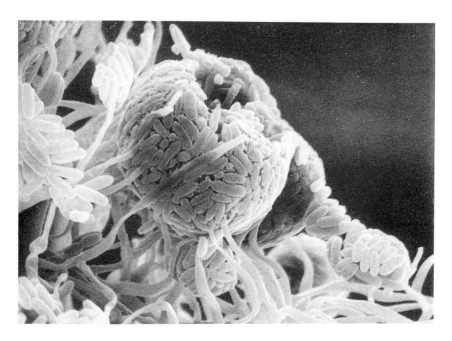

Figure 4.3. *Verticillium lecanii*: *upper*, fungus bearing spore heads shrouding the cadaver of an aphid (x 27): *lower*, sticky sporehead (x 1440) containing many spores in mucus. Photographs by P. T. Atkey and R. A. Hall.

pest's life-cycle. This permits assessment of each pathogen to decide which causes the greatest mortality and which—taking all factors into consideration—would be the most feasible microbial control agents. Field tests should follow on an increasing scale. Data, commensurate with the planned stage of investigation, should be obtained on the safety of pathogens to man, domestic animals, and wildlife. If exotic pathogens are to be imported, governmental permission must be obtained well in advance. As with chemical pesticides, governmental registration stages are normally necessary for each stage of research, development, and industrial use of a microbial insecticide.

1.4. Available pathogens

Production and use of insect pathogens is, perhaps not unexpectedly, most sophisticated in the developed world. However, one serious pest of coconut palms in developing countries, the rhinoceros beetle, although intractable to conventional control by use of chemical insecticides, was successfully curbed with a baculovirus, involving simple and inexpensive methods (section 2.2.3). Probably other such examples will be forthcoming. Meanwhile, technology and microbial products are steadily being introduced to the developing world. Available pathogens and their use are reviewed in many books (e.g. Burges and Hussey 1971; Cantwell 1974; Burges 1981*a*; Kurstak 1982; Payne and Burges 1982; Maramorosch and Sherman 1985).

Of the different types of pathogenic organisms, protozoa are most commonly found when examining live insects. This is because, like malaria in mammals, they usually kill insects slowly, although taking considerable (but often unseen) toll of insect life. Because of their relatively low pathogenicity they will not be considered further here. Nematodes will also be excluded because the principles of pest control using nematodes are similar to those described below for the three important pathogen groups, viruses, fungi, and bacteria. Viruses cause spectacular outbreaks of disease in heavy pest populations and have considerable microbial control potential though their use is handicapped by economic production being possible at present only in live insects. Among insect-pathogenic fungi, *Beauveria bassiana* (Bals.) has a unique place in the history of biological sciences, in the developed world at an early stage of its development. This fungus was the first micro-organism to be demonstrated (in 1835 by Bassi de Lodi in Italy) as the causative agent of an infectious disease (Steinhaus 1949). It causes the silkworm muscardine disease, which terminates in dead larvae shrouded in a white mat of sporulating fungus. Fungi often cause insect outbreaks to collapse dramatically, leaving foliage littered with fungus-coated cadavers. Early scientists tried to use these fungi for insect control, with erratic, unpredictable results. Modern methods have been more successful. Bacteria do not kill insects in a spectacular way, but have proved to be the easiest and most successful for commercial development. Since they are the leaders, the bacteria will be considered in detail first.

2. Present use of insect pathogens in the developed and the developing world

Since use of and research into microbial bioinsecticides in the developed world greatly exceeds and influences that in the developing world, the significance and perspective of the latter can be expressed only by a comparison of the two.

2.1. Bacteria

Bacteria are responsible for many infectious diseases, not only in man, but also in insects. A key feature of bacteria is the environmentally-resistant spore. This not only survives well in the insects' environment, but—when produced commercially—also ensures a long shelf-life. Consequently, spore-formers were the first insect pathogens to be used extensively as successful pest control agents as described below. One, *Bacillus popiliae* Dutky, after systematic introduction into a new geographical area, spread naturally in the fashion of a classical biological control agent, which was probably why it was the first to be subjected to prolonged trials. It was followed by *Bacillus thuringiensis* Berliner. Because this species was unable to spread rapidly enough to achieve economic control, it had to be applied regularly like a chemical insecticide. World sales of this microbial insecticide now total several thousand tons per annum. Later, a similar species, *Bacillus sphaericus* Neide, attracted attention because it can be used as a microbial insecticide, and will spread and persist to some extent. These species will be considered in historical order, illustrating evolution of thought about the potential of these bacterial pathogens.

2.2.1. *Bacillus popilliae* group

The first successful, well-documented, microbial control agent in the developed world was used to attack the Japanese beetle, *Popillia japonica* Newman. This scarabaeid beetle was accidentally imported into the USA, unaccompanied by its natural enemies (Klein 1981). Larvae feed for 2 or 3 years on roots, particularly those of grasses. They ravage pasture, scrubland and high quality grass, as on lawns. Adults feed on foliage and damage crops as well. *B. popilliae* and related species were introduced into infested areas and spread rapidly among larvae. Effective spread was possible because of a combination of the following factors (Klein 1981): (i) the bacteria kill larvae by slow infection, producing mainly large cadavers packed with spores; (ii) spores survive many years in soil, at high concentrations where a larva has died in untilled soil; (iii) larval populations were usually high, allowing an extensive reservoir of spores to build up; (iv) most grassland was economically unaffected by subsequent small surviving larval populations, which maintained spore reservoirs; (v) some surviving larvae become infected adults, which flew long distances and spread the bacteria widely beyond inoculated areas.

Unfortunately, production of bacteria was a limiting factor. Although the bacteria could be grown in complex media, few infective spores, resistant to environmental conditions, were formed. The only economic method for spore production was propagation in live larvae and adults. The United States Department of Agriculture developed and proved the methodology in the 1940s then handed over production to industry. Two small firms have maintained production for 45 years, limited by the availability of field-collected insects, because breeding in the laboratory is difficult. Probably, because this is the longest-used microbial control agent, recent surveys have show that the pest has resurged in some areas, indicating a need for remedial research, e.g. re-introduction of virulent strains, examination for resistance of the pest to the bacterium, use of new strains to combat resistance, etc. (Klein 1981). Related bacteria have been studied in Europe, New Zealand, and Australia (Milner 1981), but none of these has been successfully exploited to date.

In the developing world, however, no bacteria in this group have been researched although valuable species may be present.

2.1.2. *Bacillus thuringiensis*

Bacillus thuringiensis is the microbial control agent most used in the developed world. It is an aerobic spore-forming species readily cultured in media and produced commercially by conventional, liquid, stirred-tank fermentation. At sporulation, each cell produces a bipyramidal crystal of protein, a potent insect gut toxin that is the major component of commercial products in killing most susceptible species (see Fig. 4.1).

First described in 1911 from flour-moth larvae in a European flour mill, it subsequently proved common in Lepidoptera in dry protected habitats such as warehouses, food factories, silkworm farms, and beehives (Burges 1973). Although spores and crystals in dry larval cadavers survive almost indefinitely in food residues left when a store is unloaded, the bacterium does not spread enough to give useful control (Burges and Hurst 1977). In comparison it is rare in field and forest, where it spreads even less than in stores, probably because the crystal toxin is denatured in moist conditions (as in soil) and spores are killed by solar radiation (Pruett *et al.* 1980; Burges 1982; West 1984; West *et al.* 1984*a*, *b*). It must be applied regularly to crops and forests as a microbial insecticide (Burges 1981*c*; Grimble and Lewis 1984).

The crystal of protein, that is produced with every spore of *B. thuringiensis*, is an inactive protoxin, composed of molecules of mass 130 kDa. It is activated by alkaline dissolution and breakdown by gut proteases into active toxins, varying in molecular mass from 60 to 65 kDa. These toxins destroy the epithelium of the insect midgut, acting on the cell membrane. In Lepidoptera, they probably bind with glycoproteins, causing cytolysis by altering membrane permeability (Knowles *et al*. 1984; Ellar *et al*. 1985) and destroying regulation of the passage of glucose and ions such as K^+ (Fast 1981). This has interested biochemists and industry alike as a possible source of a new family of pest-specific insecticides, with the advantage of a reasonably wide range of target species among pests.

All major countries of the developed world have contributed to extensive research on the toxin (Bulla *et al*. 1980; Fast 1981; Faust and Bulla 1982; Ellar *et al*. 1985).

Thus, at first sight, the *B. thuringiensis* crystal appears to be an ideal insecticide. It is specific to and highly potent in a wide range of pest insect larvae. It is harmless to man, even if eaten in considerable quantities or injected into the body, harmless to wildlife (except some Lepidoptera) and biodegradable (WHO 1982; Burges 1982). However, its use is restricted by its being a stomach poison that must be eaten by larvae to take effect, i.e. it has no contact action and attacks no other developmental stage of the insect. Although Krieg and Langenbruch (1981) list *B. thuringiensis* as pathogenic in 328 Lepidoptera species, which comprise mainly pests, many more pest and non-pest species have not been tested. However, individual strains of the bacillus have narrower host ranges. Larval behaviour of many susceptible species, e.g. wood- and stem-borers, make them unsuitable targets for stomach poisons. Thus, the HD–1 strain used in most current industrial agricultural products is recommended for use against only about 50 lepidopteran pest species, worldwide. Consequently, the pest complex of a crop in a particular geographical area often contains species, including Lepidoptera, that are unharmed or only partially controlled by *B. thuringiensis*.

The disadvantages of *B. thuringiensis* can be alleviated by selection and genetic manipulation. Extensive selection from natural strains in the developed world, including an International Programme (Dulmage and co-operators 1981), produced a succession of better strains. These were adopted over two decades by industry, increasing potency of products by over 100-fold, and improving host range. The genes controlling the crystal toxins are borne on plasmids (extrachromosomal elements of DNA), which can be exchanged between strains with relative ease. This enables strains to be genetically tailored to the needs of particular pest complexes. What is probably the first patent of a genetically modified *B. thurginiensis* strain was taken out in 1984 in the UK by Burges and Jarrett (1984). The recent finding of a strain able to kill a number of beetles (Krieg *et al*. 1983) may herald extension of the host range to Coleoptera.

The developing world has contributed new strains and varieties of *B. thuringiensis* (de Barjac 1981), particularly from countries sited along well-established trade routes (Norris and Burges 1963; Burges 1973). The variety *ostriniae* was first found in China; var. *pakistani* in Pakistan and var. *wuhanensis* in China. From the early 1960s, commercial producers in the USA searched for markets in the developing world, but found only a limited number of opportunities. They conducted trials, as did local scientists and pest control officers. Some researchers studied mode of action in local insect species, finding results similar to those already published. Some local investigators grew their own strains in the laboratory for trial, e.g. five strains were studied by Cabrera (1983) in Mexico.

In one area of West Africa, the Upper Volta region, a new variety of *B. thuringiensis* had a great impact (Guillet 1983). This was var. *israelensis* = serovar H–14, first discovered in Israel in soil at mosquito habitats (Goldberg and Margalit 1977). It extended the *B. thuringiensis* host range not only to mosquitoes, but also to blackflies (Simuliidae) among the Diptera. In the USA and elsewhere commercial products were produced in a remarkably short time of 6 years from the date of first discovery of this new variety. It is used as a larvicide both of nuisance species and vectors of human disease. Its biggest single market and immediate application was in the river-blindness eradication programme against blackflies in the Upper Volta region. This large programme, implemented by the World Health Organization (WHO), had already been in progress for 7 years by 1982 when it was threatened with collapse by appearance of serious resistance to the organophosphate insecticide, temephos, and other acceptable chemicals in two important blackfly species over about 25 per cent of the treated area. The var. *israelensis* became commercially available just in time to prevent loss of control of the blackflies. Because it was relatively expensive, it was used only in the dry season, when water areas requiring treatment were at their lowest. In the wet season another organophosphate, chlorphoxim was used, but resistance to this chemical also became serious (Guillet 1983). Luckily, this blackfly resistance proved to be transient and was abated by the use of *B. thuringiensis* in the following summer, so that subsequently the bacterium and the chemical could be alternated seasonally. This demonstrates an important principle for successful pest control, the regular alternation in time of biological agents and chemicals, to combat appearance of resistance to chemical insecticides.

The progress of insect control by var. *israelensis* was aided by international funding. This was administered by the Biological Control Component of the Special Programme of WHO directed against six major tropical diseases in the developing world (Burges *et al.* 1981). Significantly, this programme encouraged not only search for new pathogens of all types, but also research and improvement of research personnel and facilities in the developing world. As a result new isolates of var. *israelensis* have been found, also many laboratory studies have examined host range and the requirements for successful use. These studies include formulation to keep toxin crystals suspended in the feeding zones of blackfly and mosquito larvae (WHO 1982; Lacey *et al.* 1984) and the effect of environmental conditions, such as pollution with organic particles and inorganic silt. Many field trials with mosquitoes in relatively small test plots have been undertaken (Gaughler and Finney 1982; WHO 1982) and attempts are currently being made to conduct large scale field trials with mosquitoes. For example, in a Sri Lankan malaria control project, which operated independently of WHO, trials against mosquito larvae in 80 pools have led to the start of an operational scheme, beginning with a single village and radiating outwards. The larvicidal *B. thuringiensis* was formulated in a surface active monolayer, itself insecticidal mainly against mosquito pupae and emerging adults (Roberts and Burges 1984).

Studies in developed countries have shown that the var. *israelensis* crystal toxin causes cytolysis of gut epithelial cells, but has considerable differences from that of other varieties (Fast 1981; WHO 1982). The relatively small mosquito-toxic molecule (mass about 25 kDa) probably acts by causing rearrangement of membrane phospholipids similar to that caused by detergents (Thomas and Ellar 1983; Ellar *et al.* 1985). The crystals contain endogenous proteases that may aid dissolution into active toxin in the insect gut (Chilcott *et al.* 1983).

2.1.3. *Bacillus sphaericus*

From both the developed and developing world, some strains of another spore-forming species of *Bacillus*, *B. sphaericus*, are highly potent against larvae of some mosquito species, but not against blackflies (WHO 1980*a*; Yousten 1984). Classification of strains of this species by bacteriophage and by H-antigen typing give virtually perfect agreement, all the mosquito-active pathogens occurring in both phage type 3 and serotype H–5a5b (Yousten 1984). *Bacillus sphaericus* can be produced in bulk by fermentation and formulated like *B. thuringiensis* var. *israelensis* (Vandekar and Dulmage 1983; Yousten *et al.* 1984). A proteinous toxin is located mainly in the spore coat and, in some strains, in crystals as well, serving a similar function to that of the toxin of *B. thuringiensis* H–14 (Davidson 1981, 1982). However, the bacterium multiplies freely in larval cadavers and recycles in larvae but probably insufficiently to maintain long-term effective control (Davidson *et al.* 1984). The spore persists in the environment for a long time, but tends to accumulate in bottom sediments away from feeding zones of mosquito larvae (Yousten 1984). *Bacillus sphaericus* grows saprophytically on a wide range of products that pollute water (Hornby *et al.* 1981) and in some polluted waters, e.g. some sewage installations in the USA (B. C. Hertlein, personal communication). This may be an environment in which *B. sphaericus* might have an advantage over *B. thuringiensis* var. *israelensis* especially against the main vector of Bancroftian filariasis, *Culex quinquefasciatus*, which breeds in polluted pools, worldwide. Pilot production, field experiment and small trials have occurred in many countries (WHO 1980*a*).

2.2. Viruses

While a relatively few species of bacteria cause disease in a large proportion of insect species, a large number of viruses are, in total, equally or more destructive. Martignoni (1981) has listed nearly 1000 host-virus relationships involving over 640 species of insects and mites. Probably many more await description. Among seven virus families (Payne and Kelly 1981), three have naked virus particles and others have particles protected in a proteinaceous inclusion body enabling lengthy survival in the environment. Among the latter a family of DNA viruses, the Baculoviridae, have most potential for insect control so further consideration will be limited to this group.

Morphologically and biologically, baculoviruses are unrelated to other animal viruses or to plant viruses. They are regarded as safe and ecologically acceptable (Ignoffo 1975; Burges 1981*a*). They often terminate outbreaks of important agricultural and forest pests, but not before the plants have been severely damaged. They can be grown in cell cultures (Stockdale and Priston 1981; Hink 1982), but commercial production is practical only in living insects. Although this is costly, they are sometimes so infectious that the quantities needed for use as viral insecticides can be economically competitive. Some are able to keep pests below the economic density threshold by spread from spaced application foci.

Three examples will be described in some detail to illustrate the three ways in which viruses can be utilized: (i) as viral insecticides, (ii) as introductions, and (iii) by environmental manipulations. These will involve discussions of application in both developed and developing countries.

2.2.1. Control of forest and field pests by baculoviruses

Baculoviruses (Fig. 4.2A) protect their infective DNA (Fig. 4.2D) in lipoprotein-sheathed particles (Fig. 4.2C), themselves embedded in a proteinaceous matrix to form an inclusion body (Fig. 4.2B). In Europe and North America, inclusion bodies of a baculovirus of the gypsy moth, *Lymantria dispar* L., have been produced to control this severe forest pest (Lewis 1981; Cunningham 1982). A pilot product ('Gypcheck'), using laboratory-reared insects is scheduled for commercialization in the 1985–86 season by Reuter Laboratories in the USA (Grimble and Lewis 1984). For virus production, disease-free larvae, grown for 14 days at 29°C, are transferred to agar-based artificial insect food already inoculated with pure virus under strictly hygienic laboratory conditions. After 10 days they are deep frozen and stored at −30°C. Virus inclusion bodies are harvested by blending thawed insects in water, straining through cheesecloth, centrifuging, and drying overnight under a hood fed with a laminar flow of sterile air. After grinding, quality control consists of checking potency of the powder by bioassay, and ensuring absence of dangerous bacteria by plating on selective bacteriological media and injection into mice. Field application is by conventional ground sprayers, or by aircraft spraying in carefully monitored meteorological conditions. For example, two aerial applications of 2.5×10^{11} inclusion bodies/ha with, for instance, molasses and a suitable sticker are recommended, the first timed when leaves are partly expanded and larvae in development stages 1 and 2 are present. The second is timed 5–10 days later, but before larvae grow to stage 4. This treatment should reduce gypsy moth egg masses in the next generation by 75 per cent or more and give acceptable reduction of tree defoliation. Gypcheck was produced by the United States Department of Agriculture and registered in 1978 by the Environmental Protection Agency in the USA. Costs can be greatly reduced by improved technology, e.g. an estimated US $ 24–80/ha in 1977 was reduced to $ 1.00–1.50 plus 0.90 for quality control (Shapiro 1982).

Gypcheck is used strictly as a viral insecticide. Sprays are rapidly inactivated by solar radiation and infection by virus takes a relatively long time to kill larvae. The virus neither spreads across the forest nor persists enough to give adequate control in subsequent years. In contrast, baculoviruses from a number of sawflies (*Neodiprion* spp. and *Gilpinia hercyniae* (Htg.)), produced from larvae collected from infected infestations in forests in North America and Europe, have been applied at spaced loci from which they spread to give good control lasting many years (Cunningham and Entwistle 1981).

Viral insecticides have been commercialized for use on field crops, notably against *Heliothis* spp. on cotton. These have not yet been profitable, in contrast to some products for forest use.

In the developing world, there have been attempts to use viruses in agriculture. One project by F. Moscardi won a national young scientist award in Brazil. His research on a baculovirus of the velvetbean caterpillar, *Anticarsia gemmatalis* Hübner, has led to national production and application of the virus throughout the soybean production area of Brazil. In 1984, 100 000–300 000 ha were treated and, over 3–4 years, a projected 1 700 000 ha are to be treated (Anon. 1984).

Chinese workers have used baculoviruses against, or isolated them from, *Heliothis armigera* Hübner and *Mythimna (Leucania) separata* (Walker), including use of viruses in attractant bait (Hussey and Tinsley 1981). In the Indian subcontinent there is a similar experimental interest in viruses of the latter two pests, of *Chilo infuscatellus* Sn. the sugar-cane borer (Easwaramoorthy and David 1979) and also of *Agrotis segetum* (D. and S.) (Shah *et al.* 1979). Research has included; (i) effects of sunlight and temperature on survival and infectivity of virus; (ii) virus survival in soil (Manjunath and Mathad 1978, 1981); (iii) combination with a Braconid parasitoid and the insecticide fenitrothion (Dilawari *et al.* 1981; Savanurmath and Mathod 1981); (iv) integrated use, biology, and mode of action of an Indian strain of the *Heliothis* virus (Battu and Dilawari 1978; Komolpith and Ramakrishnan 1978; Narayanan and Jayaraj 1978; Godse and Patil 1979). Islam (1973), and Islam and Sana (1974) studied a virus of *Spodoptera exigua* (Hübner) and an unidentified virus of the jute hairy caterpillar (*Diacrisia obliqua* Walker).

2.2.2. Control of a pasture pest by a baculovirus

Production of a virus in laboratory or field-reared insects is sometimes impracticable. Occasionally, the environment can be manipulated to spread virus and produce adequate pest control. Improvement of pasture land for New Zealand's major industry, sheep and cattle production, caused severe outbreaks of *Wiseana* spp. (Hepialiid moths). Ploughing, resowing grass, and cutting successive hay crops buried and prevented spread of a baculovirus that probably once reduced populations of larvae. Kalmakoff and Crawford (1982) recommended that a favourable ecological balance between virus and pests can be restored by maintaining the integrity of pastures as follows.

(1) Renovation of damaged pastures and conversion of native tussock grasslands to regular pastures by applying fertilizer to the surface and drilling seed directly into grassland, without the disruption of ploughing.

(2) Intensive rotational stocking to ensure that disturbance of the soil surface by animal activity helps to disperse virus.

(3) Winter hay feeding to enable badly damaged pastures to be renovated by drilling grass seed instead of ploughing.

(4) Avoidance of taking successive crops of hay.

2.2.3. Spread of a baculovirus to control the rhinoceros beetle in developing countries

In a range of developing countries, there has been a dramatic example of virus spread by releasing infected insects—akin to classical biological control methods (Bedford 1980, 1981).

The coconut rhinoceros beetle, *Oryctes rhinoceros* L., was accidentally introduced into West Samoa from India and South East Asia early this century. It subsequently spread to neighbouring Tonga, Fiji, and Tokelau, where it all but obliterated the coconut palm. It had also been introduced inadvertently to New Britain and Micronesia. In one year alone (1968), estimated damage to the economies of South Pacific countries totalled about US $1 000 000. In Fiji, quarantine services designed to minimize spread of the beetle, from its first discovery in 1953 to 1971, cost an estimated $1 700 000.

A full scale project from 1964 to 1975, backed by international agencies, introduced natural enemies of the beetle. By far the most effective was a baculovirus, discovered by A. Huger (Huger 1966) and exceptional in that the virus particles were not protected in an inclusion body. When introduced into Samoa, the virus spread rapidly within and between islands (Marschall 1970; Young 1974). From this evolved a whole range of technological breakthroughs in virus propagation and release. At present, a stock of virus is produced and maintained by the Ministry of Agriculture in Fiji, using simple techniques at low cost. Infected larvae are stored refrigerated. When required, they are ground in a simple mincing machine and mixed into water, in which adult beetles are dunked for 1–2 min, so that virus enters by mouth to start gut infections. On release, infected adults fly actively and spread virus mainly by depositing contaminated faeces in larval breeding sites. The method is simple and inexpensive. It can be applied anywhere that has background knowledge and safety clearance, the method being universally considered safe.

2.3. Fungi

Fungi vie with viruses in containing the largest number of insect pathogens causing diseases that naturally terminate large pest-outbreaks. Fungal spores grow on and penetrate insect cuticle. Thus, victims of fungi include insects that suck plant juices or blood, food substrates unlikely to harbour spores of other insect pathogen groups that rely only on oral transmission. In water, many fungi have motile spores with some ability to search for hosts. On land, fungi are limited by the need for very high humidity and sufficiently high temperature for spore germination and production. Water-based fungi attack only aquatic stages of hosts. The usual factors, sufficiency of inoculative stage in the environment and adequate host density, commonly limit effective pest control to the termination of major pest outbreaks. Some fungi are important natural agents that severely limit insect numbers. The general strategy used to obtain adequate pest control is application of heavy inocula early in insect population cycles. Most insect pathogens belong to the Fungi Imperfecti, the Entomophthorales and the 'water moulds'.

2.3.1. Fungi Imperfecti

This group contains the easiest fungi to produce for microbial control. Most species are readily propagated by deep liquid fermentation alone. Others will sporulate only when vegetative growth in the liquid is added to sterilised solid media, such as broken cereal grains. Species producing spores in sticky heads, on insect cuticle, are spread by rain and by contact between healthy and diseased hosts. Others that produce dusty spores are spread by air currents and contact. Their spores are washed off foliage and out of the air by rain, but a combination of rain (to produce high humidity) and dry periods (to allow spread of spores) is ideal for dusty-spored species. Fungi of both types survive in soil.

Strains of *Verticillium lecanii* (Zim.) Viegas (Fig. 4.3), are the agents in the most advanced mycoinsecticides. These are produced in the UK as the commercial products 'Vertalec' and 'Mycotal' for use in greenhouses. Vertalec contains a strain used against aphids and Mycotal contains another strain used against a scale insect, the whitefly (*Trialeurodes vaporariorum* (Westwood)). Both products are applied only to crops grown at sufficiently high humidity and temperature (Hall 1981; Hall and Papierok 1982; Burges and Hall 1982). One prophylactic spore-spray, directed at the undersides of leaves of a young crop, is usually adequate to provide crop-long protection because the fungus spreads to successive waves of invading pests. Establishment early in the development of the crop, when pests are sparse, is aided by incorporating a nutrient medium in the formulation. This allows microcolonies, virtually invisible to the naked eye, to grow and sporulate on leaves. The key to success was the realization that the aphid strain would spread effectively in populations of aphids, but not in populations of whiteflies, and *vice versa* for the whitefly strain. This emphasizes the importance among fungi of using the correct strain for each type of pest. One hectare requires a spray of 500–1000 litres, containing 2.5 g product/litre (Burges and Hall 1982). Conidiospores of most Fungi Imperfecti are difficult to keep alive for long periods and *Verticillium* products must be stored at about 4°C, except during transport. *Verticillium* fills an important gap in integrated control programmes in greenhouses, caused by resistance to key chemical insecticides. It is compatible with most pesticides, providing they are not applied simultaneously in the same spray mixture or for several days before (Hall 1981).

In the developing world, *V. lecanii* causes insect disease throughout the humid tropics and subtropics, where applications outside greenhouses are being sought (Hall and Papierok 1982). Studies have involved the guava scale, *Pulvinaria psidii* Maskell, chili aphid, *Myzus persicae* (Sulzer) (Easwaramoorthy and Jayaraj 1977*a*, *b*, 1978); brown plant hopper of rice, *Nilaparvarta lugens* (Stal), (Balasubramanian 1979); and brinjal leaf beetle, *Henosepilachna vigintioctopunctata* Fab. (Santharam *et al.* 1978) on the Indian subcontinent; *Coccus viridis* (Green) on coffee (Viegas 1939) in Brazil; *C. viridis* and *Saissetia coffeal* (Walker) cited as *hemisphaerica* Targioni on coffee in the Philippines (Gabriel 1968).

Muscardine fungi, *Beauveria bassiana* and *Beauveria brongiatti* (Sacc.) Petch, are also used for pest control to some extent. *Beauveria bassiana* is marketed in the USSR as 'Boverin' against the Colorado potato beetle (Weiser 1982; Ferron 1978; 1981; Samsinakova *et al.* 1981). In large trials in France, very heavy applications of the fungus to soil established infections which persisted for many years in the soil, giving slow, limited control of a scarabeiid beetle—the cockchafer, *Melolontha melolontha* L. (Ferron 1981).

In the developing world, the fungus used most is another muscardine, *Metarhizium anisopliae* (Metsch.), which has green powdery spores, hence its name 'green muscardine'. In Brazil, it is mass-produced in several ways, particularly on sterilized rice in autoclavable plastic bags, inoculated by injection of mycelium from liquid fermentation. The moulded rice is dried at 25°C for 72 h, ground into a powder and sold as 'Metaquino', which should be stored at about 7°C (Ferron 1981; Wassink 1983). Its extensive use includes aerial application against the spittlebug, *Mahanarva posticata* Stal, among which it appears to spread for the duration of the crop (Wassink 1983). Quality control in the 1970s was non-existent, but is now being introduced. *Metarhizium anisopliae* is also used to back up the baculovirus in a few environmental niches in the rhinoceros beetle programme (Bedford 1981).

In China, *B. bassiana* is produced using simple techniques by pest-control specialists in communes. Its use has been extensive, for example on 400 000 ha in Kirin province in 1977, against many pests, especially the European corn borer, *Ostrinia nubilalis* (Hübner), *Dendrolimus punctatus* Walker on pines, and *Nephotettix* leaf hoppers on tea (Hussey and Tinsley 1981). Records of occurrence of muscardine fungi in developing countries are too numerous to be detailed here, but the following publications may be consulted: Ferron (1981), Ramakrishnan and Kumar (1977), Pameswaran and Sankaran (1979), Israel and Padnanabhan (1980), Hirashima *et al.* (1979), Gomez (1979), Atuahene and Teyegaga (1979), and Wassink (1983).

Other Fungi Imperfecti, listed from developing countries by Wassink (1983), have been well-studied, but not successfully commercialized. These include *Nomuraea rileyi* (Farlow) (Ignoffo 1981; Ignoffo *et al.* 1975, 1978); *Hirsutella thompsonii* Fisher (on citrus, coconut and other plants; van Brussel 1975; Yen 1974; Chiang and Huffaker 1976; Ignoffo *et al.* 1973; McCoy *et al.* 1971; Hall *et al.* 1980; Hall and Espinosa 1981; McCoy 1981); *Culicinomyces clavosporus* Romney and Rao and *Tolypocladium cylindrosporum* Gams on mosquito larvae (WHO 1980*b*; Soares *et al.* 1979; Weiser and Pillai 1981; Gardner *et al.* 1982; Yu *et al.* 1980; P. O. Matewele, personal communication, 1985).

2.3.2. Entomophthorales

Many fungi in the large number of genera in the Entomophthorales are ubiquitous in nature. They occur mainly in temperate areas, principally on Hemiptera, Orthoptera, Lepidoptera, and Diptera. After spread in favourable seasons by very delicate condiospores, they commonly terminate insect outbreaks, although some fungal species can limit insect populations continuously below the levels which would cause economic loss of crops (e.g. green apple sucker in Canada; Jaques and Patterson 1962). When insect populations decline, environmental factors trigger the production of thick-walled, dormant, resting spores in the insects' bodies (Latgé *et al.* 1979;

Wilding 1981). These survive in the soil and various factors that break dormancy synchronize spore germination at a time when insect populations become active (Latgé *et al*. 1978; Wilding 1981).

Entomophthorales are difficult to use as mycoinsecticides (Wilding 1981). This is because most cannot be produced readily *in vitro*; conidiospores are so delicate that they cannot be handled, while resting spores are difficult to produce and to germinate synchronously. Effectiveness is determined by the density of fungus inoculum, density, and spatial distribution of the host, as well as the level of moisture in the insects' microenvironment (Soper and McLeod 1981; Wilding 1981). The difficulty of *in vitro* production usually prevents effective supplementation of the fungus inoculum (Gustafsson 1965; Soper *et al*. 1975; Krejzova 1973). Although they are important population-limiting agents in nature, the main hope of using them for pest control is to introduce exotic, more effective, species into a particular area or to manipulate the environment.

In developing countries records of natural occurrence of these fungi are too numerous to list here, but there has been one field application in China. *Erynia (Entomophthora) aphidis* (Hoffman), grown on soaked bran, was sprayed at 6×10^6 conidiospores/ml, resulting in 76–100 per cent aphid mortality (Hussey and Tinsley 1981). Thus, although hope of using this fungal group for pest control should not be discarded, success in producing *predictable* and *repeatable* pest control may be limited to certain very favourable cropping areas.

2.3.3. Water moulds

The genus *Coelomomyces* is the outstanding insect pathogenic group among these moulds (Mastigomycetes). There are about 50 species which, together with a few strains of *Lagenidium giganteum* (Couch), are the only fungi known to cause intense disease outbreaks among mosquito larvae, including important vector-species of human diseases (Federici 1981). These highly specialized pathogens alternate between stages that produce motile spores in mosquito larvae and a gamotophytic stage in a copepod or rarely an ostracod (Whisler *et al*. 1975; Pillai *et al*. 1976; Weiser 1976; Federici 1981; Toohey *et al*. 1982). At present, mass propagation of these fungi seems unlikely, though some may have potential for mosquito control. In the USSR, Dzerzhinskii and Deshevykh (1979) claimed successful control of the mosquito *Culex modestus* Ficabi with *Coelomomycies iliensis* Dub., Desh and Danev by transferring water, containing the whole biota, from pools naturally infected with the fungus to uninfected sites.

In the developing world, a similar approach is underway to treat Indian rice fields (Anon. 1982). Basic biology and field ecology, particularly of *Coelomomyces indicus* Iyengar, are being studied in Asia and Africa (Roberts and Strand 1977; Roberts *et al*. 1983; Anon. 1981, 1982; Chandrahas and

Rajagopalan 1979) and in the Philippines (L. Padua, personal communication). M. Laird (personal communication, 1967) established *Coelomomyces stegomyiae* Keilin :n a tree-hole mosquito in one of the atolls of the Tokelau group of islands in the South Pacific, but control was limited.

3. Potential of insect pathogens in the developing world

Pest problems are most severe in the developing world because so many of the countries have tropical and subtropical climates in which pests multiply so rapidly. The occurrence and variety of insect pathogens is correspondingly great. These features should eventually lead to increased use of insect pathogens there.

Insect pathogens have two main advantages to offer the developing world. One is specificity, enabling conservation of other natural control agents, which are often abundant and of particular value in economies that cannot afford to become increasingly dependent on costly, unspecific chemical pesticides, which are usually lethal to beneficial agents. The other advantage is safety; if properly produced, insect pathogens will not cause harm to the environment or people by overdosing or misuse, a great asset among poorly educated workers and owners of small farms.

Eventual use will depend on how effective insect pathogens prove to be in specific locations and how rapidly both technical knowledge and the insect pathogens themselves can be made available at all levels from large plantation to peasant farm. Progress should be most rapid where the various inadequacies of chemicals are most severe. Microbial agents with wide geographical application and wide host range should increase in use most rapidly; those of limited or local value may take longer.

Future control of imports and self-help in the developing world

Trends in pest control in developing countries will depend on their individual economic and political futures. Presumably, the developing world will develop and become more self-sufficient. The five areas of the developed world, listed in the first paragraph of this chapter, are fairly well-defined zones both economically and politically, arranged in rough order of expertise in use of microbial control. Africa, and Central and South America are fragmented into mainly developing countries in which the five 'developed areas' compete for influence. Microbial control in these countries will thus tend to follow the dominating political and economic influences. The Indian sub-continent, like a number of other regions, has benefited from former colonial scientific effort from which has grown local microbial control research. The People's Republic of China is exceptional in that considerable research, development, and use of pathogens has been undertaken by Chinese people, with the help of foreign literature, but not of foreign aid. Many developing countries have strict currency curbs, some forbid imports and strive correspondingly hard to encourage local products. Within this broad framework, the ability of a developing country to import microbial products and/or initiate its own microbial industry will depend on its wealth to do so. Nigeria, for instance, could afford to do both, but will probably strive to start its own indigenous industries (Okafor, 1985). Others could develop an industry for producing micro-organisms, but only with low-cost technology, although many could establish labour-intensive methodology where labour is abundant and inexpensive. Some developed countries could even operate labour-intensive processes in the developing countries to produce microbials for a world market. Certain developing countries could produce a local microbial product in excess of their internal needs and export the surplus, at least to neighbouring countries, but none do so yet.

Other factors, specific to microbial control agents, may have a dominant influence. Governments are cautious about importing exotic organisms, creating an advantage to national production of endemic species or strains. Increased knowledge about the performance of exotic organisms should reduce this problem. Even though many countries follow safety registration precedents established in the USA and Europe (WHO 1981; Burges 1981b), registration of a product for use in many different countries is time-consuming and collectively expensive. As confidence in safety tests and guidelines that control them increases, this obstacle should lessen. In particular, the time lapse between application for registration and the official response should be reduced. National requirements can be very specific, for example, countries that produce natural silk have included tests on silk worms among their registration requirements, e.g. for baculoviruses and *B. thuringiensis* in India (Dahutti and Mathad 1981).

The need for effective quality control of products is a major consideration in local production. Paramount is the safety aspect to ensure that a dangerous organism is not produced in error, or that a dangerous contaminant is not present (Burges 1981b). It is also important to ensure that the product is efficacious. Existing production of insect pathogens in developing countries has largely ignored quality control. As these countries advance towards standards required, for instance in the USA and in Europe (Burges 1981b), quality control technology will be needed and the scale of production will have to be increased to encompass the extra costs. Brazil has already sought American expertise to help formulate national safety testing and quality control guidelines for microbial insecticides (D. W. Roberts, personal communication). Some processes, such as production of mermithid nematodes in mosquito larvae or production and release of infected insects, involve virtually no risk. Others, such as production of *B. thuringiensis* and fungi by fermentation, are well documented (Vandekar and Dulmage 1983) and require defined standards of fermentation and quality control. Production on an economic scale at only one site would be adequate or even excessive for the requirements of a small or moderate-sized country.

The scale of crop production in a developing country will influence the approach taken to local production of microbial insecticides. Some crops grown in large plantations, many owned by a single large commercial company, may have a microbial-pesticide requirement sufficient to warrant local production financed and conducted by the company itself. Individual farms would have smaller requirements and would have to rely on a central supplier or co-operative effort. Peasant farming would have to rely on both centralized supply and help with application expertise; it could possibly be serviced with projects backed by government or international agencies, or with communal self-help efforts—as in the use of a *B. thuringiensis* var. *israelensis*-monolayer mixture in a project financed by OXFAM in Sri Lanka (Roberts and Burges 1984). A production facility organised by a government could eventually be passed to private industry as happened with *B. popilliae* in the USA.

Even microbial agents planned for introductions—as opposed to regular use as microbial insecticides—require sizeable production operations that may be commercialized, particularly if the introduction is required every season or at the beginning of every crop. These agents require more post-application study for efficacy than the short-term microbial insecticides.

Much research in the developed world is devoted to problems in common with, or specific to, the developing world. While this should continue, many local problems could be best investigated locally, if adequate technical and scientific resources are available. Local study is necessary for pathogens that are able to spread unaided through pest populations and thus give great benefit from the introduction of small quantities. This makes them difficult to commercialize so they will fall into the domain of non-commercial activities; their importance must be recognized.

Research on-site is desirable at all levels. An important basic research activity is collection of new isolates and species of micro-organisms as potential control agents. These may be primarily for local study, as with *B. thuringiensis* strains discovered in Mexico, or they may add to the world's joint fund of agents and the genes contained therein. Local trials of microbial agents are essential, particularly for those most sensitive to environmental conditions, with special emphasis on the possible roles of agents in relation to regional crop husbandry. Locally-orientated practical work is likely to give the quickest return for effort to the developing world, so some teaching of insect pathology should be incorporated in university curricula. A text book on insect pathology and microbial control has been written in Spanish by a South American group (Kuno *et al.* 1982).

An increasing role in fundamental research should be played by the developing world as regional scientific manpower and sophistication improve. Some fundamental studies have already been undertaken. For instance, Sosa Gomez and Alves (1983) studied the relationship between virulence and enzymes produced by the fungus *Metarhizium anisopliae* and the genetics of this fungus are being studied in Brazil (Wassink 1983); also *B. thuringiensis* strains found in Mexico are the subject of work at various levels, including attempts to produce monoclonal antibodies to the crystal toxins.

Recent highlights of fundamental work in developed countries offer prospects of far-reaching advances. These highlights include research on the mode of action of pathogens and toxins, chemistry of toxins, genetics and genetic manipulation. Microbial control has the great advantage of specificity to pests, but this also creates the problem of limiting the market size for individual microbial insecticides. Genetic recombination has created strains of *B. thuringiensis* with improved host ranges. The *B. thuringiensis* endotoxin gene has been inserted and expressed in the tobacco plant to create a plant systemically protected from caterpillar attack (Yanchinski 1985). The toxin might possibly be inserted into blue green algae to engineer mosquito-larvicidal algae that grow freely in natural waters. To protect plant roots from root-feeding caterpillars, the toxin has been transferred to soil bacteria that grow around the rhizospheres (Beardsley 1984). Some 30 per cent of the protein in the *B. thuringiensis* cell becomes crystal toxin and the polyhedrin protein in which virus particles are embedded in baculovirus inclusion bodies far exceeds in quantity the actual virus particles. Thus, both types of insect pathogen have very powerful producer sequences. These sequences can be used to amplify expression of the products of introduced genes. For instance, genes coding for human beta interferon and the enzyme β-galactosidase of *Escherichia coli* (Mig.), have been incorporated into the virus DNA of a baculovirus of *Autographa californica* (Speyer) (Smith *et al.* 1983; Pennock *et al.* 1984). When infecting insect cell cultures, the recombinant viruses produced the products of the passenger genes (interferon or β-galactosidase) at a much higher level than most other vector–gene–expression systems tested. Thus, insect virus and insect cell systems might be used industrially outside the realm of pest control.

The developing world should, indeed, increase its scientific capacity to take part in the prospects offered, not only by microbial control *per se*, but also in the wider potential use of insect pathogen genes in other applications (Kurstak and Tijsson 1982; Burges 1985, 1986).

References

Anon. (1981). *The eighth annual report of the International Centre of Insect Physiology and Ecology*, Nairobi, Kenya. October, 1981.

—— (1982). *Annual report for the Vector Control Research Centre*, Pondicherry, India.

—— (1984). Insect pathologist wins award. *Soc. Invert. Pathol. Newslett.* 16, 13–14.

Atuahene, S. K. N. and Teyegaga, A. (1979). The pathogenicity of *Beauveria bassiana* (Bals) Vuill. on larvae of *Lamprosema lateritalis* Hamps (Lepidoptera: Pyralidae) a pest of afrormosia in Ghana. Abstract of a paper presented to the Scientific Conference of the African Association of Insect Scientists, Dec 3–8, 1979. *Bull. Afr. Insect Sci.* 3, 12.

Balasubramanian, M. (1979). Pest management studies for rice brown planthopper in Tamil Nadu, India. In *Recent trends in rice brown planthopper control* pp. 9–10. Colloquim on rice brown plant hopper, 24 June 1949, Coimbatore, India. Sandoz (India), limited publication.

Barjac, H, de (1981). Identification of H-serotypes of *Bacillus thuringiensis*. In *Microbial control of pests and plant diseases 1970–1980* (ed. H. D. Burges) pp. 35–43. Academic, London.

Battu, G. S. and Dilawari, V. K. (1978). Preliminary investigations on the safety evaluation of *Spodoptera litura* (Fabricius) nuclear polyhedrosis virus (SLNPV) against a parasitoid, *Parasarcophaga misera* (Walker). *Entomologists' Newslett.* 8, 6.

Beardsley, T. (1984). Monsanto goes ahead with trials. *Nature, Lond.* 312, 686.

Bedford, G. O. (1980). Biology, ecology and control of palm rhinoceros beetle. *Ann. Rev. Ent.* 25, 309–40.

—— (1981). Control of the rhinoceros beetle by baculovirus. In *Microbial control of pests and plant diseases 1970–1980* (ed. H. D. Burges) pp. 409–26. Academic, London.

Bulla, L. A., Jr, Bechtel, D. B., Kramer, K. J., Shethna, Y. I., Aronson, A. I., and Fitz-James, P. C. (1980). Ultrastructure, physiology and biochemistry of *Bacillus thuringiensis*. In *Critical reviews in microbiology* (ed. H. D. Isenberg) Vol. 8, pp. 147–204. CRC Press, Florida.

Brussel, E. W., van (1975). Interrelations between citrus rust mite *Hirsutella thompsoni* and greasy spot on citrus in Surinam. *Landbouwproefst Suriname Bull.* 98, 66 pp.

Burges, H. D. (1973). Enzootic diseases of insects. In *Regulation of insect populations by micro-organisms* (ed. L. A. Bulla Jr). *Ann. New York Acad. Sci.* 217, 31–49.

—— (ed.) (1981a). *Microbial control of pests and plant diseases 1970–1980.* Academic, London.

—— (1981b). Safety, safety testing and quality control of microbial pesticides. In *Microbial control of pests and plant diseases 1970–1980* (ed. H. D. Burges), pp. 737–67. Academic, London.

—— (1981c). Strategy for the microbial control of pests in 1980 and beyond. In *Microbial control of pests and plant diseases 1970–1980* (ed. H. D. Burges), pp. 797–836. Academic, London.

—— (1982). Control of insects by bacteria. *Parasitology* 84, 79–117.

—— (1986). Production and use of pathogens to control insect pests. *J. appl. Bacteriol. Symposium Supplement* S 1986, 1275–375.

——, (1986). Impact of *Bacillus thuringiensis* on pest control with emphasis on genetic manipulation. *Mircen J. appl. Microbiol. Biotechnol.* 2, 101–20.

——, Fontaine, R. E., Nalim, S., Okafor, N., Pillai, J. S., Shadduck, J. A., Weiser, J., and Pal, R. (1981). World Health Organization research on the biological control of vectors of insect disease: present status and plans for the future. In *Biocontrol of medical and veterinary pests* (ed. M. Laird) pp. 1–35. Praeger, New York.

—— and Hall, R. A. (1982). Bacteria and fungi as insecticides. *Outlook on Agriculture* 11, 79–86.

—— and Hurst, J. A. (1977). Ecology of *Bacillus thuringiensis* in storage moths. *J. Invertebr. Pathol.* 30, 131–9.

—— and Hussey, N. W. (eds) (1971). *Microbial control of insects and mites*. Academic, London, 861 pp.

—— and Jarrett, P. (1984). A novel strain of *Bacillus thuringiensis* having an improved activity against certain Lepidopterous pests. British Patent 84 25487, 11 pp.

Cabrera, A. L. B. (1983). Identification bioquimica y serologica de cinco cepas de *Bacillus thuringiensis* aisladas de suelo del estado de Nuevo Leon. Tesis de Quimico Bacteriologo Parasitologo, Universidad Autonoma de Nuevo Leon, Monterrey, 41 pp.

Cantwell, G. E. (ed.) (1974). *Insect diseases* Vol. 1, 2. Dekker, New York.

Chandrahas, R. K. and Rajagopalan, P. K. (1979). Mosquito breeding and the natural parasitism of larvae of a fungus *Coelomomyces* and a mermithid nematode *Romanomermis*, in paddy fields in Pondicherry. *Indian J. Med. Res.* 69, 63–70.

Chiang, H. C. and Huffaker, C. B. (1976). Insect pathology and microbial control of insects in China. In *Proc. 1st Int. Colloq. Invertebr. Pathol. IXth Ann. Meet. Soc. Invertebr. Pathol.*, Kingston, Canada. pp. 42–6.

Chilcott, C. N., Kalmakoff, J., and Pillai, J. S. (1983). Characterization of the proteolytic activity in *Bacillus thuringiensis* var. *israelensis* crystals. *FEMS Microbiology Lett.* 18, 37–41.

Cunningham, J. C. (1982). Field trials with baculoviruses: control of forest insect pests. In *Microbial and viral pesticides* (ed. E. Kurstak) pp. 335–86. Dekker, New York.

—— and Entwistle, P. F. (1981). Control of sawflies by baculovirus. In *Microbial control of pests and plant diseases 1970–1980* (ed. H. D. Burges) pp. 379–407. Academic, London.

Dahutti, S. G. and Mathad, S. B. (1981). Effect of nuclear polyhedrosis virus of armyworm *Mythimna separata* on Eri Silkworm *Philosamiaricini*. *Entomon* 6, 115–16.

Davidson, E. W. (1981). Toxin-producing bacilli other than *Bacillus thuringiensis*. In *Pathogenesis of invertebrate microbial diseases* (ed. E. W. Davidson) pp. 269–91. Allanheld, Osmum, Towota N.J.

—— (1982). Purification and properties of soluble cytoplasmic toxin from the mosquito pathogen *Bacillus sphaericus*. *J. Invert. Pathol.* 39, 6–9.

——, Urbina, M., Payne, J., Mulla, M. S., Darwazeh, H., Dulmage, H. T., and Correa, J. A. (1984). Fate of *Bacillus sphaericus* 1593 and 2362 spores used as larvicides in the aquatic environment. *Appl. Environmental Microbiol.* 1, 125–9.

Dilawari, V. K., Mahal, M. S., and Bains, S. S. (1981). Role of a Braconid parasite and a viral disease in the population decline of armyworm *Mythimna separata* (Noctuidae: Lepidoptera) during an outbreak. *Ind. J. Ecol.* 8, 65–73.

Dulmage, H. T. and co-operators (1981). Insecticidal activity of isolates of *Bacillus thuringiensis* and their potential for pest control. In *Microbial control of pests and plant diseases 1970–1980* (ed. H. D. Burges) pp. 193–223. Academic, London.

Dzerzhinskii, V. A. and Deshevykh, I. D. (1979). The possibility of introduction of entomopathogenic fungus *Coelomomyces iliensis* in south-east Kazakhstan. (In Russian). *Parazitologiya* (Leningrad) 13, 166–8.

Easwaramoorthy, S. and David, H. (1979). A granulosis virus of sugar cane shoot

borer, *Chilo infuscatellus* Snell. (Lepidoptera: Crambidae). *Current Sci.* 48, 685–6.
—— and Jayaraj, S. (1977*a*). Control of guava scale *Pulvinaria psidii* and chilli aphid *Myzus persicae* with *Cephalosporium lecanii* and insecticides. *Ind. J. Agric. Sci.* 47, 136–9.
—— and —— (1977*b*). The effect of temperature, pH and media on growth of the fungus *Cephalosporium lecanii*. *J. Invert. Pathol.* 29, 399–400.
—— and —— (1978). Effectiveness of the white halo fungus, *Cephalosporium lecanii* against field populations of coffee green bug *Coccus viridis*. *J. Invert. Pathol.* 32, 88–96.
Ellar, D. J., Thomas, W. E., Knowles, B. H., Ward, E. S., Todd, J., Drobniewski, F., Lewis, J., Sawyer, T., Last, D., and Nicholas, C. (1985). Biochemistry, genetics and mode of action of *Bacillus thuringiensis* δ endotoxins. In *Molecular biology of microbial differentiation* (eds J. Hoch and P. Setow) 230–40. A.S.M. Publications, Washington.
Fast, P. G. (1981). The crystal toxin of *Bacillus thuringiensis*. In *Microbial control of pests and plant diseases 1970–1980* (ed. H. D. Burges) pp. 223–48. Academic, London.
Faust, R. M. and Bulla, L. A., Jr (1982). Bacteria and their toxins as insecticides. In *Microbial and viral pesticides* (ed. E. Kurstak) pp. 75–208. Dekker, New York.
Federici, B. A. (1981). Mosquito control by the fungi *Culicinomyces*, *Lagenidium* and *Coelomomyces*. In *Microbial control of pests and plant diseases 1970–1980* (ed. H. D. Burges) pp. 555–72. Academic, London.
Ferron, P. (1978). Biological control of insect pests by entomogenous fungi. *Ann. Rev. Ent.* 23, 409–42.
—— (1981). Pest control by the fungi *Beauveria* and *Metarhizium*. In *Microbial control of pests and plant diseases 1970–1980* (ed. H. D. Burges), pp. 465–83. Academic, London.
Fine, P. E. M. (1984). Vertical transmission of pathogens of invertebrates. In *Comparative pathobiology* (ed. T. C. Cheng) Vol. 7, pp. 205–41. Academic, London.
Fuxa, J. R. and Tanada, Y. (eds) (1986). *Epizootiology of insect diseases*. Wiley, New York.
Gabriel, B. R. (1968). Entomogenous micro-organisms in the Philippines: New and past records. *Philippine Entomologist* 1, 97–130.
Gardner, J. M., Chang, C. M., and Pillai, J. S. (1982). The effects of salinity and temperature on the growth of three strains of the mosquito pathogenic fungus, *Tolypocladium cylindrosporum* Gams. WHO Mimeographed Document WHO/VBC/82.849.
Gaughler, R. and Finney, J. R. (1982). A review of *Bacillus thuringiensis* var. *israelensis* (serotype H–14) as a biological control agent of blackflies (Simuliidae). *Misc. Publ. Ent. Soc. Am.* 12, 1–17.
Godse, D. B. and Patil, R. B. (1979). Persistence of nuclear polyhedrosis virus of tobacco caterpillar *Spodoptera litura* Fabr. in Karnatak soil. *Ind. J. Exp. Biol.* 17, 1286–7.
Goldberg, L. J. and Margalit, J. (1977). A bacterial spore demonstrating rapid larvicidal activity against *Anopheles sergentii*, *Uranotaenia unguiculata*, *Culex univitattus*, *Aedes aegypti* and *Culex pipiens* (Dipt., Culicidae). *Mosq. News* 37, 355–8.
Gomez, A. L. (1979). Incidencia del hongo *Beauveria bassiana* en Mexico y su reproduccion en laboratorio para control microbiolegico inducido. In *VII Reunion*

nacional de control biologico, Veracruz, Mexico (Comité Organizador de la VII Reunion Nacional de Control Biologico), pp. 107–10.

Grimble, D. G. and Lewis, F. B. (1984). *Microbial control of Spruce budworms and gypsy moths.* United States Department of Agriculture, Broomall, Pennsylvania.

Guillet, P. (1983). Biological control of onchocerciasis in West Africa. *News Lett. Soc. Invert. Pathol.* 15, 21–2.

Gustafsson, M. (1965). On species of the genus *Entomophthora* Fresnius in Sweden. II. Cultivation and physiology. *Lanthbrukshogkolans Annalar* 31, 405–57.

Hall, R. A. (1981). The fungus *Verticillium lecanii* as a microbial insecticide against aphids and scales. In *Microbial control of pests and plant diseases 1970–1980* (ed. H. D. Burges) pp. 483–98. Academic, London.

—— and Espinosa, A. (1981). The coconut mite, *Eriophyes guerreronis* with special reference to the problem in Mexico. *Proceedings 1981 British Crop Protection Conference – Pests and Diseases*, 113–20.

——, Hussey, N. W., and Mariau, D. (1980). Results of a survey of biological control agents of the coconut mite, *Eriophyes guerreronis. Oleagineux* 35, 395–400.

——, and Papierok, B. (1982). Fungi as biological control agents of arthropods of agricultural and medical importance. *Parasitology* 84, 205–40.

Hink, W. F. (1982). Production of *Autographa californica* nuclear polyhedrosis virus in cells from large-scale suspension cultures. In *Microbial and viral pesticides* (ed. E. Kurstak) pp. 492–506. Dekker, New York.

Hirashima, Y., Aizawa, K., Miura, T., and Wongsiri, T. (1979). Field studies on the biological control of leafhoppers and planthoppers (Hemiptera: Homoptera) injurious to rice plants in South-West Asia. Progress report for the year 1977. *Esakia* 13, 1–20.

Hornby, J., Hertlein, B., Levy, R., and Miller, T. (1981). Persistent activity of mosquito larvicidal *Bacillus sphaericus* 1593 in fresh water and sewage. WHO mimeographed document WHO/VBC/81.8.30. World Health Organization, Geneva.

Huger, A. M. (1966). A virus disease of the Indian rhinoceros beetle, *Oryctes rhinoceros* (Linnaeus), caused by a new type of insect virus, *Rhabdionvirus oryctes* gen. n., sp. n. *J. Invert. Pathol.* 8, 38–51.

Hussey, N. W. and Tinsley, T. W. (1981). Impressions of insect pathology in the People's Republic of China. In *Microbial control of pests and plant diseases 1970–1980* (ed. H. D. Burges), pp. 785–95. Academic, London.

Ignoffo, C. M. (1975). Evaluation of *in vivo* specificity of insect viruses. In *Baculoviruses for insect pest control: safety considerations* (eds M. Summers, R. Engler, L. A. Falcon, and P. Vail) pp. 52–62. American Society of Microbiologists.

—— (1981). The fungus *Nomuraea rileyi* as a microbial insecticide. In *Microbial control of pests and plant diseases 1970–1980* (ed. H. D. Burges), pp. 513–38. Academic, London.

——, Barker, W. M., and McCoy, C. W. (1973). Lack of *per os* toxicity or pathogenicity in rats fed the fungus *Hirsutella thompsonii. Entomophaga* 18, 333–5.

——, Hostetter, D. L., Biever, K. D., Garcia, C., Thomas, G. D., Dickerson, W. A., and Pinnell, R. E. (1978). Evaluation of an entomopathogenic bacterium fungus and virus for control of *Heliothis zea* on soybeans. *J. Econ. Ent.* 71, 165–8.

——, Putler, B., Marston, N. L., Hostetter, D. L., and Dickerson, W. A. (1975). Seasonal incidence of the entomopathogenic fungus *Spicaria rileyi* associated with Noctuid pests of soybeans. *J. Invert. Pathol.* 25, 135–7.

Islam, M. A. (1973). Some studies on the natural enemies of *Spodoptera exigua*,

(Lepidoptera: Noctuidae) in Bangladesh. *Bangladesh J. Biol. Agric. Sci.* 2, 25–7.

—— and Sana, D. L. (1974). Virus disease of the jute hairy caterpillar *Diascaria obliqua* (Lepidoptera: Arctiidae) in Bangladesh. *Bangladesh J. Zool.* 2, 83–7.

Israel, P. and Padnanabhan, S. Y. (1980). Biological control of stem borers of rice in India. Final report (U.S.P.L. 480 Project) Central Rice Research Institute, Cattack, Orissa.

Jacques, R. P. and Patterson, N. A. (1962). Control of the apple sucker, *Psylla mali* by the fungus *Entomophthora sphaerosperma*. *Can. Ent.* 94, 818–25.

Kalmakoff, J. and Crawford, A. M. (1982). Enzootic virus control of *Wisseana* spp. in the pasture environment. In *Microbial and viral pesticides* (ed. E. Kurstak) pp. 435–48. Dekker, New York.

Klein, M. G. (1981). Advances in the use of *Bacillus popilliae* for pest control. In *Microbial control of pests and plant diseases 1970–1980* (ed. H. D. Burges) pp. 183–92. Academic, London.

Komolpith, U. and Ramakrishnan, N. (1978). Joint action of a baculovirus of *Spodoptera litura* and insecticides. *J. Ent. Res. (New Delhi)* 2, 15–19.

Knowles, B. H., Thomas, W. E., and Ellar, D. J. (1984). Lectin-like binding of *Bacillus thuringiensis* var *kurstaki* lepidopteran-specific toxin is an initial step in insecticidal action. *FEBS Lett.* 168, 197–202.

Krejzova, R. (1973). The preparation of the reproductive stages of *Entomophthora exitialis* Hall et Dunn. *Vestnik Ceskoslovenske Spolecnosti Zoologicke* 37, 21–2.

Krieg, A., Huger, A., Langenbruch, G. A., and Schnetter, W. (1983). *Bacillus thuringiensis* var *tenebrionis*: ein neuer, gegenüber larven von Coleopteren wirksomer pathotyp. *Zeitschrift angewandte entomologie* 96, 500–8.

—— and Langenbruch, G. A. (1981). Susceptibility of arthropod species to *Bacillus thuringiensis*. In *Microbial control of pests and plant diseases 1970–1980* (ed. H. D. Burges) pp. 837–96. Academic, London.

Kuno, G., Mulett, J., and de Hernandez, M. (1982). *Patologia de insectos con enfasis en las enfermedades infecciosas y sus aplicaciones en el control biologico.* Universidad del Valle, Columbia, 212 pp.

Kurstak, E. (1982). *Microbial and viral pesticides.* Dekker, New York.

—— and Tijsson, P. (1982). Microbial and viral pesticides: modes of action, safety and future prospects. In *Microbial and viral pesticides* (ed. E. Kurstak) pp. 3–32. Dekker, New York.

Lacey, L. A., Urbina, M. J., and Heitzman, C. M. (1984). Sustained release formulations of *Bacillus sphaericus* and *Bacillus thuringiensis* (H–14) for control of container-breeding *Culex quiquefasciatus*. *Mosquito News* 44, 26–32.

Laird, M. (1967). A coral island experiment. *WHO Chronicle* 21, 18–26.

Latge, J.-P., Perry, D., Papierok, B., Coremans-Pelseneer, J., Remaudiere, G., and Reisinger, O. (1978). Germination des azygospores d'*Entomophthora obscura* Hall et Dunn, role du sol. *Compt. Rend. Acad. Sci., Serie D*, 943–6.

——, ——, Reisinger, O., Papierok, B., and Remaudiere, G. (1979). Induction de la formation de spores de resistance d'*Entomophthora obscura* Hall et Dunn. *Compt. Rend. Acad. Sci. Serie D* 288, 599–601.

Lewis, F. B. (1981). Control of the gypsy moth by a baculovirus. In *Microbial control of pests and plant diseases 1970–1980* (ed. H. D. Burges) pp. 363–77. Academic, London.

Manjunath, D. and Mathad, S. B. (1978). Temperature tolerance thermal inactivation and UV light resistance of nuclear polyhedrosis virus of the armyworm *Mythimna separata* (Lepidoptera: Noctuidae). *Zeits. ang. entomol.* 87, 82–90.

—— and —— (1981). Stability of nuclear polyhedrosis virus of the armyworm *Mythimna separata* (Lepidoptera: Noctuidae) in soil. *Entomon* 6, 1–6.

Maramorosch, K. and Sherman, K. E. (1985). *Viral insecticides for biological control.* Academic Press, New York.

Marschall, K. J. (1970). Introduction of a new virus disease of the coconut rhinoceros beetle in Western Samoa. *Nature, Lond.* 225, 288–9.

Martignoni, M. E. (1981). A catalogue of viral diseases of insects, mites and ticks. In *Microbial control of pests and plant diseases 1970–1980* (ed. H. D. Burges) pp. 899–911. Academic, London.

McCoy, C. W. (1981). Pest control by the fungus *Hirsutella thompsonii*. In *Microbial control of pests and plant diseases 1970–1980* (ed. H. D. Burges) pp. 499–512. Academic, London.

——, Selhime, A. G., Kanavel, R. E., and Hill, A. J. (1971). Suppression of citrus mite populations with application of fragmentation mycelia of *Hirsutella thompsonii*. *J. Invert. Pathol.* 17, 270–6.

Milner, R. J. (1981). Identification of the *Bacillus popilliae* group of insect pathogens. In *Microbial control of pests and plant diseases 1970–1980* (ed. H. D. Burges) pp. 45–9. Academic, London.

Narayanan, K. and Jayaraj, S. (1978). Factors responsible for the mode of action of polyhedral inclusion bodies of nuclear polyhedrosis virus in the tobacco caterpillar *Spodoptera litura* F. *Curr. Sci.* 47, 310–11.

Norris, J. R. and Burges, H. D. (1963). Esterases of crystalliferous bacteria pathogenic for insects: epizootiological applications. *J. Insect Pathol.* 5, 460–72.

Okafor, N. (1985). Present state and future prospects of the fermentation industry in the developing countries of Africa: the Nigerian case. *Process Biochem.* 1985, 23–5.

Pameswaran, G. and Sankaran, T. (1979). Record of *Beauveria bassiana* (Bals) Vuill. on *Linschcosteus* sp. (Hemiptera: Reduviidae: Tritominae) in India. *J. Ent. Res.* 1, 113–14.

Payne, C. C. and Burges, H. D. (eds) (1982). Invertebrate pathology and microbial control. *Proc. IIIrd International Colloquium on Invertebrate Pathology; XVth Annual Meeting of the Society for Invertebrate Pathology, Brighton 1982.* Glasshouse Crops Research Institute, Littlehampton.

—— and Kelly, D. C. (1981). Identification of insect and mite viruses. In *Microbial control of pests and plant diseases 1970–1980* (ed. H. D. Burges) pp. 61–91. Academic, London.

Pennock, G. D., Shoemaker, C., and Miller, L. K. (1984). Strong and regulated expression of *Escherichia coli* ß-galactosidase in insect cells with a baculovirus vector. *Mol. Cell. Biol.* 4, 399–406.

Pillai, J. S., Wong, T. L., and Dodgshun, T. J. (1976). Copepods as essential hosts for the development of *Coelomomyces* parasitizing mosquito larva. *J. Med. Ent.* 13, 49–50.

Pruett, C. J. H., Burges, H. D., and Wyborn, C. H. (1980). Effect of exposure to soil on potency and spore viability of *Bacillus thuringiensis*. *J. Invert. Pathol.* 35, 168–74.

Ramakrishnan, N. and Kumar, S. (1977). Biological control of insects by pathogens and nematodes. *Pesticides Ann.* pp. 32–47.

Roberts, D. W., Daoust, R. A., and Wraight, S. P. (1983). Bibliography on pathogens of medically important arthropods: 1981. W.H.O. Document VBC.83.1. 324 pp.

—— and Strand, M. A. (1977). Pathogens of medically important arthropods. *Bull. WHO.* Vol. 55 Supplement No. 1.

Roberts, G. M. and Burges, H. D. (1984). Combination of *Bacillus thuringiensis* and a surface active monolayer for mosquito control. *1984 British Crop Protection Conference—Pests and Diseases* 4A, 287–92.

Samsinakova, A., Kalalova, S., Vicek, V., and Kybal, J. (1981). Mass production of *Beauveria bassiana* for regulation of *Leptinotarsa decemlineata* populations. *J. Invert. Pathol.* 38, 169–74.

Santharam, G., Easwaramoorthy, E., and Jayaraj, S. (1978). Preliminary laboratory evaluation of *Cephalosporium lecanii* (Zimm.) as a pathogen of brinjal leaf beetle Henosepilachna vigintioctopunctata (Fabr.). *Curr. Sci.* 47, 477.

Savanurmath, C. J. and Mathud, S. B. (1981). Efficacy of fenitrothion and nuclear polyhedrosis virus combinations against the armyworm *Mythimna separata* (Lepidoptera: Noctuidae). *Zeit. ang. entomol.* 91, 464–74.

Shah, B. H., Zethner, O., Gul, H., and Chaudhry, M. I. (1979). Control experiments using *Agrotis segetum* granulosis virus against *Agrotis ipsilon* (Lep: Noctuidae) on tobacco seedlings in northern Pakistan. *Entomophaga* 24, 393–401.

Shapiro, M. (1982). *In vivo* mass production of insect viruses for use as pesticides. In *Microbial and viral pesticides* (ed. E. Kurstak), pp. 463–92. Dekker, New York.

Smith, G. E., Summers, M. D., and Fraser, M. J. (1983). Production of human beta interferon in insect cells infected with a baculovirus expression vector. *Mol. Cell. Biol.* 3, 2156–65.

Soares, G. G., Jr., Pinnock, D. E., and Samson, R. A. (1979). *Tolypocladium* a new fungal pathogen of mosquito larvae with promise for use in microbial control. *Proc. Calif. Mosq. Vector Contr. Ass.* 47, 51–4.

Soper, R. S., Holbrook, F. R., Majchrowicz, I., and Gordon, C. C. (1975). Production of *Entomophthora* resting spores for biological control of aphids. Life Sci. Agric. Exp. Stn. Univ. Me. Orono. Tech. Bull. No. 76.

—— and McLeod, D. M. (1981). Descriptive epizootiology of an aphid mycosis. USDA Sci. Education Administration Tech. Bull. No. 1632.

Sosa Gómez, D. R. and Alves, S. B. (1983). Caracterización de once aislamientos de *Metarhizium anisopliae* (Metsch.) Sorok. 1. Estandarizacion, virulencia y actividad enzimática. *CIRPON, Rev. Invest.* 1, 83–102.

Steinhaus, E. A. (1949). *Principles of insect pathology*. McGraw-Hill, New York. Reprinted 1967, Hafner, London.

Stockdale, H. and Priston, R. A. J. (1981). Production of insect viruses in cell culture. In *Microbial control of pests and plant diseases 1970–1980* (ed. H. D. Burges) pp. 313–28. Academic, London.

Thomas, W. E. and Ellar, D. J. (1983). *Bacillus thuringiensis* var *israelensis* crystal endotoxin: effects on insect and mammalian cells *in vitro* and *in vivo*. *J. Cell Sci.* 60, 181–97.

Toohey, M. K., Prakash, G., Goettel, M. S., and Pillai, J. S. (1982). *Elaphoidela taroi*: the intermediate copepod host in Fiji for the mosquito pathogenic fungus *Coelomomyces*. *J. Invert. Pathol.* 40, 378–82.

Vandekar, M. and Dulmage, H. T. (eds) (1983). Guidelines for production of *Bacillus thuringiensis* H–14. World Health Organization, Geneva.

Viegas, A. B. (1939). Um amigo do fazendeiro *Verticillium lecanii* (Zimm.) n. comb., o causador do halo branco do *Coccus viridis* Green. *Revista do instituto du cafe do estado de Sao Paulo* 14, 754–72.

Wassink, H. (1983). Insect pathology in South America. *Soc. Invert. Pathol. Newslett.* 15, 33–4.

Weiser, J. (1976). The intermediate host for the fungus *Coelomomyces chironomi*. *J. Invert. Pathol.* 28, 273–4.

—— (1982). Persistence of fungal insecticides: influence of environmental factors and present and future application. In *Microbial and viral pesticides* (ed. E. Kurstak), pp. 531–58. Marcel Dekker, New York.

—— and Pillai, J. S. (1981). *Tolypocladium cylindrosporum* (Deuteromycetes, Moniliaceae) a pathogen of mosquito larvae. *Entomophaga* 26, 357–61.

West. A. W. (1984). Fate of the insecticidal, proteinaceous parasporal crystal of *Bacillus thuringiensis* in soil. *Soil Biol. Biochem.* 16, 357–60.

——, Burges, H. D., White, R. J., and Wyborn, C. H. (1984a). Persistence of *Bacillus thuringiensis* parasporal crystal insecticidal activity in soil. *J. Invert. Pathol.* 44, 128–33.

——, ——, and Wyborn, C. H. (1984b). Effect of incubation in natural and autoclaved soil upon potency and viability of *Bacillus thuringiensis*. *J. Invert. Pathol.* 44, 121–7.

Whisler, H. C., Zebold, S. L., and Shemanchuk, J. A. (1975). Life history of *Coelomomyces psorophorae*. *Proc. Nat. Acad. Sci. USA* 72, 693–6.

Wilding, N. (1981). Pest control by Entomophthorales. In *Microbial control of pests and plant diseases 1970–1980* (ed. H. D. Burges) pp. 539–54. Academic, London.

WHO (1980a). Data sheet on the biological control agent *Bacillus sphaericus*, strain 1593. Mimeographed document WHO/VBC/80.777.VBC/BCDS/80.10, World Health Organization, Geneva, 16 pp.

—— (1980b). Data sheet on the biological control agent *Culicinomyces clavosporus*. Mimeographed document WHO/VBC/80.775, Rev. 1. World Health Organization, Geneva, 7 pp.

—— (1981). Mammalian safety of microbial control agents for vector control: a WHO memorandum. *Bull. Wld. Hlth Org.* 59, 857–63.

—— (1982). Data sheet on the biological control agent *Bacillus thuringiensis* serotype H–14 (de Barjac 1978). Mimeographed document WHO/VBC/79.750 VBC/BCDS/79.01, Rev. 1, World Health Organization, Geneva.

Yanchinski, S. (1985). Plant engineered to kill insects. *New Scientist* 14, Feb., 25.

Yen, H. (1974). Isolation of *Hirsutella thompsonii* from *Phyllocoptruta oleivora*. *Acta Entomol. Sinicia* 17, 225–6.

Young, E. C. (1974). The epizootiology of two pathogens of the coconut palm rhinoceros beetle. *J. Invert. Pathol.* 24, 82–92.

Yousten, A. A. (1984). *Bacillus sphaericus*: microbiological factors related to its potential as a mosquito larvicide. *Adv. Biotechnol. Processes* 3, 315–43.

——, Madhekar, N., and Wallis, D. A. (1984). Fermentation conditions affecting growth, sporulation and mosquito larval toxin formation by *Bacillus sphaericus*. *Dev. Indust. Microbiol.* 25, 757–62.

Yu, H. S., Cho, H. W., and Pillai, J. S. (1980). Infection studies of mosquito pathogen *Culicinomyces* sp. against *Aedes* and *Culex* larvae. *Korean J. Entomol.* 10, 62–3.

5

Marine biotechnology and the developing countries

Rita R. Colwell

1. Introduction

Marine pharmaceuticals, genetic engineering of marine and estuarine animals and plants for food production, and marine specialty chemicals offer prospects for both immediate and long-term rewards for island and riparian nations (Colwell 1984a, b).

In the short-term, practical results can be anticipated in improving stocks of fish and shellfish for aquaculture and in developing compounds of pharmaceutical value from marine sources for human and domestic animal application. Also anticipated to be of immediate benefit are vaccines produced by genetic engineering methods for prevention and control of diseases of fish and shellfish in aquaculture.

Marine biotechnology, with an inherent industrial potential, holds the greatest promise for developing countries where fish and shellfish are a major source of food. Furthermore, developing countries possess abundant marine and estuarine natural resources that are ideally suited for exploration of marine biotechnology. This is because access to unusual and/or novel marine life permits direct utilization of new and/or unique products by genetic engineering and ensures continuous production in the laboratory, eliminating fluctuations caused by weather and climate.

Genetic engineering, i.e. applied molecular biology, represents a giant, technological 'leap' forward, a major accomplishment in the structural development of modern civilization. All institutions are being affected, not the least of which is the influence of this new technology on the role of universities in society (van Hemert et al. 1983; Sasson 1984).

Island and riparian countries should explore the potential of marine biotechnology unique to their regions. Existing marine laboratories should be linked with molecular genetic laboratories, if any exist within the country. If sufficient financial resources are available, investment in a genetic engineering laboratory with sufficient staff to accomplish the necessary work for genetic engineering of marine systems should be established. If financial constraints limit development, linkage of the marine facility with a molecular genetic/genetic-engineering laboratory in a developed country can be a cost-effective mechanism for establishing a marine biotechnology centre.

Workshops, seminar programmes, and short-course training for molecular biologists to become familiar with the workings of marine systems can also provide a means of technology transfer, i.e. to abbreviate the route to establishing a marine biotechnology capability.

Direct linkage of a marine field station with molecular genetic facilities and staff can be established to develop a marine biotechnology capability. However, for technology transfer to industry, further direct linkage to the marine industries of the country concerned will ensure that the results of marine biotechnology research are made available to relevant and interested industries.

2. Aquaculture

Historically, aquaculture has long been established in China, where for example, approximately 2 million metric tons of fin fish are produced every year, mostly in the form of carp grown in ponds, lakes, reservoirs, and ditches. However, the rate of knowledge-gathering in marine biology and advances made in technical expertise for applying discoveries in marine biology to aquaculture has increased significantly over the past decade. Research is underway on marine shrimp, freshwater prawn, crayfish, blue crab, brine shrimp, salmon and other fin fish, oysters, clams, abalone, and scallops.

Gene manipulation, i.e. the introduction of new combinations of heritable materials by insertion into any virus, bacterial plasmid, or other vector system of nucleic acid molecules (the basic genetic material of all organisms except some viruses), allows for the incorporation of heritable material into a host organism in which it does not naturally occur, but in which it becomes capable of continued propagation. These methods are being applied to aquaculture, employing larvae of fish and shellfish, including oysters, clams, abalone, and other molluscan species.

In vitro manipulations, such as cloning, cell fusion, production of chimeras, and other recombinant DNA techniques applied to marine animals provide an impetus for major advances in fish and shellfish genetics. Successful aquaculture of many species of invertebrate animals has been achieved and the stage set for the realization of the potential of genetic engineering, since very large populations of shellfish—in the form of larvae and intermediate stages—can be manipulated and their genes cloned. Work at the University of Maryland has recently yielded a 'genomic library' for *Crassoastrea virginica*, a commercially important shellfish species in the Chesapeake Bay.

As one of several examples, the reproduction and growth of gastropod molluscs of the genus *Haliotis* can now be controlled. It has been shown that spawning is normally regulated by prostaglandins (hormones regulating reproduction in humans and other animals) with the rate-limiting process in the mollusc being enzymatic synthesis of the prostaglandins. Synthesis of the latter in reproductive tissue is controlled by very small amounts of hydrogen peroxide. Incidentally, enzymatic synthesis of prostaglandins occurs not only in the reproductive tissues of *Haliotis*, but in that tissue of many other molluscs as well. Spawing in a large number of these species can be induced conveniently, reliably and inexpensively simply by adding a low concentration of peroxide to the surrounding seawater. In fact, thirteen species of abalone, including *Haliotis gigantea* from Japan, four species of *Crassostrea*, three species of *Mytilis*, two species each of *Tridacna*, *Cellana*, and *Trochus*, and several other genera, have been found to spawn in response to stimulation with hydrogen peroxide.

After spawning and developing of the fertilized eggs to the larval stage, settlement and metamorphosis has been shown to be under similarly stringent biochemical control, dependent upon larval recognition of specific molecular signals deduced early on from the patterns of substrate-specific recruitment of larvae; an observation made with many species.

The inducer required for induction of *Haliotis* larval settlement and metamorphosis has been identified and characterized. Intriguingly, the inducer is a chemical uniquely available to the larvae only at the surface of crustose red algae, including species of *Lithothamnium*, *Lithophyllam philippi*, and *Hildenbrandia nardo*. Larval contact with the inducers at the algae surface triggers rapid settlement and metamorphosis, accounting for the substrate-specific recruitment of *Haliotis* larvae to crustose red algae in the benthic environment.

Inducer molecules extracted from the natural recruiting algae and active for the abalone are derivatives of γ-aminobutyric acid (GABA), a simple amino acid neurotransmitter. The GABA-related molecule appears to be recognized by stereochemically specific chemosensory receptors on the larval epithelium. GABA alone will induce rapid and complete settlement and metamorphosis when an available substrate is provided.

Interestingly, binding of the neurotransmitter-like, GABA-related, inducing molecule to larval receptors activates the behavioral and development sequence, resulting in settlement and metamorphosis. Morse and Morse (1984) at the University of California at Santa Barbara have described the stereochemical specificity and binding properties of larval receptors. Also described is the regulation of the receptors by endogenous and exogenous factors. Amongst these are the seawater-borne amino acids and the induced ionic flux resulting in depolarization of the chemosensory membrane with subsequent transduction of the inducing signal. The early sequence of developmental changes, including larval secretions, abscission of the velum, internal organogenesis, and shell growth, all of which lead to irreversible commitment to the benthic habitat, have been elucidated, as well as the subsequent initiation of growth of the attached juvenile.

Along with the ability to control spawning, larval settlement and induction of metamorphosis, enhancement of growth will affect aquaculture of the abalone significantly, since the animal grows relatively slowly. In fact, abalone require several years to mature, with significant heterogeneity in growth parameters. Post larval abalone growth can be accelerated significantly by addition of specific, exogenous peptide hormones. The growth-regulating hormones, insulin and growth hormone isolated from mammals have proven effective, and both act in a concentration-dependent manner to accelerate early growth, yielding accelerations of approximately 25 per cent over the mean growth rate within the first few days following metamorphosis, while also reducing hetereogeneity in growth rates and sizes. Apparently, the active hormones increase efficiency of nutrient assimilation and utilization, rather than increase feeding activity and ingestion.

To scale-up the production and provide for the use of homologous, molluscan growth-regulating peptide hormones, encoded by the *Haliotis* DNA cloned in microbial plasmid vectors, DNA from abalone sperm has been purified, treated with DNA-restriction enzymes and ligated or recombined with the resulting DNA fragments of the *Haliotis* genome (both *en masse* and as electrophoretically purified, separate genes) with the DNA of genetically modified, autonomously replicating plasmids. The plasmids used as cloning vectors were selected for proven ability to amplify the production of peptide hormones from cloned DNA templates when introduced into cells of rapidly dividing producer strains of bacteria or yeast.

The abalone gene bank and its counterpart in the Eastern oyster, *Crassostrea virginica*, and individually cloned genes of these animals will prove useful for economical production of safe, homologous (molluscan) peptide hormones for the enhancement of nutrient assimilation, increased synthesis of protein and glycogen (meat constituents of greatest value), and acceleration of growth.

A role of prostaglandin has also been shown for barnacles in egg hatching, with the major marine invertebrate source of prostaglandins being the gorgonian, *Plexaura homomalla*. The larvae of the coral-eating nudibranch *Phestilla siboqae* settle and metamorphose specifically in response to a soluble, coral-produced substance. They undergo slow, but complete metamorphosis, in response to choline, GABA and related compounds. For the chiton *Tonicella lineata*, a relatively high molecular weight factor (60 000–100 000 daltons) associated with food of the chiton, the coralline alga *Lithothamnion*, induces settlement of chiton larvae.

Clearly, major advances have been made since the late 1920s, at which time, and earlier, it was generally assumed that metamorphosis of marine larvae was simply a function of the developmental state of the animal, i.e. once larvae developed the ability to metamorphose, they would do so, and for the few which by chance 'fell upon good ground' and survived, the vast majority would 'fall by the wayside and be lost'. With the most recently gained new information it is evident that planktonic larvae of many benthic invertebrates settle and metamorphose in response to specific substances or conditions in their environment, and may delay metamorphosis indefinitely in the absence of those substances. Metamorphosis-stimulating factors have, in almost every case, been shown to originate from, or be related to, some feature of the preferred adult environment. These features include presence of other individuals of the same species, algal or bacterial films, specific types of substrata, or certain plant or animal species, often those upon which the metamorphasing species will feed as an adult. For any particular species, two or more of these factors may act together as the appropriate metamorphic stimulus. Thus, once the factor(s) can be identified, the genes cloned and the production of the factors exogenously amplified, controlled aquaculture is possible.

At the University of Maryland, studies on bacterial films implicate bacterial by-products as factors responsible for induction of metamorphosis of the Eastern oyster, *Crassostrea virginica*. Serotonin (5-hydroxytryptamine), succinylcholine chloride, or acetyl-beta-methylcholine chloride have routinely been used to 'artificially' induce the metamorphosis of larvae of the gastropod *Ilyanassa obsoleta*. Veitch and Hidu (1971) demonstrated that thyroxine and related iodinated compounds stimulated setting of *Crassostrea virginica*.

L-DOPA has been found to be active in promoting oyster attraction and attachment to a surface, i.e. 'set', by Weiner *et al.* (1986). *Crassostrea virginica* larvae become competent to metamorphose 1–2 days following the appearance of pigmented eyespots and this developmental stage is typically reached at a specific shell size (about 260 μm). Interestingly, it was discovered that component larvae have only a few days during which they can metamorphose and then only if presented with an attractive substratum. Once competence is acquired, the ability of a group of larvae to metamorphose declines, so that after eight days, only a small percentage (approximately 11 per cent) of the larvae are still capable of metamorphosis. The unmetamorphosed larvae show increasingly reduced activity and eventually die.

Metamorphosis will occur on a suitable substratum, and evidence shows that oyster larvae have a predilection for surfaces coated with periphytic microbiota. Consistently present in this primary film is a bacterium, LST (Lewes, Delaware, Spat Tank isolate, i.e. LST) which has been repeatedly isolated from the substratum surface film to which oyster larvae set.

Invertebrate species, about which the effect of periphytic organisms on induced metamorphosis has been studied, include the sea urchin *Lytechinus pictus*, cnidarians *Hydractinina echinata* and *Cassiopea andromeda*, and the annelid *Janus brasiliensis*. For *Lytechinus*, the responsible factor is a low molecular weight (less than 5000 daltons) bacterial by-product, very likely a protein. It is known that planulae larvae of *Hydractinia* metamorphose in response to a product released by certain marine, Gram-negative bacteria at the end of their exponential growth phase. If the bacterial cultures are subjected to osmotic shock, the activity shows up in the supernatant, suggesting it to be a soluble factor rather than a bound one.

In addition to the well-documented case involving LST and *Crassostrea virginica* larvae cited above, preliminary evidence points to other potential examples of bacterial-invertebrate symbioses. Melanin has been reported to protect organisms in the marine environment, and in addition, it has been shown that some procaryotes, *viz.* vibrios, survive longer when they are associated with invertebrate chitin. Therefore, it is concluded that associations between bacteria and invertebrates are strongly mutualistic, and bacterial products can function as mediators.

The bacterium LST adheres very strongly to 'cultch', i.e. shell and other solid surface for spat set and other hard surfaces, it forms micro-colonies on cultch, and when present in sufficient numbers during the decline phase of its growth, produces a high concentration of pigment sufficient to attract *Crassostrea virginica* larvae. The larvae can feed on LST and, as noted earlier, induce reproduction of the bacterium, much as a lectin produced by *Halochondrea panicea* stimulates the bacterium *Pseudomonas insolita*. Thereby the larvae are able to disseminate the bacterium, and reciprocally, the bacterial metabolite can, by production of a hormone-like compound, stimulate larval development and metamorphosis.

It is significant that the natural molecular inducer required for *Haliotis* recruitment, settlement, and metamorphosis is an amino acid-derived, neurotransmitter-related, small molecule linked to a large (protein) polymer. As stated above, this class of molecular structure has been documented to be involved significantly in the induction of larval settlement and metamorphosis. *Crassostrea virginica* larvae are induced to settle and undergo metamorphosis by contact with melanin-like polymers of dihydroxyphenlyalanine (DOPA) produced by marine bacteria and by various analogues of the amino acid-derived, neurotransmitter-related compound, DOPA.

Application of genetic engineering and modern biotechnology permits cloning of genes controlling production of these attractants and inducers. Recombinant DNA probes and templates can be used to analyse and control the life-cycle processes of benthic invertebrates (Colwell 1984*c*).

3. Vaccine production

Another important area in which genetic engineering and advances in applied molecular biology can be applied to aquaculture is in the control of microbially-mediated diseases. *Vibrio* disease, for example, is widespread amongst fish. Viral agents, including infectious pancreatic necrosis virus (IPN), and other viruses, as well as *Aeromonas* app. and a variety of other bacteria, cause diseases and loss of hatchery stocks. Production of vaccine strains employing genetic engineering for excision of virulence factors, as has been done for *Vibrio cholerae* and other agents of human diseases, should be equally effective for controlling *Vibrio* diseases of fish and shellfish.

An extract from *Ecteinascidia turbinada* (Ete) has been shown to enhance haemocyte function of invertebrates, e.g. the blue crab (*Callinectes sapidus*), crayfish (*Procambarus clarkii*), and prawn (*Macrobrachium rosenbergii*), possibly rendering the animals more resistant to infection. Interestingly, intraperitoneal injection of Ete renders eels strongly resistant to *Aeromonas hydrophila* and appears to potentiate phagocytic activity. Ete also causes changes in the concentration of peripheral blood leucocytes.

Thus, biotechnology offers opportunities for control of diseases occurring in aquaculture of many species of shellfish and fin fish. Furthermore, it offers a means for obtaining presently 'recalcitrant' species in culture. Obviously, such organisms represent excellent opportunities for gene selection, manipulation, and amplification.

Production of vaccines employing both hybridoma technology and genetic engineering can advance aquaculture significantly, especially in increasing productivity and improving success in maintaining animals from the egg through larval stages, presently a high-risk portion of the life cycle of a significant number of cultured species.

4. Seaweeds

Marine plants offer special opportunities. Genetic engineering of osmo-regulation, for example, is being studied. Plants which are halophytes can be introduced to agricultural areas where the soil has become too salty for conventional agriculture. Halophytes, as well as selected stocks of marine and estuarine grasses, can be beneficial in managing erosion and shoreline losses. This represents another unique opportunity for island and riparian countries with arid regions.

Seaweeds are far more economically important than generally realized. They are used as human and animal food, in medicine and agriculture, and as a source of raw materials for many industries. The *Porphyra*, or nori, industry in Japan alone is estimated to involve over 60 000 hectares in cultivation area and to be worth more than $730 million annually. In fact, *Porphyra* is the most important mariculture crop in Japan at the present time.

In the western hemisphere, seaweeds are principally utilized as a source of phycocolloids, which include agar, carrageenan, and alginate. These three phycocolloids have a combined current world market value in excess of $250 million annually.

Many workers have been successful in cultivating agar and carrageenan-producing seaweeds on a small, experimental scale in both the USA and Canada. Commercial seaweed cultivation is currently being conducted in both enclosed systems and ocean farms in other parts of the world as well.

Altogether, there are approximately 11 genera (less than 20 species) of seaweeds being cultivated commercially to a significant extent on a worldwide basis. For some developing countries, seaweeds represent an important food source.

In general, the application of genetic modification/improvement techniques to seaweeds is recent and is as yet somewhat limited. The most widely used approach has been that of simple strain selection, i.e. the screening of wild plants for desirable traits such as rapid growth. Strain selection experiments have been conducted on several economically important seaweeds, including *Chondrus*.

Perhaps the most notable success to date in genetic improvement of seaweeds has been that of Chinese researchers working with the kelp *Laminaria japonica*, a plant not native to Chinese waters. Through the use of a variety of techniques, including intensive inbreeding and selection, X-ray induced mutations, and colchicine treatment, new and improved strains have been produced that have resulted in higher yields and extensive geographical expansion of the *Laminaria* culture industry in China. *Laminaria* provides a major food product in China.

Some of the recent research employing protoplast fusion-somatic hybridization techniques is aimed at producing new, cultivatable strains of high quality, agar-producing seaweeds. A major advantage is that genetic traits can be transferred from one species to another without involving (or requiring) sexual reproduction. Thus, it is theoretically possible to hybridize individuals from different (sexually incompatible) species (or genera), as well as from the same species (or genus). Furthermore, somatic hybridization offers the potential of hybridizing sterile individuals as well as species in which male and female reproductive structures are rare and/or difficult to synchronize. Therefore, with such methods it should be possible to hybridize species of *Gracilaria* and *Gelidium* which does not occur via a sexual hybridization (Cheney 1983).

Interestingly, a large body of literature is available concerning protoplast isolation and fusion in 'higher' (i.e. seed) plants, supporting the notion that with increased emphasis on marine plants, breakthroughs can be expected within the near future. In fact, protoplast isolation has already been accomplished in approximately 28 genera of algae.

5. Marine products

5.1. Marine pharmaceuticals

Perhaps one of the most dramatic examples of marine biotechnological applications is in marine pharmaceuticals. At a conference held in September 1977 at Norman, Oklahoma, USA, cardiotonic polypeptides isolated from sea anemones were described, as were an adrenergic compound from the sponge, *Verongia fistularis* and potential anti-cancer agents from Caribbean gorgonians and soft corals. An extensive literature is now available on marine natural products; many appear to have potential pharmacological value. The Porifera (invertebrates), algae, and coelenterata were studied and compounds extracted from sponges, coelenterates, algae, and seaweed, most of which have biological activity. In general, a high proportion of extracts studied have proven to be cytotoxic in preliminary experiments with many of the anti-bacterial, anti-fungal, and anti-viral compounds that have been isolated. In many cases, biological activities have been confirmed in more extensive assays employing tumour cells, pathogenic mico-organisms and viruses. Of special interest are the didemnins (depsipeptides isolated from a didemnid tunicate) inhibiting several RNA and DNA viruses, and exhibiting potent cytotoxicity against tumour cell lines.

Extracts prepared from the Caribbean tunicate, an ascidian or sea squirt, of the family Didemnidae, inhibit growth of DNA and RNA viruses, as well as L1210 leukemic cells. These depsipeptides—termed didemnins after the name of the tunicate family, Didemnidae from which they are isolated—are closely related, but vary in activity. The discovery indicates that the subphylum Tunicata or Urochordata (phylum Chordata) may be an abundant source of bioactive compounds of pharmaceutical interest. The tunicate of the *Trididemnum* genus, when extracted with methanol:tolulene (3:1) showed activity against type 1 *Herpes simplex* virus grown in CV–1 cells (monkey kidney tissue), indicating that it inhibited the growth of the virus. This antiviral activity may also involve anti-tumour activity. When tested against other viruses, essentially all extracts of the tunicate collected at a number of sites showed activity in inhibiting both RNA and DNA viruses. The suggestion that the extracts might also have anti-tumour properties was evidenced from their high potency against L1210 murine leukemic cells. The novelty of the didemnins results from a new structural unit for depsipeptides, hydroxy-isovalerylpropionate, and a new stereoisomer of the higher, unusual amino acid, statine.

In the literature, a variety of compounds from the sea are described which act on the cardiovascular and central nervous systems. Marine animals and plants have yielded cardiovascular-active substances including histamine and N-methylated histamines of sponges, vis. *Verongia fistularis*, asystolic nucleosides from the sponge *Dasychalina cyathina*, and the nucleoside spongosine isolated from *Cryptotethya crypta*.

Several marine organisms provide useful drugs: liver oil from some fish provides an excellent source of vitamins A and D; insulin extracted from whales and tuna fish; and the red alga *Digenia simplex*, long used as an antihelmintic. Bacteriologists have for many years incorporated agar and alginic acids into laboratory media. In general, it has been uneconomical to extract and purify a drug from an organism which has to be collected in large quantities from remote corners of the world. Thus, only a few marine organisms are currently sources of useful drugs. Genetic engineering can change this situation dramatically, opening up a vast and diverse range of marine life to probing for valuable pharmacological compounds, if the genes coding for production of the compounds can be cloned into laboratory strains of micro-organisms. In the long run these opportunities will open as the tools for gene cloning are sharpened and the applications broadened.

5.2. Marine toxins

Of particular interest in marine systems are toxins produced by marine organisms. A toxin is a substance possessing a specific functional group arranged in the molecule(s) and showing strong physiological activity. A toxin has the potential of being applied as a drug or pharmacological reagent. Furthermore, even if direct use as a drug is not feasible because of potent or harmful side effects, the toxin can serve as a model for synthesis or improvement of other drugs. Many attempts have been made to develop useful drugs from the sea by screening for anti-carcinogenic, antibiotic, growth-promoting (or inhibiting), haemolytic, analgetic, antispasmodic, hypotensive, and hypertensive agents.

With increased interest in marine toxins (or bioactive substances in marine organisms), research on these substances has increased in recent years, with a number of monographs and reviews appearing in the literature. The burgeoning research work in this area has provided the focal topics of symposia and conferences since 'Drugs from the Sea', the first one, which was held during August 1967 in Rhode Island, USA.

Marine toxins show great promise as pharmacological reagents, viz., tetrodotoxin, and as models for development of new synthetic chemicals. Recently, ciguatoxin, palytoxin, and halitoxin have also been investigated and provide interesting new information. However, it must be emphasized that, for the moment, applications of marine toxins are limited, to say the least. It is mainly in the area of understanding the structure and function of neurological systems that the toxins are of interest.

Applications of hybridoma technology in marine pharmacology are practically unlimited, including study of the structure and function of the toxins, as well as for production of anti-toxins for treatment. The latter is an especially important application, since there are no antidotes for many of the toxic venoms of marine animals at the present time. The occurrence of ichthyosarcotoxism and toxic effects of poisonous fish and invertebrates is not uncommon, especially in native populations of island communities.

The need at the moment in marine biotechnology is for strategies for collecting, culturing, and screening marine organisms from which bioactive agents can be isolated and characterized. Probably the immediate successes will occur in discoveries of novel anti-bacterials or antibiotics produced by marine bacteria. However, the potential for engineering the production of the more complex pharmaceuticals and polysaccharides of commercial value exists. Ingenuity will certainly provide the means, and profit the initiative.

5.3. Industrial chemicals

Marine toxins are fascinating from a scientific point of view, but it is more likely in the short term that the marketable products will come from marine polysaccharides, carotenoids, and specialty chemicals, such as unusual sugars, enzymes, and algal lipids. These represent products having possibilities for short-term, perhaps immediate pay-off.

In fact, carrageenan is a major product from the red seaweeds and is widely used as an extender in foods and related products, ranging from evaporated milk to toothpaste. Agarose is widely employed in electrophoresis and chromatography analyses in the laboratory. Because of its significant economic value, seaweed culture offers an opportunity for gene cloning and transfer into microbial systems that can extend the presently profitable market by providing a stable, bioengineered source of the polysaccharides for production.

Specialty chemicals from salt-tolerant microbial systems, notably poly-saccharides and lipids, offer the greatest potential in the immediate future (Colwell 1985).

Besides toxins and biologically active substances and those substances already exploited commercially, such as carrageenin, chitin, and agarose, a variety of interesting compounds and metabolites not yet observed from terrestial sources have also been reported, including spatane diterpenoids from the tropical marine alga *Stoechospermum marginatum*.

Sponges and gorgonians have been useful sources of biologically active metabolites because they are frequently abundant, permitting pursuit of trace metabolites. These unusual compounds may be pathway intermediates and offer potential sources of new chemicals. Sea hares provide the advantage of being rich sources of interesting metabolites, but the ultimate source of the latter is not always the animal itself, often proving to be algae on which it feeds or is associated with. Extracts of the sea hare, *Aplysia dactylomela* show both cytotoxicity and *in vivo* anti-tumour activity.

6. Biodegradation in the marine environment

In contrast to natural products, man-made compounds are relatively refractory to biodegradation, creating special problems for waste treatment and environmental protection. Usually, this happens because organisms naturally present in the environment often cannot produce enzymes necessary for transformation of the original compound. The resulting accumulation of intermediate metabolites can be toxic and/or refractory to further metabolism, i.e. catabolism.

Required steps to initiate biodegradation are reasonably well understood. Halogenated compounds are known to be persistent because of the location of the halogen atom, the halide involved and the extent of halogenation. Selective use of micro-organisms, including actinomycetes, fungi, bacteria, phototrophic micro-organisms, anaerobic bacteria, and oligotrophic bacteria, is not new, but represents a common practice in certain applications, such as waste water treatment for biological removal of nitrogen via sequential nitrification and denitrification. Controlled mixed cultures comprised of heterotrophic bacteria, photosynthetic bacteria and algae are already in use in Japan for treating selected industrial wastes in reactors. Population selection based on the use of various methods of genetic engineering to develop optimized proliferation and maintenance of selected populations will certainly become widely used.

What has not yet been widely applied however, is the engineering of micro-organisms to be added to wastes that are to be discharged into the marine environment. It is obvious that with increased use of the oceans as a depository for mankind's waste, attention must be paid to the problems of marine pollution. Pollutants entering the ocean that can interfere with the integrity of ecosystems include synthetic organics, chlorination products, dredged spoils, litter, artificial radionuclides, trace metals, and fossil fuel compounds. Toxaphene, a group of slightly under 200 compounds, i.e. chlorinated hydrocarbons produced by chlorination under ultra-violet light of wood waste products, contain carcinogenic and mutagenic members, and may be more persistent in the environment than DDT and its degradation products. A concerted effort should be made to engineer marine micro-organisms that can be added to waste effluent prior to discharge in order to ensure degradation of the recalcitrant species of compound.

The problems of *in situ* degradation are much greater than for contained application. The modifications of genetic information resident in micro-organisms that are useful in pollution control are: (1) amplification of enzyme concentrations in an organism, either by selection of constitutive mutants, increase in the number of copies of the gene coding for the enzyme, or both; (2) rearrangement of regulatory mechanisms controlling the expression of specific genes in response to specific stimuli; (3) introduction of new enzymatic functions into organisms not possessing them; and (4) alteration of the characteristics of specific enzymes, viz. substrate specificity, kinetic constants (K_n and V_{max}) or factors such as pH optimum. To achieve these modifications, it is possible to employ *in vitro* recombinant DNA manipulation, *in vitro* modification via transposon mutagenesis or other transposon-mediated gene manipulation, genetic exchange via transduction, transformation, or conjugation, protoplast fusion, specific site mutagenesis, and specialized selection procedures to enrich for mutants. What has not been considered to date is the engineering of micro-organisms capable of flourishing in the marine environment. Ability to grow at low temperatures, in a high saline environment (35 parts per thousand), at a relatively high pH (8.2), and in the deep sea under elevated hydrostatic pressure are characteristics of organisms which should be engineered for use in treatment of recalcitrant wastes that are dumped into the ocean.

Of the achievements in marine biotechnology cited above, those of particular relevance to developing countries are in marine pharmacology, aquaculture, and marine plants. The products that can be expected include new drugs, food products and specialty chemicals. Of particular interest, however, is the possibility of establishing aquacultured species which have enhanced genetic traits, i.e. resistance to disease, more rapid growth, etc., achievable by genetic engineering.

Looking at the future, i.e. in projecting trends for the field of marine biotechnology, one can predict an increasing effort being placed on search and discovery of pharmacologically active substances, notably anti-microbial and anti-neoplastic agents. These products offer immediate economic reward and profits may be used to underwrite research in other, more long-term projects.

Close upon the pharmacologically active substances will be the development of vaccines and/or agents for the treatment of diseases of aquacultured fish and shellfish species. In the US and the United Kingdom vaccines against *Vibrio* disease (also called vibriosis, a devastating disease of salmonids in culture) are being 'engineered' i.e. genetic engineering is being done to modify agent(s) for use as live vaccines, while new methods of vaccination are also being tested. Control of fish diseases will move conventional aquaculture significantly forward, carrying the industry from the precarious to the predictable, i.e. ensuring profit-making. Culture of fin fish in Japan provides evidence of what can be done with currently available methods. Extrapolating benefits expected from the application of genetic engineering permits the prediction of significantly increased production, if the diseases are also fully controllable.

Natural product research is moving forward impressively and the opportunities for applying strains of genetically engineered bacteria to carry genes for the production of alginates, carrageenins, and/or xanthans, are enormous. This line of research is being pursued by at least one company in the US and very likely other companies in countries such as Japan and Western Europe will soon follow, if they have not already begun to do so. Thus, specialty chemicals, notably those useful for food manufacture and processing, offer possibilities for short-term pay-off.

Over the long range, it can be predicted that basic research in aquaculture, seaweed culture, and mariculture will lead to closed-system, controlled production of fish and shellfish. Ultimately, tissue culture of fish and shellfish will provide fish protein for food in many countries, notably developing countries. For island and riparian countries, especially those falling into the category of developing nations, the opportunity for self-sufficiency in protein should be very attractive.

Also long-term, but achievable, is return from investment in bioengineering of marine bacteria for controlled, closed system biodegradation. This can be both cost-effective and environmentally protective. For developing nations this aspect of marine biotechnology holds the advantage that an economic yield can be attained from by-product industries that can be developed, while the natural resources of the country, e.g. unusual species of animals and plants (a *major* resource in the years ahead), are protected.

In short, it is clear that future trends can be predicted which show possibilities for both short- and long-term pay-offs from investment in marine biotechnology (Colwell 1984*d*).

7. Conclusion

In order for developing countries to achieve progress from applied marine biotechnology, a cadre of trained marine biotechnologists must be produced and the necessary research and development facilities provided. It would be most efficient to link existing marine technology laboratories with molecular genetic laboratories, if both exist in the developing countries. If this is not the case, it would be effective to link with a developed country and to establish an exchange of students and faculty.

Technology transfer can be achieved effectively by workshops, training courses and visiting lecture series. However, practical experience must be included, so that the procedures and methods of genetic engineering can be instituted within the developing country (Sinskey 1985).

Because of the diversity and abundance of unusual marine animals and plants in island and riparian countries, the opportunities for discovery of new drugs, food sources, and specialty chemicals are great. Thus, the attraction of marine biotechnology for such countries is significant.

To ensure results from research in marine biotechnology are made available to the relevant industries, a close linkage between the research laboratory and industry should be encouraged. In fact, investment by industry in marine biotechnology ventures will catalyse development, allowing industry access to new products with the result being immediate development and market exploitation.

A marine biotechnology centre located in one country may focus on research questions quite different than if it were located in another country, because of climate, available natural resources, and diversity of animal and plant species unique to each locale. For example, a tropical country could focus on marine pharmaceuticals because it possesses an abundance of animal and plant species, coupled with the occurrence of toxic or poisonous species of fish and/or marine plants; whereas another country highly dependent on fish as a food staple would choose to emphasize aquaculture and genetic engineering of fish, with the objective of becoming protein self-sufficient. Diverse interests would not be an obstacle and would in fact, argue in favour of the establishment of several centres, as for example, the MIRCENs (Microbiological Resource Centres) which are sponsored by UNESCO, but on a larger, more focussed scale of operation. Linkage would then be very effective, with all areas of marine biotechnology being addressed, though not necessarily at a single location.

The retraining of scientific personnel is a possibility to consider when staffing a centre(s). It may be relatively easy to retrain a molecular biologist to work with marine systems, but this may not necessarily always be the case. For example, marine physiologists can be encouraged to learn the methods of marine biotechnology. On the other hand, aquaria curators and lower level technicians will be required. Retraining of these people would not be necessary because their duties and responsibilities for marine biotechnology remain essentially the same, i.e. to maintain cultures for laboratory research and development.

In moving from the ideal described above to the more pragmatic possibility, i.e. exploiting existing facilities in building a marine biotechnology centre, it is possible to move in either of two directions, i.e. towards upscaling a marine facility or re-directing a molecular biology facility to marine biotechnology. As stated above, linkage of existing facilities is a reasonable alternative to construction and staffing of a completely new facility. The options are ranked as follows: a new facility being ideal; linkage of existing facilities a reasonable substitute; adding a marine facility to a molecular biology facility a possibility; adding a molecular biology laboratory to a marine facility also a possibility; linkage with a facility in another region or country an alternative to consider when funds are limiting. None of these options, it should be emphasized, precludes or prevents successful development of a marine biotechnology centre. Only the rate of progress and extent of success over the long term will be influenced, not whether or not a biotechnology centre is feasible.

Having established the will to move into marine biotechnology, an assessment of the strengths and/or weaknesses of the country with respect to molecular biology and marine sciences should be done. This can be accomplished by engaging consultants, i.e., either internal or external experts. Once the potential of the country is known, viz. marine pharmacology or aquaculture, a plan for development of a centre can be initiated.

Goals of the marine biotechnology effort should be determined. For example, if the goal is to attract new industry, expertise in marine biotechnology research and development will be needed. If local industry is to be strengthened, linkages with universities, especially with molecular genetic and genetic engineering talent should be established. Projects to be undertaken should be a mix of short-term and long-term efforts.

Linkage of developing nation centres with counterparts in developed countries should allow more rapid achievement of goals if the individual country does not internally have the combination of molecular genetic and marine biology facilities and intellectual resources.

The potential of marine biotechnology from developing countries is especially great. The opportunities now waiting should not be lost.

References

Cheney, D. R. (1983). *Proceedings of the Sea Grant Conference.* M.I.T., Cambridge (in press).

Colwell, R. R. (1984*a*). Biotechnology in the Marine Sciences. In *Biotechnology in the Marine Sciences* (eds R. R. Colwell, E. R. Pariser, and A. J. Sinskey) *Proc. First Annual MIT Sea Grant Lecture and Seminar.* John Wiley, New York.

——, (1984*b*). Biotechnology in the marine sciences. *Science* 222, 19–24.

——, (1984*c*). Microbial ecology of biofouling. In *Biotechnology in marine sciences* (eds) R. R. Colwell, E. R. Pariser, and A. J. Sinskey) Proceedings of the First Annual MIT Sea Grant Lecture and Seminar. John Wiley, New York.

——, (1984*d*) The industrial potential of marine biotechnology. *Oceanus* 27, 3–12.

——, (1985) Marine polysaccharides for pharmaceutical and microbiological applications. In *Biotechnology of marine polysaccharides* (eds R. R. Colwell, E. R. Pariser, and A. J. Sinskey). Proceedings of the Third Annual MIT Sea Grant College Program Lecture and Seminar. Hemisphere Publishing Corporation, New York.

Morse, A. N. C. and Morse, D. E. (1984). GABA-mimetic Molecules from *Porphyra* (Rhodophyta) Induce Metamorphosis of *Haliotis* (Gastropoda) Larvae. *Hydrobiologia* (in press).

Sasson, A. (1984). *Biotechnologies: Challenges and promises,* United Nations Educational, Scientific and Cultural Organization, Paris.

Sinskey, A. J. (1983). The effects of development in biotechnology on the third world: current status, transfer of technology, future of biotechnology, National Policies Workshop, November 8–12, 1983, Port of Spain, Trinidad and Tobago.

van Hemert, P. A., Lelieveld, H. L. M., and la Riviere, J. W. M. (eds) (1983). *Biotechnology in Developing Countries.* Delft University Press, Delft.

Veitch, F. P. and Hidu, H. H. (1971). *Chesapeake Sci.* 12, 173.

Weiner, R., Bonaw, D. B., and Colwell, R. R. (1986) (in press).

6

Fermented foods: tradition and current practice

Dr K. H. Steinkraus

1. Introduction

There are many indigenous food fermentations in the world. Some of these such as Chinese soy sauce/Japanese shoyu, Japanese miso, Indonesian/ Malaysian tempe, Indonesian tape/Malaysian tapuy, Japanese sake, Indian idli/dosai, Mexican pulque, Nigerian ogi, South-East Asia fish sauces and pastes, and African kaffir (Bantu) beer have been intensively studied to determine the optimum conditions for fermentation, the essential micro-organisms, the biochemical, nutritive, and flavour/texture changes that occur and the possible toxicological problems that can arise (Steinkraus 1983).

This rather intensive study is required before the food can be produced by modern, large-scale processing plants. All of the indigenous fermented foods were originally produced on a small scale in the home or by small cottage industries. Many are still produced in the home and commercialization is still a long distance in the future.

The huge international enzyme industry today can be traced directly to the indigenous oriental soy sauce/shoyu/miso/sake fermentations. The large monosodium glutamate industry and the nucleotide flavour-enhancing industry also are outgrowths of the soy sauce fermentation. The soy sauce, shoyu, miso, and sake industries are very large in Japan and other parts of the Orient. However, these and related products are also made in small, primitive factories or even in the home in some cases. Fish sauces and pastes are produced industrially in Burma, South Vietnam, Thailand, and the Philippines and on a smaller scale in a number of other countries. African kaffir (Bantu) beer is produced as a home industry in many parts of Africa, but is produced as a huge commercial industry for the mine workers in South Africa. Mexican pulque has been a primitive fermentation for probably centuries, but has now been upgraded through scientific study and engineering to a modern controlled industrial fermentation. Nigerian ogi, a staple in the diet of several countries of Africa and produced in relatively small quantities by numerous small producers, can also be produced by large commercial plants designed by the Federal Institute of Industrial Research (Oshodi, Nigeria).

Most indigenous food fermentations still require much additional scientific study before they can be commercialized. In the meantime, they are produced by the primitive traditional methods (Steinkraus 1983).

This chapter will concentrate on those indigenous food fermentations that have been upgraded and become modernized.

2. Indonesian tempe kedele

Indonesian tempe kedele is a white, mould-covered cake produced by fungal fermentation of dehulled, hydrated (soaked), and partially cooked soybean cotyledons. The mould grows throughout the bean mass knitting it into a compact cake that can be sliced thin, deep-fat fried or cut into chunks, and used as a protein-rich meat substitute in soups. *Rhizopus oligosporus* Saito is the species identified as most characteristic and best adapted for production of tempe (Steinkraus *et al.* 1960; Hesseltine 1961).

Tempe has been produced for centuries as a traditional cottage industry. Research on tempe that was begun in laboratories at Cornell University and in the Northern Regional Research Laboratory (NRRL) of the US Department of Agriculture yielded enough basic information so that small factory production became possible and within a 25-year interval, 53 factories were producing tempe for the American market (Shurtleff and Aoyagi 1984). Results of the American studies were quite rapidly transferred to and adapted to tempe production in Indonesia.

The essential steps in the production of tempe are the following: (i) cleaning the soybeans; (ii) hydration and bacterial acid fermentation; (iii) dehulling dry or following hydration; (iv) partial cooking; (v) draining, cooling, surface drying; (vi) placing soybean cotyledons in suitable fermentation containers; (vii) inoculating with tempe mould before or after placing in fermentation container; (viii) incubating until the cotyledons are completely covered with mould mycelium; (ix) harvesting and selling; (x) cooking for consumption, deep fat frying, or using in soups in place of meat.

2.1. Traditional tempe fermentation

The soybeans are washed and soaked in water overnight during which time the beans undergo bacterial acid fermentation under tropical conditions reducing the pH to 5.0 or lower. An alternate process is to place the soybeans in water, bring it to a boil, and then allow the beans to soak overnight. The general purpose of the boil is to facilitate hull removal. The hulls are removed by rubbing the soaked beans between the hands or by stamping them with the feet. The loosened hulls are then floated away with water. The cotyledons are then given a short boil, cooled, surface dried, and inoculated with tempe mould either from a previous batch of sound tempe or the mould grown and dried on leaves. Traditionally, the inoculated cotyledons are then wrapped in small packets using wilted banana or other large leaves, incubated in a warm place for 2 or 3 days during which time the cotyledons are completely overgrown by the mould mycelium. The tempe is then ready for cooking (Steinkraus *et al.* 1960).

2.2. Industrial production of tempe

Twenty years ago most tempe was prepared for sale using the traditional process described above. Then tempe research in the United States resulted in some improvements in medium-scale processing of tempe. Steinkraus *et al.* (1965) described a pilot-plant process in which the soybeans were dehulled dry by passing them through a properly adjusted burr mill. Preceding the burr mill, the soybeans were given a short heat treatment at 104°C (220°F) to shrivel the cotyledons. The hulls were then removed from the cotyledons by passing them through an aspirator or over an Oliver gravity separator. Alternatively, the beans were soaked and dehulled wet by passing them through an abrasive vegetable peeler. Acidification of the beans, considered to be an essential step, particularly in large-scale processing where invasion by food spoilage organisms could ruin large batches, was accomplished by adding 1 % v/v lactic acid to the soak and cook waters. The partially cooked beans were then drained, cooled and inoculated with powdered pure culture tempe mould grown on sterilized soybeans and freeze-dried. The inoculum was mixed with the cooked cotyledons in a Hobart mixer. The inoculated beans were then spread on dryer trays (35 × 81 × 1.3 cm) covered with a layer of wax paper and incubated at 37°C (98.6°F) and 90 per cent relative humidity. By this procedure, fermentation was complete in less than 24 hours. The tempe was cut into 2.5-cm squares and the dryer trays were placed in a circulating hot air dryer at 104°C (220°F), dehydrated to less than 10 per cent moisture, and packaged in polyethylene bags for distribution.

Within a few years, the commercial tempe industry in Indonesia had adopted wooden trays with dimensions similar to those used above. They lined the trays with plastic sheeting perforated to allow access of air to the mould.

Martinelli and Hesseltine (1964) developed a new method of incubating the tempe in plastic bags with perforations at 0.25–1.3-cm intervals to allow access of oxygen. By this method the soybean cotyledons are inoculated with the mould and placed in the plastic bags or in plastic tubes similar to sausage casings. They can be immediately incubated or stored in a refrigerator until fermentation is desired. The plastic bag process has been widely adopted in Indonesia and is also being used commercially in new tempe factories in the United States.

According to Shurtleff and Aoyagi (1984), the largest tempe factory in the United States produces 3182 kg (7000 lbs) of tempe per week. The largest operation in Indonesia produces 795 kg (1750 lbs) of tempe per day (Shurtleff and Aoyagi 1980). Thus, tempe production is still a relatively small commercial operation. Looking toward the future, if tempe should become a staple in the diet of Americans, it is conceivable that tunnel reactors would be used in which soybeans would be cleaned and dry dehulled, soaked and cooked continuously, cooled, inoculated and passed continuously through a tunnel fermenter. The fermented tempe would emerge from the tunnel fermenter following an approximate 20-hour processing time.

3. Soy sauce—Japanese shoyu (Yokotsuka 1983; Steinkraus 1983)

Soy sauce is light brown to black liquid with a meat-like, salty flavour manufactured by hydrolysing soybeans, with or without the addition of wheat or other starchy carbohydrate in a strong salt brine (approximately 18 per cent w/v using enzymes produced by *Aspergillus oryzae* (*A. soyae*). A two-stage fermentation is used. An aerobic, solid-state mould fermentation is followed by a mixed *Lactobacillus* yeast submerged fermentation (Yokotsuka 1960, 1972).

3.1. Traditional manufacture of soy sauces

The most ancient process of soy sauce manufacture (and also miso manufacture) is the miso-dame process in which soybeans are soaked, cooked, mashed, and formed into a ball, suspended in a straw bag under the eaves, where it becomes overgrown with mould prevalent in the environment, and then is mixed with salt and some water to provide the proper consistency. As enzymes from the mould hydrolyse the proteins, lipids, pectins, and carbohydrates, the substrate becomes more soluble and a liquid (tamari) soy sauce separates. The remainder of the semi-solid substrate is consumed as a miso (Shurtleff and Aoyagi 1983). The miso-dame process requires that the cooked soybeans be mashed to provide the surface and nutrients required for growth of the moulds.

A subsequent modification of the miso-dame process is to coat the soaked, cooked soybeans with ground toasted wheat or wheat flour. This enables the moulds to grow on the whole intact soybeans, facilitating growth over the entire soybean and thereby producing sufficient enzyme activity for the subsequent fermentation.

While the ultimate product in the miso process is a semi-solid paste, the end product in soy sauce is a liquid. Placing the soybeans overgrown with the mould in an 18–23 per cent salt brine is most advantageous for production of a liquid sauce.

Soy sauces can be produced by simple traditional fermentation in relatively small quantities using woven bamboo mats for the initial solid state fermentation in which the mould overgrows the soaked, cooked soybeans coated with ground, toasted wheat and crocks of suitable size enabling the mould-covered soybeans to be covered with 18 per cent salt brine for the subsequent submerged fermentation. It is common practice to stir such crocks at least once a day and expose the surface of the substrate to the sun to inhibit continuing growth of the mould.

3.2. Modern soy sauce—Japanese shoyu production (Yokotsuka 1972, 1983)

Clean soybeans are soaked for 10–15 hours in water, which is changed every few hours to prevent acidification by bacteria. The weight of the beans should increase 2.1–2.15 times (Nakaya 1934). The hydrated beans are cooked for 1 hour in steam at 10–14 psi (6.9×10^4–9.7×10^7) Pascals in a NK-rotary cooker (capacity 1 ton). The cooked beans are then cooled rapidly. Continuous cookers that operate at higher temperatures for shorter times have been developed.

Whole wheat or wheat flour is essential for production of typical Japanese shoyu flavour. Usually, low-protein (soft) flour is used. The wheat is roasted in sand for several minutes at 170–180°C (338–356°F). A rotary cylinder about 0.7 m in diameter and 2 m in length (capacity about 0.5 ton/hour), rotating at 25–30 rev/min is used. The sand is recycled and the wheat grains are crushed into four or five pieces in a roller mill. A slightly charred flavour is desirable. If wheat flour or bran is used, it is usually only steamed (Yokotsuka 1977).

The procedure starts with the preparation of koji *Aspergillus orzae* (Ahlberg) Cohn is grown on the cooked soybeans coated with ground roasted wheat. The best shoyu soy sauce is made from a soybean–wheat mixture of 50:50 w/w or 52:48 v/v. The soybean–wheat mixture is inoculated with 1–2 per cent w/w of seed koji-*A. oryzae* grown on polished rice. In earlier times, the mixture was inoculated and placed in wooden trays 30 by 70 cm and incubated in a room with temperature and humidity control. Now tons of the soybean–wheat mixture at 25–35°C (77–95°F) are placed to depths of 30–40 cm on porous stainless steel plates several metres in length and width. Careful temperature, aeration, and moisture control allow completion of mold growth in 45 hours, help prevent development of contaminants and enhance development of proteolytic enzymes. The end product is called koji, a mixture of fungal hydrolytic enzymes and the substrate.

The procedure continues by the fermentation of the solvent koji. The koji is mixed with 1.2–1.5 volumes of salt brine (23 per cent w/v salt) to make mash which ferments in a wood, concrete, or steel tank with a capacity of 10–20 m^3. Pure cultures of *Pediococcus soyae* Sakaguchi and *Zygosaccharomyces* (*Saccharomyces*) *rouxii* Stelling-Dekker are added to the moromi at the start and after 1 month, respectively, to promote the desirable fermentations. The high-salt content ensures the development of flavour-enhancing yeasts (Yong 1971). Traditional fermentation continues for 1–3 years at ambient temperatures, as the colour and flavour becomes more intense.

The fermented moromi is then filtered by pressing in a hydraulic filter press at 1379 psi (100 kg/cm^2 or 9.5 × 10^6 Pascals) for 2 or 3 days. Although such high pressure incurs added expense, lower pressures, and longer filtration times result in loss of flavour due to oxidation.

The liquid soy sauce is subsequently heated to a pasteurization temperature of 70–80°C (158–176°F) either in a kettle or in a heat exchanger, cooled, filtered, and then stored. Pasteurization helps remove heat-coagulable material as well as preserve the product.

About 60 per cent of Japanese shoyu is packed in 2-litre glass bottles and another 30 per cent is packaged in 1-litre plastic bottles (Yokotsuka 1983). Benzoic acid or propyl- or butyl-*p*-hydroxy benzoate is sometimes added as a preservative.

4. Japanese miso (Ebine 1983; Steinkraus 1983; Shurtleff and Aoyagi 1983)

Miso is a salty, meat-flavoured smooth to chunky paste with a high protein content. It is made by fermentation of cooked soybeans generally with the addition of a rice or barley koji overgrown with moulds belonging to species *Aspergillus oryzae (soyae)* and *Lactobacilli* and a yeast *Zygosaccharomyces* (*Saccharomyces rouxii*) in the presence of 5.5–13 per cent salt. Although miso is a seasoning, it is also a traditional dietary staple used by the Japanese in preparation of soups for breakfast. Miso is a product which can serve the consumer's craving for meat-like flavours such as bouillon cubes, Maggi cubes, marmite, and vegemite do in various parts of the world.

Miso has been manufactured in Japan for at least 1000 years. There are several major types (Shibasaki and Hesseltine 1962; Shurtleff and Aoyagi 1983). White miso, preferred in western Japan, has a light colour, a very sweet flavour, a low concentration of salt (5–6 per cent), and a short fermentation about 1 week at 23–33°C (73–91°F). *Edo* miso, preferred in the region around Tokyo, is light reddish-brown has a low salt content, and requires a 2-week fermentation. *Sendai* miso has a high salt content (12–13 per cent), requires a 1-year fermentation and has excellent keeping quality. *Shinshu* miso, popular in Tokyo and central Japan, is light yellowish-brown, has a high-salt content, and requires more than a year to ferment. *Mame miso*, preferred in Nagoya and central Japan, is made solely from soybeans. It is deep reddish-brown, requires 2 years to ferment, and has excellent keeping quality.

4.1. Traditional miso fermentation

We have already discussed the most primitive miso process, the miso-dame process which leads to a primitive soy sauce, 'tamari' as the soybeans are hydrolysed. Most miso is produced using a koji in which polished rice or pearl barley is used as the substrate for the mould content. The koji is then ground and mixed thoroughly with cooked soybeans, salt, and water which following fermentation becomes miso.

The essential steps in modern miso production include the following (Shurtleff and Aoyagi 1980, 1983). Polished rice or pearl barley is washed, soaked, steamed, cooled, inoculated 0.1 per cent w/w with selected strains of *Aspergillus oryzae (soyae)*, placed in fermentation pans or chambers suitable for rapid growth of the fungus and incubated at a favourable temperature 28–32°C (82–90°F) for about 3 days, sufficient time for the mould mycelium to grow throughout the rice or barley koji. The soaking is done in large fibreglass or epoxy-lined steel tanks holding a ton or more of material.

Soaked rice or barley is transferred continuously on stainless steel mesh belts 9.1–6.1 m (3 × 20 feet) through tunnel steamers in which steam under low pressure (4.3 psi, 3.0×10^4 Pascals) is injected into the rice or barley in a first chamber and at 3.5 psi (2.4×10^4 Pascals) in the second chamber. The partially steamed rice may be rinsed with water as it passes from the first to the second chamber to remove free starch.

The rice/barley is then cooled on another conveyor belt and transferred to the koji fermenter. The rice or barley is inoculated with pure culture strains of *Aspergillus oryzae* selected for their ability to produce the required proteases, amylases, lipases, and other enzymes in proper proportions and quantities.

Koji fermenters are of four main types. The first are large rotating drums, about 1.8 m (6 feet) in diameter and 3.6 m (12 feet) long with temperature, humidity, and air flow controls. The drums contain finger projections inside that break up the developing koji as the drum is periodically rotated (Shurtloff and Ayoaji 1983).

The second type of koji fermenter is a room approximately 9 × 15 × 2 m (30 × 50 × 7 feet) lined with insulated stainless steel with floors of stainless steel mesh covered with fine-mesh nylon with temperature, humidity and air-flow controls. The developing koji is mixed manually with shovels. Depth of the koji is about 20 cm (8 inches).

The third type of koji fermenter is a horizontal, circular koji fermenter 50 feet (15 m) in diameter in which mechanical fingers from a revolving crane stir the developing koji. The fourth type of koji fermenter is a large rectangular chamber with a floor 12 × 21 m (40 × 70 feet) where an overhead crane with metal projections passes along the length of the fermenter periodically stirring the developing koji.

Cleaned soybeans are hydrated to approximately double their dry weight by soaking in large tanks. If the soybeans are to be dehulled, they can be dehulled dry by passing the size-graded soybeans through a properly spaced Burr mill or they can be dehulled wet following soaking by passing them through an abrasive mechanical vegetable peeling machine. Dry hulls are separated by aspiration or winnowing. Wet hulls must be removed by flotation.

Following soaking and dehulling, the soybeans must be thoroughly cooked in large batch-type or continuous soybean cookers which heat the beans either in steam or in water to temperatures of about 250°F (121°C) for 30–40 minutes or higher temperatures for shorter times. The autoclaves (retorts) used for this purpose generally will hold a ton or more of soaked soybeans, and often are mounted so that they can be rotated during cooking to decrease cooking time and also can be tilted so that the beans can be emptied onto conveyor belts. As soon as the soybeans have been cooked for the necessary time, they are rapidly cooled to prevent any further darkening (Shurtleff and Aoyagi 1987).

The cooked soybeans must then be ground and mixed with the koji, the required amounts of salt and water added, and the mixture inoculated with either miso (from a previous batch) or pure cultures of selected *Lactobacilli* such as *Pediococcus halophilus* Mees which produces the necessary acidity and *Zygosaccharomyces* (*Saccharomyces*) *rouxii* which produces alcohol contributing to the formation of esters, and other flavour and aromo compounds. Machines similar to large sausage grinders, mash the soybeans, koji, salt, and inocula together, and mixing continues in large mixing vats with heavy paddles. Five-and-a-half to thirteen per cent w/w salt is added, determined by the type of miso to be prepared.

The paste is then conveyed to temperature-controlled tanks or fermentation tanks, vats, or rooms where the fermentation is completed. The tanks or vats may hold as much as 12 tons. Spigots allow the removal of liquified 'tamari' sauce as the fermentation progresses.

The lower the salt content and the higher the proportion of rice–barley koji to soybeans used, the sweeter the resulting miso, since the cereal grain contributes starch which is hydrolysed to sugars by the mould. The more thoroughly the soybeans are cooked, the darker the colour and the higher the salt content, the longer the fermentation, and also the more robust the meat-like flavour.

If the finished miso is to be packaged in plastic bags, it must be pasteurized or preserved with the addition of preservatives. Automatic packaging machines are used to fill the plastic bags.

5. African kaffir (kaffircorn or sorghum) beer (Novellie 1968; Steinkraus 1983)

Kaffir beer is an alcoholic, effervescent, pinkish-brown beverage with a sour flavour resembling yogurt, the consistency of a thin gruel, and an opaque appearance due to its content of undigested starch residues, yeasts, and other micro-organisms. Kaffir beer is not hopped or pasteurized and is consumed while still actively fermenting. Kaffir beer, also known as Bantu beer, is the traditional beverage of the Bantu tribes of South Africa (Novellie 1963, 1977).

The art of kaffir beer brewing goes back to prehistoric times. In the villages, kaffir beer is made by the women, and every girl learns to brew kaffir beer before she marries (Platt 1964; Platt and Webb 1946, 1948).

Because of the great importance of kaffir beer in the nutrition of Bantu tribesmen who work the diamond and gold mines, and who are removed from their tribal homes for long periods of time, the brewing of kaffir beer in South Africa has become big business, with production three times that of ordinary modern beers (Novellie 1968). Thus, kaffir beer is of unusual interest to those studying indigenous fermented foods and the effects modern processing methods can have on them. Kaffir beer production is the only large modern industry founded on tribal art of African origin (Novellie 1968).

The major part of the sorghum crop (*Sorghum caffrorum* or *S. vulgare*) is malted and used for brewing beer. Maize is frequently substituted for sorghum depending upon the relative cost (Schwartz 1956). The millets, *Eleusine coracana* and *Pennisetum typhoides*, are also malted and used in place of sorghum. Because of the small size of the millet seed, however, it is not malted by modern plants (Platt 1964).

5.1. Small-scale brewing

The essential steps in small-scale brewing are malting, mashing, souring, boiling, conversion, straining, and alcoholic fermentation. In the traditional village processes, kaffir beers are made in large drumlike pots in 115–180-litre batches (Platt 1964). Each litre of beer requires 180–360 g of grain. Sorghum malt is produced by soaking the grain in water for 1 or 2 days, draining, and then allowing the seed to germinate for a few days. The sprouted grain is sun-dried and allowed to mature for several months. Next, the malt is pulverized, slurried to a thin gruel, boiled, and cooled, and a small amount of fresh, uncooked malt is added, probably both for its amylolytic action, and also as a source of yeast and other micro-organisms. The mixture is held 1 day. On the second day, it is boiled in the cooking pot and returned to the brewing pot. On the third and fourth days, more pulverized uncooked malt is added, and on the fifth day, the brew is strained through a coarse basket to remove the husks. The beer is then ready to drink (Platt 1964).

About equal quantities of malted and unmalted grains are mashed in cold and boiled water, respectively, at the time they are combined so that the temperature of the mixture turns out to be 37°C (98.6°F). Platt and Webb (1946) concluded that the major saccharification was not due to malt amylases, but to moulds growing on the grain. Novellie (1968) concluded, however, that the ability of the malt to produce sufficient diastatic power is the most critical factor in brewing kaffir beer. He distinguishes native trade malts and malt for municipal beer brewing, but both are made by similar processes. Grain is steeped for 8–24 hours (frequently 16–18 hours). Germination proceeds for 5–7 days. Municipal malts are more thoroughly precleaned, and carefully washed and watered during germination (Novellie 1962). Sorghum requires a warm temperature of 25–30°C (77–80°F) for optimum production of amylases in a reasonable length of time. For optimum malting, the grains must be kept moist and aerated and turned to prevent overheating. The plumules should be 2.5–5.0 cm (1–2 inches) in length following germination. Gentle drying in the sun or in hot air at 50–60°C (122–140°F) preserves the enzymes. At best, the diastatic power does not go much beyond 20° Lintner under commercial conditions. In the laboratory it is possible to reach approximately 70° Lintner (Novellie 1968).

5.2 Large-scale brewing of Kaffir beer

The municipal brewing process involves two distinct fermentations (Novellie 1966a, b) a lactic acid and an alcoholic fermentation. Souring (lactic acid production) is achieved by holding a mixture of sorghum malt and water at 48–50°C (118–122°F) for 8–16 hours until the proper degree of acidity is attained. The 'sour' is about one-third the final volume of beer. The souring step controls the course of the remaining fermentation, mashing, body, and alcoholic content of the beer. Thus, it is very important. Although pure culture inoculation of lactic acid bacteria is not used, about 10 per cent of each batch of sour is used to inoculate the next batch (van der Walt 1956). The pH at the end of souring should be 3.0–3.3 and total acidity should be 0.3–1.6 per cent (average 0.8 per cent) as lactic acid.

The soured malt mixture is pumped to the cooker and diluted with 2 volumes of water. An adjunct (usually maize grits) is added and the whole mixture is boiled 2 hours. Most boiling is carried out at atmospheric pressure, but slight pressure cooking may be used at high altitudes where boiling temperature is low. The thick, cooked sour mash is cooled to 60°C (140°F). A small amount of malt may be added when the temperature reaches 75–80°C (167–176°C) to reduce the viscosity. At 60°C (140°F), the conversion malt is added and the temperature is held for 1.5–2 hours. The mash is now thinner and sweet. It is cooled to 30°C (86°F) and pitched (inoculated) with a top-fermenting yeast strain of *S. cerevisiae*. The yeast is produced locally and distributed as a dry yeast which is slurried before pitching. No yeast is recovered because it is sold as part of the beer.

The pitched mash is passed through coarse strainers to remove husks. Both screw presses and basket centrifuges are used. The wort then goes to fermentation tanks for 8–24 hours fermentation at 30°C (86°F). Fermentation continues in the carton in which it is distributed.

Starch is a very important component in Bantu beer. The beer contains a considerable amount of both gelatinized and ungelatinized starch (Novellie 1966*b*; Novellie and Schutte 1961). The gelatinized starch helps keep the ungelatinized starch in suspension, makes the beer creamy, and adds body. The starches provide calories to the consumer during the day when kaffir beer is a major dietary component.

Traditional kaffir beers contain 2–4 per cent w/v ethanol, 0.3–0.6 per cent acid (as lactic), and 4–10 per cent total solids. The final pH is generally between 3.3 and 3.5. Municipal kaffir beers have pH ranges from 3.2 to 3.7 with an average of 3.4. Total solids range from 3.0 to 8.0 per cent with an average of 5.4 per cent. Alcohol ranges from 1.8 to 3.9 per cent w/v with an average of 3.0 per cent. Total acidity (as lactic acid) ranges from 0.16 to 0.25 per cent with an average of 0.21 per cent (Novellie 1968).

6. Mexican pulque (Sanchez-Marroquin 1983; Herrera *et al*. 1983; Steinkraus 1983)

Pulque is the national drink of Mexico where it was inherited from the Aztecs (Concalves de Lima 1975). Pulque is a white, viscous, acidic alcoholic beverage made by fermentation of the juice of *Agave*, mainly *Agave atrovirens* or *Agave americana* (the century plant). By official definition, its refractometer reading is 25–30° Brix at 20°C (68°F). Additional characteristics include: pH 3.5–4.0; 4.0–6.0 per cent v/v alcohol; 400–700 mg total acidity (as lactic)/100 ml; 200–400 mg fixed acid (as lactic)/100 ml; density (20°C): 0.9960–1.000; 200–500 mg reducing sugars (as glucose)/100 ml; 300–500 mg protein (N × 6.25)/100 ml; 2.0–3.0 g total solids/100 ml; 20–30 g esters (as ethylacetate)/100 ml; up to 2.5 mg aldehydes (as acetaldehyde)/100 ml; and 80–100 mg higher alcohols (fusel oils)/litre.

Agave spp. grow on very poor soil, making consumption of pulque particularly important in the diets of the low-income people. Children at or under school age receive pulque three times a day. This provides 2.2–12.4 per cent of their calories and 0.6–3.2 per cent of their protein requirements each day (Institute Nat. de Nutricion 1976).

Pulque is distributed commercially in 250-litre wooden or fibreglass barrels by car, truck, and railroad to the pulquerias, which are special bars where the beverage can either be consumed immediately or taken home for consumption.

Consumption is similar to that of draft beer. In Mexico City, with a population of 10 000 000 people, about 1000 kilolitres of pulque are consumed each day. Drinkers consume from 3 to 10 glasses to several litres per day, mainly at lunch and during leisure periods. Peasants are the prime consumers, but the middle-class also consumes pulque on birthdays, at weddings, on picnics, and as an accompaniment to local foods.

6.1. Traditional production of pulque

The substrate for pulque is *Agave* juice called *aguamiel*. It is extracted from mature (8–10-year-old) plants. The floral stem (primordium) is cut by means of a special knife, an operation described as 'castration' of the floral bud. Juice gradually accumulates in the caviety left by its removal. The accumulated juice is removed daily by oral suction through a large, dried gourd (or squash) plant called an *acocote*. The juice is then carried to the *tinacales*, where it is fermented in open wooden, leather, or fibreglass tanks whose capacity is usually about 700 litres.

In the usual fermentation, uncontrolled natural inoculum from a previous pulque fermentation is added to the tank. The fermentation lasts 8–30 days, depending upon the temperature, seasonal changes, and other uncontrolled factors. Tank volume is kept constant by removing quantities equal to those being added to allow a semi-continuous production. Because neither the optimum amount of fresh juice that should be added each time nor the desirable time interval between additions is ever determined, the product quality, the degree of fermentation obtained, and other characteristics vary. Therefore, pulque producers frequently resort to adulteration of the final product. It is impossible to bottle the spontaneously-fermented pulque untreated because it never ferments completely under the above conditions and the bottles burst.

6.2. Modern production of pulque

Sanchez-Marroquin (1953, 1957, 1967, 1970; Sanchez-Marroquin and Hope 1953) studied the microbial flora of pulque and developed a pure-culture process. Using a mixed inoculum (5–10 per cent v/v) containing *Saccharomyces cervisiae*, a homo-fermentative *Lactobacillus* sp., *Zymomonas mobilis* (Lindner) Kluyver Van Niel a *Leuconostoc* sp., pasteurized *Agave* juice with 8° Brix sugar concentration, a pH of 6.0–7.0, and a temperature in the range of 15–28°C (59–82°F) (optimum temperature is 28°C), fermentation time was markedly shortened to 48–72 hours. Settling, racking, and blending required another 12–24 hours.

A pilot plant with a capacity of 1500 litres per day followed by a commercial 50 000 litres per day plant were built. A series of 5000-litre fibreglass fermentation tanks assure semi-continuous operation. Settling blending tanks hold 10 000 litres. Pasteurization of the fresh *Agave* juice is accomplished in plate heat exchangers, and the juice then passes through pre-steamed lines to the fermentation tanks. The tanks are inoculated and fermentation continues 48–72 hours or until the sugar has been metabolized. Next, the pulque is piped to the settling, blending, and viscosity adjusting tanks. The product is then bottled or canned. This process markedly shortens fermentation time, allows optimization of operational parameters, and yields a standardized product.

6.3. Recent improvements in pulque-processing technology

Sanchez-Marroquin (1983) studied the effects of mechanically pressing the whole *Agave* plant, including stems and leaves, to improve the recovery of the sugary juice used for the fermentation. Once the juice is filtered and clarified, it can be adjusted for optimum sugar concentration and pH prior to fermentation.

Although the viscosity of traditional pulque is one of the unique and typical characteristics of this beverage, the younger generation prefers a low viscosity product. Thus, in the modern pilot plant, pulque is manufactured without the use of *Leuconostoc* spp., which produces the dextran responsible for the viscosity. *Leuconostoc* is grown in separate tanks and the viscous culture is added to a portion of the pulque for those consumers who prefer the viscous beverage. The final product is also enriched with additional yeast which raises both the B vitamin and protein contents.

In addition to the above modifications and improvements of the process, pure starters and concentrated *Agave* syrup are being distributed to manufacturers in order to shorten the fermentation and improve the quality of the product. Yeast that settles from the fermenters is recovered and is either added to the pulque as a supplement or used for cattle feed.

7. Japanese sake (Kodama and Yoshizawa 1977; Steinkraus 1983)

Although Japanese sake manufacture today is highly advanced technology, it started as an indigenous fermentation in which in ancient times, soaked, steamed rice was chewed to introduce the necessary amylolytic enzymes (Kodama and Yoshizawa 1977).

Japanese sake is a clear, pale yellow, rice wine with an alcoholic content of 15–16 per cent or higher, a characteristic aroma, little acid, and slight sweetness (Murakami 1972). Originally, the wines were probably made from millet and the beverage was consumed unclarified, cloudy with the micro-organisms and residual solids still in suspension, much as primitive rice wines are consumed in South-East Asia today.

The sake fermentation today utilizes a koji in which steamed rice is overgrown with selected strains of *Aspergillus oryzae* (Ahlberg) Cohn high in amylolytic enzymes. The koji is then combined with additional steamed rice. Saccharification and fermentation with the yeast *Saccharomyces cerevisiae* Hansen (*Saccharomyces sake*) proceed simultaneously. This situation leads to very high populations of yeast cells (2.5×10^8 cells/g) and ethanol concentrations as high as 20 per cent v/v.

Under traditional fermentation conditions, the first organisms to develop are bacteria such as *Pseudomonas*, *Achromobacter*, *Flavobacterium*, or *Micrococcus*, all of which are nitrate-reducing (Murakami 1972). These organisms are followed or possibly accompanied by *Leuconostoc mesenteroides* tsenkovskii var. *sake* and Lactobacillus sake (Kodama and Yoshizawa 1977). These bacteria reach populations of 10^7–10^8 cells/g, acidify the mash and then tend to disappear. Then *Saccharomyces cerevisiae* type yeasts take over. The wild yeasts present cannot withstand the nitrate produced by the nitrate reducing organisms. *Hansenula* type yeasts cannot survive the anaerobiosis in the mash.

In the modern industrialized sake process, the mash is acidified by addition of lactic acid. This suppresses the growth of non-essential micro-organisms and shortens the total fermentation 7–15 days. Traditional fermentations are ordinarily conducted at low temperatures (maximum 18°C or 64°F). Using acidified mash, the fermentation temperature can be raised to 18–22°C (64–72°F). However, with artificial acidification, wild yeasts can play more of a role since no nitrite is present to control their development.

In contrast to traditional rice wines, Japanese sake is noted for its clarity and absence of any residual insoluble solids or micro-organisms.

7.1. Details of sake production

Water used for sake should be colourless, tasteless, odourless, neutral or weakly alkaline, and substantially free of iron (less than 0.02 ppm), nitrate, ammonia, organic substances, and harmful micro-organisms (Kodama and Yoshizawa 1977).

Rice of the short-grained varieties is considered best for sake manufacture and large grains are considered desirable. The rice is polished to remove proteins, lipids, and minerals that are in excess in the bran and germ. Generally, 25–30 per cent of the original weight of the rice kernels is removed. About 280 kl of sake can be produced from 150 tons of brown rice (Kodama and Yoshizawa 1977).

Rice is washed and steeped in water before steaming. During washing, the grains absorb 9–17 per cent of their weight of water. During steeping, water uptake increases to 25–30 per cent of the original rice weight. Steeping requires from 1 to 20 hours, depending upon the hardness of the rice. Following steeping, the rice is drained for 4–8 hours before steaming.

During steaming, starch is gelatinized, the protein is denatured, and the grains are also sterilized. Steaming generally requires 30–60 min, during which time the rice grains absorb an additional 7–12 per cent moisture, resulting in a total uptake of 35–40 per cent water during washing, steeping, and steaming based upon original dry weight. The steamed rice is cooled to 35°C (95°F) for koji manufacture and to 10°C (50°F) for preparation of moromi (Kodama and Yoshizawa, 1977).

For the production of koji, rice is soaked, drained, steamed, cooled, and then inoculated with spores from a tane-koji. Tane-koji is produced by culturing *A. oryzae* on soaked, steamed, polished rice at 28–30°C (82–86°F) for 5 or 6 days or until there is abundant sporulation. In typical koji manufacture, 60–100 g of tane-koji may be used to inoculate 1000 kg of steamed, cooled rice. The inoculated rice is heaped on the floor of a room with controlled humidity and held at 26–28°C (79–82°F). The temperature is about 31–32°C (88–90°F) within the heaped rice. After 10–12 hours, the mould spores have germinated and have begun to grow, so the rice is mixed to maintain uniformity of fungal mycelial growth, temperature, and moisture content. When the temperature rises to 32–34°C (90–93°F) after 20–24 hours, the rice is placed in wooden boxes, each of which holds 15–45 kg of the developing koji. The rice grains are mixed every 6–8 hours to prevent overheating. The grain layers may be reduced from 8 cm deep at the start, to 6 cm deep at the second, and 4 cm deep at the third mixing. After 40 hours incubation, the temperature of the developing koji reaches 40–42°C (104–108°F), mycelium covers the grains and they contain sufficient enzymes so that the koji can be used for saccharification of starch in the mash.

According to Kodama and Yoshizawa (1977), koji can be made in special boxes in which inoculated rice is aerated by forced air at the optimum temperature and humidity. By optimizing conditions, koji manufacture can be shortened by 6–8 hours.

Koontoka-moto is a hot-mash, rapid-saccharification procedure used for the preparation of yeast starter (moto). After a 6 hours starch hydrolysis step carried out at 56–60°C (133–140°F) with koji amylases, the mash is cooled, acidified, and filtered. The filtrate is used to grow a pure sake culture.

More recently, aerobically propagated compressed sake yeast has become available commercially and can be inoculated directly as 7 per cent w/w of the total rice used in a moromi mash acidified with lactic acid, eliminating the necessity of preparing moto.

For the main fermentation mash (moromi), unsterilized koji, steamed rice, and water are fermented with moto in 6–20 kl-tanks, each containing from

1.5–10 tons of rice. The yeast population in *moromi* is built up stepwise over a period of 3 days. Moto mash is combined with equal quantities of rice and water, reducing the yeast count by two-thirds. After 2 days at 12°C (54°F), the yeast population rises again to 10^8 organisms/g and the mash is diluted again by about one-half. The rice-koji-water mixture is added at 9–10°C (50°F) to suppress the growth of contaminating organisms. The following day, a third addition is made at 7–8°C (45–46°F), again reducing the yeast by one-half. In this way, the yeast population, 2.5×10^8 cells/g, is reached after about 1 week of fermentation. Such stepwise addition permits careful temperature control, important in balancing saccharification and fermentation rates. Low temperatures (below 18°C, 64°F) and gradual release of sugars due to starch hydrolysis permit equally gradual ethanol production which is important in maintaining yeast viability (Nagodawithana and Steinkraus 1976). With such controlled fermentations, ethanol concentration approaches 20 per cent (v/v) in 20–25 days.

The moromi tends to form a rather viscous foam that may occupy one-third of the fermenter volume, a problem that has lead to development of non-foaming yeasts in order to increase fermenter capacity. Further developments include acidification of the mash with lactic acid to decrease the amount of bacterial activity and allowing a higher fermentation temperature (15°C, 59°F) thus shortening the fermentation (Murakami 1972).

The processes of pressing out the mash, settling for 5–10 days, filtering, blending, and final settling (30–40 days) all at low temperature are followed by pasteurization at 55–65°C (131–149°F). Interestingly, pasteurization of sake has been practised since the 16th century, long before Pasteur's time. Aging at 13–18°C (55 to 64°F) with or without activated carbon is followed by final blending, diluting with water, filtering, and bottling.

One ton of polished rice yields about 3 kl of sake (20 per cent ethanol v/v and 200–250 kg of residue (sake-kasu).

Summary

Several indigenous food fermentations have been commercialized. These include Indonesian tempe, Chinese soy sauce, Japanese shoyu, Japanese miso, African kaffir beer, Mexican pulque, and Japanese sake. The contrast between the primitive and the commercial processes is rather dramatic. But the scientific principles behind both processes are basically the same. The Japanese shoyu, miso, and sake processes have lead directly to the international enzyme, mono-sodium glutamate and flavour enhancing nucleotide industries. Detailed studies of the microbiology, biochemistry, and nutritional implications of other indigenous fermented foods important in other countries of the world will likely lead to whole new industries and improvement in the quality and nutritional value of the food presently consumed by the world's masses of people. These same foods are likely to be of even greater importance as world population reaches 8–12 billion in the 21st century.

References

Ebine, H. (1983). Japanese miso. In *Handbook of Indigenous Fermented Foods* (ed. K. H. Steinkraus) Marcel Dekker Inc. New York.

Goncalves de Lima, W. (1975). *Pulque, Balche E. Pajuaru*. Universidade Federal de Pernambuco. Recife, Brasil.

Herrera, T., Ulloa, M., and Taboada, J. (1983). Microbiological studies on pulque. In *Handbook of Indigenous Fermented Foods*, (ed. K. H. Stein Krams), Marcel Dekker Inc. New York.

Hesseltine, C. W. (1961). Research at Northern Regional Research Laboratory on fermented foods. Proc. Conf. *Soybean Products for Protein in Human Foods*, USDA Peoria, Ill. September 13–15, pp. 67–74. United States Dept. of Agriculture, Agricultural Research Service.

Instituto Nal. de Nutricion. (1976). Encuestas Nutricionales en Mexico, Mexico City, II: 91–11.

Kodama, K. and Yoshizawa, K. (1977). Sake. In *Economic Microbiology* (ed. A. H. Rose) Vol. 1, pp. 432–75. Academic Press, New York.

Martinelli, A., and Hesseltine, C. W. (1964). Tempeh fermentation: Package and tray fermentations. *Food Technol.* 18, 167–71.

Murakami, H. (1972). Some problems in sake brewing. In *Fermentation Technology* (ed. G. Terui) pp. 639–43. Proc. IVth. Int. Fermentation Symp. Kyoto, Japan.

Nagodawithana, T. W. and Steinkraus, K. H. (1976). Influence of the rate of ethanol production and accumulation on the viability of *Saccharomyces cerevisiae* in 'Rapid Fermentation'. *Appl. Env. Microbiol.* 31, 158–62.

Nakaya, K. (1934). Studies on the treatment of soybean for the soy sauce fermentation. *Rep. Noda Soy Sauce Co. Ltd.* Nodashi, Chiba-ken, Japan, 4, 83–108.

Novellie, L. (1962). Kaffircorn malting and brewing studies. XI. Effect of malting conditions on the diastatic power of kaffircorn malt. *J. Sci. Food Agric.* 13, 115–20.

—— (1963). Bantu beer: food or beverage? *Food Industries, SA* 16, 28.

—— (1966a). Bantu beer-popular drink in South Africa. *Int. Brewer Distiller* 1, 27–31.

—— (1966b). Kaffircorn malting and brewing studies. XIV. Mashing with kaffircorn malt. Factors affecting sugar production. *J. Sci. Food Agric.* 17, 354–61.

—— (1968). Kaffir beer brewing, ancient art and modern industry. *Wallerstein Lab. Comm.* 31, 17–29.

—— (1977). Beverages from sorghum and millets. *Proceedings Symposium on Sorghum and Millet for Human Food* (ed. D. A. V. Dendy) Tropical Products Institute, London. pp. 73–7.

—— and Schutte, R. H. (1961). Kaffircorn malting and brewing studies. X. The susceptability of sorghum starch to amylolysis. *J. Sci. Food Agric*, 12, 552–9.

Platt, B. S. (1964). Biological ennoblement: Improvement of the nutritive value of foods and dietary regimes by biological agencies. *Food Technol.* 18, 662–70.

—— and Webb, R. A. (1946). Fermentation and human nutrition. *Proc. Nutr. Soc.* 4, 132–40.

—— and —— (1948). Microbiological protein and human nutrition. *Chem. Ind.* Feb. 7, 88–90.

Sanchez-Marroquin, A. (1953). The biochemical activity of some micro-organisms of pulque. *Mem. Congr. Cient. Mex. IV Centenario Univ. Mex.* 2, 471–84.

—— (1957). Microbiology of pulque. XVIII. Chemical data on the fermentation of agave juice with pure microbiol cultures. *Rev. Soc. Quim. Mex.* 1, 167–74.

—— (1967). Estudios sobre la microbiologia del pulque, XX. Proceso industrial papa la elaboracion tenica de la bebida. *Rev. Latamer. Microbiol. Parasitol.* 9, 87–90.

—— (1970). Investigaciones realizadas en al Facultad de Quimica de al Universdad Nacional Autonoma de Mexico, tenientes a la industrializacion del agave. *Rev. Soc. Quim. Mex.* 14, 184–8.

—— (1983). Mexican pulque-a fermented drink from Agave juice. In *Handbook of Indigenous Fermented Foods* (ed. K. H. Steinkraus), Marcel Dekker, New York.

—— and Hope, P. H. (1953). Agave juice fermentation and chemical composition studies of some species. *Agric. Food Chem.* 1, 246–9.

Schwartz, H. M. (1956). Kaffircorn malting and brewing studies. I. The kaffir beer brewing industry in South Africa. *J. Sci. Food Agric.* 7, 101–5.

Shibasaki, D. and Hesseltine, C. W. (1962). Miso fermentation. *Econ. Botany*, 16, 180–95.

Shurtleff, W. and Aoyagi, A. (1980). *Tempeh production.* New Age Foods. Lafayette, California.

—— and —— (1983). *The Book of Miso*, 2nd edn. Ten Speed Press, Berkeley, California.

—— and —— (1984). *The Soyfoods Industry and Market.* The Soyfoods Center. Lafayette, California.

Steinkraus, K. H. (1983). *Handbook of Indigenous Fermented Foods.* Marcel Dekker Inc. (New York). 672 pages.

——, Hwa, Y. B., Van Buren, J. P., Provvidenti, M. I., and Hand, D. B. (1960). Studies on tempeh. An Indonesian fermented soybean food. *Food Res.* 25, 777–88.

——, Van Buren, J. P., Hackler, L. R., and Hand, D. B. (1965). A pilot-plant process for the production of dehydrated tempeh. *Food Technol.* 19, 63–8.

van der Walt, J. P. (1956). Kaffircorn malting and brewing studies. II. Studies on the microbiology of kaffir beer. *J. Sci. Food Agric.* 7, 105–13.

Yokotuska, T. (1960). Aroma and flavor of Japanese soy sauce. *Adv. Food Res.* 10, 75–134.

—— (1972). Some recent technological problems related to the quality of Japanese shoya. In *Fermentation Technology Today* (ed. G. Terui), Proc. Fourth int. Fermentation Symp., Kyoto. March 19–25. Society of Fermentation Technology, (Japan).

—— (1983). Japanese shoyu. In *Handbook of Indigenous Fermented Foods*, (ed. K. H. Steinkraus) Marcel Dekker, New York.

Yong, F. M. (1971). Studies on soy sauce fermentation. MSc Thesis, University of Strathclyde, Glasgow.

7

Improvement of industrially important Streptomycetes by protoplast fusion and regeneration

W. Kurylowicz, W. Kurzatkowski, and J. Solecka

1. Introduction

In recent years a method for the improvement of *Streptomyces* strains by regeneration of protoplasts was described. Though this method is important from the industrial point of view, the principle of increased antibiotic productivity by the regenerated protoplasts is still not well known.

Conditions suitable for the formation and regeneration of protoplasts from *Streptomyces* were described previously by Okanishi *et al.* (1974). These procedures were later modified by Baltz and Matsushima (1981), Shirahama *et al.* (1981), and Ogawa *et al.* (1983).

2. Strain improvement by protoplast regeneration

Ikeda *et al.* (1983) demonstrated an improvement in macrolide-antibiotic productivity by regeneration of protoplasts of *Streptomyces* strains. When compared with the original strains *Streptomyces ambofaciens* KA–1029 and *Streptomyces fradiae* KA–427 producing spiramycin and tylosin, respectively, the strains selected after regeneration of protoplasts showed changes in antibiotic productivity.

The results demonstrated that about 90 per cent of the regenerated protplasts of *S. ambofaciens* KA–1029 exhibited increased activity of spiramycin production. In particular, some strains selected from the re-generated protoplasts synthesized twice as much. Furthermore, in comparison with the original strain *S. fradiae* KA–427, tylosin production was increased by about 70 per cent of the regenerated protoplasts, some of the regenerated protoplast strains showing a three-fold higher activity of the process of antibiotic biosynthesis.

The strains selected after regeneration of protoplasts exhibited stability in both streptomycin and tylosin production. The biosynthetic pattern of products synthesized by the regenerated protoplasts was the same as in the original strains *S. ambofaciens* KA–1029 and *S. fradiae* KA–427.

In the macrolide-antibiotic producing strains some progeny after protoplast regeneration were different in morphology from the original strains, though most of them produced more aerial mycelium. Ikeda *et al.* (1982) and Okanishi (1982) suggested that the increased activity of antibiotic production by the regenerated protoplasts originates from genetic variation, as also do the 'restoration' of antibiotic productivity (Ikeda *et al.* 1982; Okanishi 1982); changes in morphology (Ikeda *et al.* 1982); increases in antibiotic resistance (Yamashita *et al.* 1982); and the elimination of extrachromosomal elements (Kieser *et al.* 1982).

In this context, genetic variation during the protoplast formation can be explained by the liberation of plasmids attached to the membranous proteins causing changes in the genetic material of cells. These changes can be expressed in the variation of antibiotic productivity.

On the other hand, Araujo *et al.* (1984) demonstrated that the activity of cyclamycin, an anthracycline type of antibiotic, production by some regenerated protoplasts of *Streptomyces capoamus* was about five times higher in comparison with the original strain. The authors assumed that the increased cyclamycin productivity may be caused by protoplast fusion and recombination in the genetically diversified population. Again, in the protoplast medium bivalent cations such as Ca^{2+} and Mg^{2+} may function also as protoplast fusogenic agents.

Several authors underlined the important role of calcium and magnesium as fusogenic agents. According to Ferenczy (1981) the fusion of protoplasts starts during the process of protoplast formation in the protoplast medium containing calcium and magnesium.

The role of calcium in the fusion of vesicles has been studied Ingola and Koshland (1978). Artificial phospholipid vesicles fused upon addition of calcium, and the subsequent addition of valinomycin or calcium ionophore (A 23187), showed further fusion of liposomes. Lansman and Haynes (1975) described the triggering of aggregation of phospholipid vesicles by bivalent cations.

Papahadjopolous *et al.* (1974) examined the effect of divalent cations on the interaction and the mixture of components of vesicular membranes produced from acidic phospholipids. For this purpose the authors used freeze-fracture electron microscopy. The ions calcium and magnesium induced extensive mixing of components of vesicular membranes and caused drastic structural rearrangements leading to formation of new membranous structures. The calcium and magnesium ions induced formation of vesicles when the lipids are in fluid state.

Kurzatkowski *et al.* (1984) investigated the productivity of antibiotics by the protoplasts and strains selected from the regenerated protoplasts. The activity of actinomycin D biosynthesis was estimated by incorporation of radioactive precursors into the antibiotic molecule. In comparison with mycelium of *Streptomyces melanochromogenes* no. 4, its protoplasts showed similar activity of L-/U/^{14}C/-valine incorporation into actinomycin D. The mycelium regenerated from the protoplasts of *S. melanochromogenes* no. 4 yielded a ten times higher incorporation yield (data not shown).

Different results were obtained with *Streptomyces griseus* As–4 and *S. griseus* LS–1–3. The activity of their protoplasts in ^{14}-C-myo-inositol incorporation into streptomycin was ten times higher in comparison with that of the mycelium. However, the regenerated protoplasts of both compared strains were only five times more active than the mycelium.

Variation in the yield of actinomycin D and streptomycin production by protoplasts may be explained by changes in the activity of the transport through protoplast membranes. It is possible that a number of membranous proteins involved in the transport are lost from the cell membrane during the digestion of cell wall by lysozyme. In other organisms cell wall removal has been shown to cause loss of enzymes and proteins associated with the transport (Barash and Halpern 1971; Gerdes *et al.* 1977; Heppel 1967).

In this context, D-valine, *cis*-3-methylproline, and α-methyl-DL-tryptophan were previously found to inhibit the biosynthesis of actinomycin by strains of *Streptomyces antibioticus* and *Streptomyces parvulus* ATCC 29651. However, an accumulation of 4-methyl-3-hydroxy-anthranilic acid, which is a direct precursor of the actinomycin chromophore production occurred (Sivak and Katz 1962; Yoshida *et al.* 1966; Katz and Weissbach 1962; Golub *et al.* 1969; Salzman *et al.* 1969; Hitchcock and Katz 1978). The activity of the protoplast system is much less sensitive to inhibition by D-valine, *cis*-3-methylproline, and α-methyl-DL-tryptophan. This may be due to reduced uptake of these compounds as the rate of transport of several amino acids such as methionine, sarcosine, and L- and D-valine into protoplasts. The transport of amino acids with protoplasts is known to be much lower than that with mycelium.

The leakage of soluble nucleotides and nucleic acids from protoplasts during preparation has been reported (Okanishi *et al.* 1974). Weibull (1956) suggested that the effect of magnesium on the stabilization of spheroplasts might be ascribed to the prevention of lipid release from the cell membrane. Muschel (1968) reported that magnesium has been employed by several researchers for stabilizing spheroplasts or protoplasts. Tabor (1962) reported that calcium as well as spermine prevented lysis of lysozyme induced protoplasts of *Escherichia coli* and *Micrococcus lysodeikticus* in hypertonic media. The liberation of the vesicles of the mesosomes from protoplasts of *Bacillus licheniformis* was controlled by the magnesium concentration present during the removal of the walls by lysozyme. Moreover, functioning of the cytolasmic membranes is also controlled by magnesium concentration (Reaveley and Rogers 1969).

Another explanation of variation in antibiotic productivity can be ascribed to aminoglycosides such as streptomycin and kanamycin, which appear to be shunt products diverging from a pathway leading to cell wall biosynthesis (Nara 1977; Nara *et al.* 1977). Indeed, the inclusion of cell wall inhibitors in growth medium has been found, to increase production of aminoglycoside antibiotics.

Actinomycin D can be also a shunt product of cell wall biosynthesis. The antibiotic is accumulated in the cell wall (Kurzatkowski *et al.* 1976) and, it has been possible to stimulate or to inhibit the production of actinomycin D by different inhibitors of cell wall biosynthesis. Gentian violet inhibited both the cell wall biosynthesis of protoplasts *S. melanochromogenes* no. 4 and actinomycin production. On the other hand, bacitracin, novobiocin, and vancomycin stimulated production of actinomycin. In the light of these results, we suggest a common pathway for cell wall and antinomycin biosynthesis which branches off after the production of uridine-5'-pyrophosphate N-acetylmuramic acid, which is an intermediate of cell wall biosynthesis (unpublished data).

3. Strain improvement by protoplast fusion

Protoplast fusion in bacteria, streptomycetes, yeasts, and fungi is a recent successful technique which has contributed not only to the improvement of industrially important strains, but also, to a better understanding of the secondary metabolic pathways in antibiotic biosynthesis. It has facilitated elucidation of several genetic consequences of microbial recombination, such as the elaboration of numerous linkage maps bearing accurately located genes. However, in most cases their biochemical expression is unknown.

The polyethylene glycol (PEG) induced protoplast fusion in order to obtain recombinants in *Streptomyces* consists of: (i) selection of two or more parent strains to be recombined; (ii) obtaining marked mutants (preferably auxotrophs) resistant to antibiotics; (iii) formation of stabilized protoplasts able to regenerate into normal wall-bearing cells and to fuse with high efficiency; and (iv) identification and selection of recombinants, and their analysis.

Selection of strains for recombination should be conducted based on data of biosynthetic pathways of secondary metabolites with antimicrobial action overproduced by both strains to be crossed. The aim of the fusion of protoplasts of two or more parent strains producing different antibiotics, but structurally belonging to similar groups is to transfer genes coding for antibiotic synthesizing enzymes. This would probably enhance the chance that an enzyme introduced into a new host would find a substrate chemically different from its normal substrate, but released to it, and would be capable of conversion to an analogue of a 'natural antibiotic' (Normansell 1983).

Though recombination in *Streptomyces* is widespread, it has nevertheless been applied to a rather limited extent. According to Hopwood and Wright (1978) the reason for its neglect has probably been the necessity to introduce selectable markers into the strains to be crossed in order to identify recombinants. Marking is time consuming.

The auxotrophic mutants of *Streptomyces* can easily be selected after N-methyl-N'-nitro-N-nitrosoguanidine (NTG) treatment. It induces primarily base transition mutations of GC-AT types, although AT-GC transitions, transversions and even frameshifts arise at low frequencies. It is highly mutagenic at low killing levels, induces multiple mutations in localized regions. It is particularly effective for the isolation of auxotrophic mutants (Carlton and Brown 1981). Several methods such as the filtration method (Woodward *et al.* 1954), the penicillin method (Davis 1948) or the replica plating technique of Lederberg and Lederberg (1952) can be used for the selection of auxotrophs in *Streptomyces*. Delić *et al.* (1970) described the conditions of effective mutagenesis by NTG in *Streptomyces*, which requires a longer and more intensive treatment than in some other micro-organisms. With treatment at a concentration of 1 mg/ml of NTG at pH 9.0 for 30 mins 5 per cent of the survivors were auxotrophs. After incubation of *Streptomyces coelicolor* for 2 hours in 0.05 M Tris-maleic acid buffer containing 3 mg/ml of NTG about 11 per cent of auxotrophs were amongst the survivors obtained. With *Streptomyces rimosus* after incubation for 1 hour in the presence of NTG at the same concentration at pH 9.0, more than 21 per cent of the survivors were auxotrophs.

One of the possible difficulties of working with NTG is the likelihood of inducing multiple mutations in the small region of the bacterial chromosome (Guerola *et al.* 1971). As a consequence, the mutants could have complex growth requirements on account of mutations affecting more than one type of metabolic function and failure to revert to the wild stage at expected frequencies. It is thus necessary to check NTG-induced mutants carefully for reversion before attempting to use them for further analysis (Carlton and Brown 1981).

In order to obtain auxotrophic mutants of *Streptomyces* strains resistant to antibiotics, the gradient plate technique of Szybalski (1952), or the serial dilutation method incorporating repeated transfers is commonly used.

Formation of protoplasts depends on several factors that have to be established for each strain separately. Here belong the composition of the medium to produce mycelium in form of homogenous growth. The mycelium to be protoplasted has to be harvested in the middle of the exponential growth phase (average 6–10 doublings). With the addition of glycine to the medium in a concentration range from 0.5 to 1.5 per cent, these requirements can be met after 18–24 hour growth of mycelium in shake flasks. It should be noted that some mycelia need sonication before lysozyme treatment (Baltz 1980).

The mycelium, after washing with 0.3 M sucrose, is treated with lysozyme at a concentration of 1–2 mg/ml of the protoplast medium.

Ogawa *et al.* (1983) described an improved method for preparation of *Streptomyces* and *Micromonospora* protoplasts using the combination of cell-wall decomposing enzymes such as lysozyme, lytic enzyme no. 2, and achromopeptidase. In such manner the protoplasts of *Streptomyces virido-chromogenes*, *Steptomyces wedmonensis*, *Streptomyces kanamyceticus* and *Streptomyces mycerofaciens* were formed and regenerated. As a result it was found that lysozyme combined with achromopeptidase was 10–100 times more effetive in protoplast formation than the combination of lysozyme with lytic enzyme no. 2.

On an average, the appearance of protoplasts delivered after a couple of minutes. Frequent control by phase contrast microscopy showed that the time phase needed to convert the cells into the protoplasts could easily be established.

The effect of sodium, magnesium and calcium on the formation and leakage of protoplasts was tested by Okanishi *et al.* (1974). In hypertonic media containing 0.3–0.5 M sucrose, $MgCl_2.6H_2O$ of a concentration of 0.01–0.05 M and $CaCl_2.2H_2O$ of a concentration of 0.025–0.05 M, the protoplasts are well established.

The regeneration of protoplasts is a crucial step in strain improvement and in recombination in *Streptomyces*. The recombination of protoplasts of *S. griseus* ISP 5236 and *Streptomyces venezuelae* ISP 5230 was 41 and 51 per cent, respectively, depended to a great extent on the presence of bivalent cations in the regeneration medium containing phosphate and casamino acids (Okanishi *et al.* 1974). Similar efficiency of protoplast regeneration has been obtained in two auxotrophic mutants of *S. melanochromogenes* no 4 (Kurzatkowski *et al.* 1979). Again, a 50 per cent regeneration efficiency in *Streptomyces rimosus* protoplasts was reported by Pigac *et al.* (1982) by optimizing the osmotic stabilizer concentrations and by modifying the plating procedure through the use of the soft agar overlay technique.

Moreover, Shirahama *et al.* (1981) found that the regeneration frequency of protoplasts varies according to the species used, and that the high regeneration rate is limited only to some strains of *Streptomyces*. The authors suggested that improved conditions of protoplasts regeneration resulted from the use of modified regeneration medium (R_3), of embedding the protoplasts in soft, low melting point agarose (with ingredients of R_3 medium) and of seeding the embedded protoplasts on a dried surface of underlay R_3 agar (Shirahama *et al.* 1981). The protoplasts were incubated at 20°C. The regeneration frequency with an auxotrophic mutant of *Streptomyces kasugeansis* MB 273–18a was 70 per cent, with *S. fradiae* and *Streptomyces niveus* about 90 per cent, and with *Streptomyces ambofaciens* KA–1028, *S. fradiae* KA–427 and *Streptomyces cirratus* KA–412 about 100, 92, and 100 per cent, respectively. On the other hand, the regeneration frequently with an auxotrophic mutant of *Streptomyces kanamyceticus* was 30 per cent. Only *Streptomyces humidus* showed a low regeneration frequency of about 10 per cent.

According to Baltz and Matsushima (1981) efficient protoplast regeneration strongly depends upon the incubation temperature.

The influence of plasma expanders such as polyvinyl pyrrolidine K–90 MW 360 000 in amount about 0.3 per cent and bovine serum albumin (fraction V) at about 0.02 per cent on the regeneration frequency of protoplasts from three macrolide producing strains of *Streptomyces* were observed by Ikeda *et al.* (1983). Amongst the various reviews dealing with protoplast fusion, Ferenczy (1981) recently covered the whole field from bacteria to algae and discussed the various consequences of genetic transfer. The most rapid progress in protoplast fusion was achieved in microbiology. Significant advances have been made in industrial research, but it seems that one is rather at the beginning stage in the exploitation, for both theoretical and practical purposes, of the method of protoplast fusion (Ferenczy 1981).

Protoplast fusion is possible not only between two or more mutants (Hopwood and Wright 1978), but between protoplast of strains belonging to the same or different species, i.e. intra or interspecific fusion. Recent data showed that the protoplast fusion is possible between cells belonging to different genera. Even inter-kingdom protoplast fusion has been reported.

The fusion induced by polyethylene glycol (PEG) between protoplasts of different organisms seems to be correlated neither with their taxonomic position nor with immunochemical specificity. Since structure of the cell membrane involved in fusion is fairly similar throughout the whole of the living world, the fusion processes are likely to be similar (Ferenczy 1981).

The mechanism of protoplast fusion induced by PEG is not well understood. According to Ahkong *et al.* (1975*a*) fusion starts with agglutination of protoplasts caused by intensive dehydration and formation of aggregates. The protoplast shrinks and becomes highly distorted. Large area of adjacent protoplast membranes come into very close contact (Ferenczy *et al.* 1975). First a perturbation of bilayer structure of membrane lipid may occur which increases fluidity of the lipid region. The consequence of this event is the aggregation of the intramembranous glycoprotein particles. The interaction and inter-mixing of the disturbed molecules of closely opposed membranes in the region denuded of intramembranous proteins and glycoproteins may allow the adjacent cells to fuse through small cytoplasmic bridges (Ferenczy 1981).

High frequency of fusion of plant protoplasts by electric fields has been described by Zimmermann and Scheurich (1981), and Zimmermann (1982). The protoplasts of *Vicia faba* were collected by dielectrophoresis in highly unhomogenous alternating field (sine wave, 5 to 10 V peak to peak value, 500 kHz, electrode distance 200 μm). Zimmermann (1982) found that the cells formed aggregates on the electrodes, which aggregates were stable for the duration of the applied field. By additional application of a high single field pulse it was possible to introduce cell fusion within the aggregates. The fused aggregates formed spheres within a few minutes. From these results it can be concluded that in addition to close contact between membranes, the prerequisites for an electrically-stimulated fusion is a dielectric breakdown that leads to changes in the membranes conductance, permeability and probably fluidity.

4. Conclusion

The PEG-induced protoplast fusion has recently been applied in several cases to yield more data on its mechanism. It consists of the following steps: (i) selection of two or more parent strains to be recombined; (ii) formation in suspension of young protoplasts; (iii) protoplast fusion; (iv) fusion of 'active' protoplasts with heat inactivated protoplasts; (v) identification and selection of colonies regenerated from fused protoplasts, through means of high selection 'pressure'; (vi) isolation of colonies with high antibiotic yield with the overlay bioassay; and (vii) comparison of chromatographically established patterns of metabolites produced by both parental and selected strains. This model of recombination of *Streptomyces* strains by protoplast fusion can be applied for the improvement of *Streptomyces* strains and for isolation of strains producing new secondary metabolites some of which may possess new antimicrobial activity.

In order to obtain a recombinant by fusion, there must be considerable homology between the DNA of the two interacting parents (Hopwood 1981). However, it should be stressed that there is much evidence of intra- and inter-specific recombination, nutritional supplementation and gene transfer. This has been achieved in auxotrophic mutants of *S. parvulus* and *S. antibioticus* by Ochi *et al.* (1979) and proved by the production of a new hybrid antibiotic type of anthracycline through recombination by mating of strains belonging to different species (Fleck 1979). Godfrey *et al.* (1978) found that recombination by protoplast fusion is 10^4 times more efficient than recombination by mating of cells bearing the cell wall.

Intergeneric protoplast fusion has been obtained in *Saccharomyces cerevisiae* and *Saccharomyces pombe* by Svoboda (1980). Even the fusion between erythrocytes and *Saccharomyces cerevisiae* protoplasts induced by PEG has been successful (Ahkong *et al.* 1975b). Although the results of this experiment were with no immediate genetic consequences they open ways for research on membrane structure and fusion.

Normansell (1983) reported that the *Streptomyces genome* has 1.5×10^7 base pairs and the average size of a gene is taken to be 1500 pairs, i.e. there are about 1000 genes in the genome. If there are 100 genes which affect the yield of a given antibiotic, then this is equivalent to 1.5×10^5 bases pairs. The wealth of genetic information for antibiotic production in *Streptomyces* makes the protoplast fusion in these organisms very promising.

References

Ahkong, Q. F., Fischer, D., Tampion, W. and Lucy, J. A. (1975a). Mechanism of cell fusion. *Nature* 253, 194–5.

——, Howell, J. I., Lucy, J. A., Safwat, F., Davey, M. R. and Cocking, E. C. (1975b). Fusion of hen erythrocytes with yeast protoplasts induced by polyethylene glycol. *Nature* 255, 66–7.

Araujo, J. M., de A. Lyra, F. D. and Kurylowicz, W. (1984). Recombination in antibiotic producing *Streptomyces* by protoplast fusion. *Rev. Microbiol.* São Paulo 15, 67–81.

Baltz, R. H. (1980). Genetic recombination by protoplast fusion in *Streptomyces*. *Dev. Indust. Microbiol.* 21, 43–54.

—— and Matsushima, P. (1981). Protoplast fusion in *Streptomyces*: conditions for efficient genetic recombination and cell recombination. *J. Gen. Microbiol.* 127, 137–46.

Barash, H. and Halpern, Y. S. (1971). Glutamate-binding protein and its releation to glutamate transport in *Escherichia coli* K–12. *Biophys. Res. Commun.* 45, 681–8.

Carlton, B. C. and Brown, B. J. (1981). Gene mutation. In *Manual of methods for general microbiology* (ed. P. Gerhardt *et al.*) pp. 222–42. A. S. M., Washington D.C.

Davis, B. D. (1948). Isolation of biochemically deficient mutants of bacteria by penicillin. *J. Am. Chem. Soc.* 70, 4467–71.

Delić, V., Hopwood, D. A. and Friend, E. J. (1970). Mutagenesis by N-methyl-N′-nitro-N-nitrosoguanidine (NTG) in *Streptomyces coelicolor*. *Mutation Res.* 9, 167–82.

Ferenczy, L. (1981). Microbial protoplast fusion. In *Genetics as a tool in microbiology* (eds S. W. Glover and D. A. Hopwood) 31 Symp. Soc. Gen. Microbiol., pp. 1–34. Cambridge University Press.

——, Kevei, F. and Szegedi, M. (1975). High frequency fusion of fungal protoplasts. *Experientia* 31, 117–22.

Fleck, W. F. (1979). Genetic approaches to new streptomycetes products. In *genetics of industrial micro-organisms* (eds O. K. Sebek and A. J. Laskin) pp. 117–22. Am. Soc. Micribiol., Washington.

Gerdes, R. G., Strickland, K. P. and Rosenberg, H. (1977). Restriction of phosphate transport by the phosphate binding protein in spheroplasts of *Escherichia coli*. *J. Bacteriol.* 131, 512–8.

Godfrey, O., Ford, L. and Huber, M. L. B. (1978). Interspecies matings of Streptomyces fradiae with *Streptomyces bikiniensis* mediated by conventional and protoplast techniques. *Can. J. Microbiol.* 24, 994–7.

Golub, E. E., Ward, M. A. and Nishimura, J. S. (1969). Biosynthesis of the actinomycin chromophore:incorporation of 3-hydroxy-4-methylanthranilic acid into actinomycins by Streptomyces antibioticus. *J. Bacteriol.* 100, 997–84.

Guerola, N., Ingraham, L. and Cerda-Olmeda, E. (1971). Induction of closely-linked multiple mutations by introso-guanidine. *Nature* 230, 122–5.

Heppel, L. A. (1967). Selective release of enzymes from bacteria. Treatments affecting the bacterial wall remove; certain enzymes and transport factors from living cells. *Science* 156, 1451–5.

Hitchcock, M. J. and Katz, E. (1978). Actinomycin biosynthesis by protoplasts derived from *Streptomyces parvulus*. *Antimicrob. Agents Chemother.* 13, 104–14.

Hopwood, D. A. (1981). Possible application of genetic recombination in the discovery of new antibiotics in Actinomycetes. In *Future of antibiotherapy and antibiotic research*, pp. 407–18. Academic Press, London.

—— and Wright, H. M. (1978). Bacterial protoplast fusion: recombination in fused protoplasts of *Streptomyces coelicolor*. *Mol. Gen. Genet*. 162, 307–17.

Ikeda, H., Inoue, M. and Omura, S. (1983). Improvement of macrolide antibiotic-producing streptomycetes strains by regeneration of protoplasts. *J. Antibiot*. 35, 283–8.

——, Tanaka, H., and Omura, S. (1982). Genetic and biochemical features of spiramycin biosynthesis in Streptomyces ambofaciens. Curing protoplast regeneration and plasmid transfer. *J. Antibiot*. 35, 507–16.

Ingola, T.and Koshland, D. E. I. (1978). The role of calcium in fusion of artificial vesicles. *J. Biochem. Chem*. 253, 3821–9.

Katz, E. and Weissbach, H. (1962). Biosynthesis of the actinomycin chromophore; enzymatic conversion of 4-methyl-3-hydroxyanthranilic acid to actinocin. *J. Biol. Chem*. 237, 882–5.

Kieser, T., Hopwood, D. A., Wright, H. W. and Thompson, C. J. (1982). pIJ 101, a multi-copy broad host-range Streptomyces plasmid. Functional analysis and development of DNA cloning vesicles. *Mol. Gen Genet*. 185, 223–38.

Kurzatkowski, W., Gorzkowski, B., Woznicka, W., Paszkiewicz, A. and Kurylowicz, W. (1976). The site of synthesis of actinomycin D in *Streptomyces melanochromogenes*. In *Proceedings Int. Symp. Nocardia* and *Streptomyces* (ed. G. Fischer) Verlag, Stuttgart, New York.

——, Kurylowicz, W., Woznicka, W., Paszkiewicz, A., Polowniak-Pracka, H., Wawrzyniak, K. and Gumpert, J. (1979). Formation, stabilization and fusion of protoplast of two auxotrophic mutants of *Streptomyces melanochromogenes*. In *Proceedings of Congress of Hungarian Soc. Chemotherapy*, Hajduszoboszló, 129–42.

Lansman, J. and Haynes, D. H. (1975). Kinetics of a Ca^{2+}-triggered membrane aggregation reaction of phospholipid membranes. *Biochem. Biophys. Acta* 394, 335–47.

Lederberg, J. and Lederberg, E. M. (1952). Replica plating indirect selection of bacterial mutants. *J. Bacteriol*. 63, 399–408.

Muschel, L. H. (1968). The formation of spheroplasts by immune substances and the reactivity of immune substances against diverse rounded forms. In *Microbial protoplasts, spheroplasts and L-forms* (ed. I. B. Guze). Baltimore.

Nara, T. (1977). Aminoglycoside antibiotics. In *Annual reports on fermentation processes* (ed. D. Perlman) Vol. 1, pp. 299–326. Academic Press, New York.

——, Kawamoto, I., Okachi, R., and Oka, T. (1977). Source of antibiotics other than *Streptomyces*. Jap. *J. Antibiot*. 30, 174–89.

Normansell, J. D. (1983). Strain improvement in antibiotic producing micro-organisms. *J. Chem. Techn. Biotechnol*. 32, 296–303.

Ochi, K., Hitchcock, M. J. and Katz, E. (1979). High-frequency fusion of *Streptomyces parvulus* and *Streptomyces antibioticus* protoplasts induced by polyethylene glycol. *J. Bacteriol*. 139, 984–92.

Ogawa, H., Imai, S., Satch, A. and Kojima, M. (1983). An improved method for the preparation of streptomycetes and *Micromonospora* protoplasts. *J. Antibiot*. 36, 184–6.

Okanishi, M. (1982). Function of plasmid in aureothricin production. In *Tends in Antibiotic Research* (ed. H. Umezawa *et al*.) pp. 32–41. Japan Antibiotics Res. Assoc., Tokyo.

——, Suzuki, K. and Umezawa, H. (1974). Formation and reversion of streptomycete protoplasts: cultural condition and morphological study. *J. Gen. Microbiol.* 80, 389–400.

Papahadjopoulos, D., Poste, G., Schaeffer, B. E. and Vail, W. J. (1974). Membrane fusion and molecular segregation in phospholipid vesicles. *Biochim. Biophys. Acta* 352, 10–28.

Pigac, J., Hranueli, D., Smokvina, T. and Alacaric, M. (1982). Optimal cultural and physiological conditions for handling *Streptomyces rimosus* protoplasts. *Appl. Environmental Microbiol.* 44, 1178–86.

Reaveley, D. A. and Rogers, H. J. (1969). Some enzymic activities and chemical properties of the mesosomes and cytoplasmic membranes of *Bacillus licheniformis* 6346. *Biochem J.* 113, 67–69.

Salzman, L. Weissbach, H. and Katz, E. (1969). Enzymatic synthesis of actinocinyl peptides. *Arch. Biochem. Biophys.* 130, 536–46.

Shirahama, T. Furumai, T. and Okanishi, M. (1981). A modified regeneration method for streptomycete protoplasts. *Agric. Biol. Chem.* 45, 1271–3.

Sivak, A. and Katz E. (1962). Biosynthesis of the actinomycin chromosphore. Influence of α-4,5-, and 6-methyl-DL-tryptophan on actinomycin synthesis. *Biochim. Biophys. Acta* 62, 80–90.

Szybalski, W. (1952). Microbial selection, Gradient plate technique for study of bacterial resistance. *Science* 116: 46–8.

Svoboda, A. (1980). Intergeneric fusion of yeast protoplast: *Saccharomyces cerevisiae* plus *Schizosaccharomyces pombe*. In *Advances in protoplast research* (eds L. Ferenczy and G. L. Farkas) pp. 118–24. Budapest, Akademiai Kiado, Oxford.

Tabor, C. W. (1962). Stabilization of protoplasts and spheroplasts by spermine and other polyamines. *J. Bacteriol.* 83, 1101–11.

Weibull, C. (1956). The nature of the ghosts obtained by lysozyme lysis of Bacillus megaterium, *Exp. Cell Res.* 10, 214–21.

Woodward, V. W., De Zeeur, J. R. and Srb, A. M. (1954). The separation and isolation of particular biochemical mutants of *Neurospora* by differential germination of conidia, followed by filtration and selective plating. *Proc. Natl. Acad. Sci. USA* 40, 192–200.

Yamashita, F., Hotta, K., Kurasawa, S., Okami, Y. and Umezawa, K. (1982). Antibiotic formation by interspecific protoplast fusion in streptomycete and emergence of drug resistance by protoplast regeneration. In *Abstracts 4th Int. Symp. Genet. Ind. Micro-organisms. Abstr. No.* P–11–20, p. 108, Kyoto.

Yoshida, T., Mauger, A., Witkop, B. and Katz, E. (1966). Influence of methylproline isomers upon actinomycin biosynthesis. *Biochim. Biophys. Res. Comm.* 25, 66–72.

Zimmermann, U. (1982). Electric field mediated fusion and related phenomena. *Biochim. Biophys. Acta* 694, 227–77.

—— and Scheurich, P. (1981), High frequency fusion of plant protoplasts by electric fields. *Planta* 151, 26–32.

8

Biogas production in China: an overview

D. Nianguo Li

1. History of biomethanation in China

The study of biogas fermentation has a history of over a hundred years (McCarty 1981). In China, the practical application of biogas technology dates back to 1920 when Luo Guorui from the Guangdong province constructed the first methane digesters, then commercialized them by establishing the China Guorui General Firm of Gas in Shanghai (Chen 1982). Zhou Peiyuan, the ex-chairman of the State Science and Technology Association, built some power generating biogas facilities in the Zhejiang province in 1936. Twin 46 m^3 rectangular methane digesters, built in 1937 in the Hebei province, are still operational with only a minor amendment made in 1976 (Li 1984).

Wars retarded the massive construction of biogas digesters in China. After the foundation of the People's Republic, following a first national conference of experience exchange on biogas utilization, held in Guangdong by the end of 1953, a peak occurred in 1958 in methane digester building. Owing to insufficient technical assistance provided in too vast an area, the dissemination was not successful. In 1968, the construction technique for a water-pressure type digester was developed and, as it was recommended by the Chinese government, rapid implementation took place in the mid-1970s. In the eighties, although rural digester building came to a halt during a period of readjustment, industrial application of biogas is now rising.

2. Potential of biomethanation in China

China has a tremendous population of over 1 billion inhabitants in a vast territory of 9.6 million square kilometres which occupies some 1/15 of the globe's terrestrial area. The annual total energy consumption in China approaches 830 million tce (ton coal equivalent; one tce is equivalent to 29 271 MJ; Wu and Chen 1982). Hence, the average *per capita* energy consumption is equal to 0.83 tce, which is less than half of the world's average.

Biomass supplies 310 million tce of energy, with firewood, crop stalk, and straw accounting for some 224 million tce, which represents 61 per cent of their total production. The calorific value of firewood is 16 748 kJ/kg, while that of straw and stalk is 14 236 kJ/kg, taking into account the fact that the thermal efficiency of stoves is as low as 8.8–13.2 per cent. If half the amount of the straw and stalks presently burnt were fed into methane digesters as raw material for biogas production, 32.8 Gm^3 of biogas would be generated annually, producing 0.285 litres of biogas from every kg of straw or stalk.

In addition, human and animal manures are also a major source for biogas production. More than 260 million tonnes of dry matter originate from manure every year, 213 million tons of which can be used to generate biogas. Presently, the manure is used mainly as fertilizer, but if it was all used for biogas production, more than 40 Gm^3 of biogas would be produced, while the effluent remaining would still provide a high quality fertilizer (Li 1985).

Some 80 per cent of the Chinese live in the countryside. Hence, rural energy supply is a matter of particular importance. Yearly effective energy consumption by a farming family may be estimated at 0.56 tce. As a Chinese family averages five persons, the yearly *per capita* requirements is 0.11 tce, which is a very low level; this includes energy requirements for cooking, lighting, hot water supply, and heating in winter.

There is a national requirement for 3.9 million of gasoline and diesel oil for farming, plus 37.4 million of coal and 33 TWh of electricity for rural production, making a total of 55 million tce. Following the rise in living standards of farmers and the modification of agricultural practices, an estimate of 0.186 tce *per capita* yearly consumption and a 1.4 factor for rural production is closer to the present needs (Wu and Chen 1982).

This huge increase could hardly be expected from commercial energy supplies in the foreseeable future, so China is looking for a renewable energy in rural areas, mainly from biogas. Hence, some 6–7 million family-size rural methane digesters have been built in China, of which at least two-thirds are operational. Furthermore, nearly 600 biogas power plants with a total capacity exceeding 6 MW were established, as well as some 1200 biogas electricity-generating stations with an installed capacity of over 16 MW (Li 1985).

In line with pollution control, in a number of breweries, sugar refineries, textile mills, slaughterhouses, gourmet powder factories, and chemical and pharmaceutical plants, anaerobic digestion processes are being applied for the disposal of organic waste water such as distillery-spent liquor, molasses, black cooking waste water, or furfuraldehyde-containing waste water from pulp mills, producing biogas for energy.

3. Design of Chinese methane digesters

Various types of methane digester have been investigated in China. Besides the three major types in practice, namely the domed cut-spherical type, the floating-cover type, and the bag-storage type, the Chinese tested elliptical, arch-shape, flat circular, jar-shape, and global-type methane digesters. Furthermore, a rectangular type is applied for most of the larger-size community methane digesters.

3.1. Domed cut-spherical, 'Water-Pressure' methane digester (Fig. 8.1)

This is the most common type chosen for Chinese rural domestic methane digesters and has been given the name 'Chinese Digester' by foreigners. Following the increase in gas produced, the internal pressure in the methane digester goes up and a pressure difference is built between the liquid levels in the digesting and effluent chambers, pressing the supernatant to flow into the latter till an equilibrium is established. As biogas is being used, the internal gas pressure drops and the supernatant flows back into the digesting chamber, reaching a new equilibrium. This is the reason why this type of digester is called a 'water-pressure' methane digester.

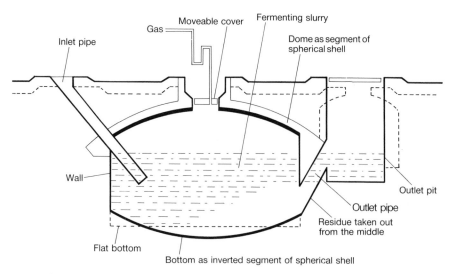

Fig. 8.1 Water-pressure digester

This type of digester has a force-bearing structure, and a size 20 per cent smaller than a similar rectangular type. Being a spherical structure, it exhibits less dead spaces which limits the activity of methanogens. Thus, it is simple in construction, convenient to manage, and requires lower investment. However, the structure is unfavourable to steady gas production owing to the variable gas pressure, and the need to prevent leakages and seepages is high, as it calls for high pressure stress.

3.2. Bag-storage type methane digester

In this type of methane digester, the gas-storage chamber is not located in the digesting chamber itself. Instead, the gas produced is fed into an external storage bag. The resulting low internal gas pressure calls for minor consideration for gas leakage and liquid seepage, and favours gas production as well. However, the system requires an additional bag and asks for fire protection measures.

The bag was formerly made of polyvinylchloride (PVC) or other kinds of plastic film, generally 0.25–0.3 mm thick, but this was found to deteriorate with age. Currently in China, 'Red Mud Plastic' (RMP) material, developed by the Chinese in Taiwan with a waste from aluminium metallurgy, is used as the raw material and mixed with plastics for reinforcement. It is presently being more and more extensively employed as bag-storage material. RMP is flexible and does not deteriorate under the effect of uv light from the sun, its use looks promising as a long-life plastic.

Taking advantage of the low pressure from this type of methane digester, some rectangular digesters were designed which are located underneath pig-barns and use the floor of the pig-barn as the roof of the digester. In this way, construction and labour cost have been further reduced.

3.3. Floating cover-type methane digester

The floating cover-type methane digester is actually a modification of the Indian 'Gobar' biogas plant. In view of the high cost of metal, the Chinese have tried to replace the metal cover of the methane digester by wire-reinforced thin concrete or an asphalt-plastered bamboo frame. The cover is installed either on top of a well beside the methane digester, or directly on top of it. Consequently, the floating cover-type methane digester has a constant gas pressure and subsequently a higher gas production rate is achieved. However, the cover results in a higher cost for materials, bringing about a limitation in its massive use in developing countries.

4. Process parameters for Chinese methane digesters (Xu 1981)

4.1. Pressure within the methane digester

Methanogens may withstand up to 40 m static water pressure, but are extremely sensitive to pressure variation. Hence, the internal pressure of a methane digester should be kept stable, and preferably at a lower rather than a higher level. The biogas pressure for cooking and lighting is best within the range of 20–60 cm water column (P = 0.02–0.06 kg/cm^2). Too low a gas pressure requires a larger effluent chamber diameter. Hence, the biogas pressure process value for domestic methane digesters is fixed at 0.06 kg/cm^2.

4.2. Biogas production rate

The daily gas production rate is best expressed relatively to the unit volume of mixed liquor (working volume of the methane digester), namely in m^3/m^3 per day. As most rural domestic methane digesters work at an ambient temperature, the temperature of the mixed liquor varies with the atmospheric temperature, within the range of 8–30°C. With a dry matter concentration of 7–10 per cent w/v, the gas production rate averages 0.10–0.30 m^3/m^3 per day under normal running conditions, higher in summer and lower in winter. Thus, the design gas production rate is set at 0.15 m^3/m^3 per day. For mesophilic methane digesters, the gas production rate is set at 1.0 m^3/m^3 per day, while for thermophilic methane digesters, the gas production rate is set at 2.5 m3^3/m^3 per day.

4.3. Size of methane digesters

The size of a methane digester deserves due consideration in its design, depending on raw material supply, gas requirements as well as gas production rate. The working volume of the digesting chamber in a rural domestic methane digester is based on the daily need value of 0.2–0.3 m^3 biogas per person. This requires a total working volume for the digester of 4–10 m^3.

For power-generating methane digesters, the total working volume of the methane digester may be as high as 60, 100, 150, or even 200 m^3, and may result from groups of methane digesters connected parallely.

4.4. Loading volume of methane digesters

The loading volume of Chinese methane digesters varies with the design of the gas storage device. For water-pressure methane digesters, the initial load occupies 70 per cent of the working volume of the digesting chamber, with 85 per cent as a maximum, in order to spare room for the biogas produced. For floating-cover or bag-storage types, the design maximum load volume may be that of the whole digesting chamber.

4.5. Biogas storage

This parameter applies to the volume of the storage bag or floating cover. In the water-pressure digester, this parameter applies to the volume of effluent chamber. In designing rural domestic digesters, the storage bag should hold gas for a 12-hour use period, which amounts to 50 per cent of the gas produced daily.

5. Advanced technologies for biomethanation in China

The simple conventional process with no temperature control or effective mechanical stirring, adopted in Chinese rural area, saves labour and excessive investment. Yet the gas production rate remains as low as 0.1–0.25 m^3 biogas per m^3 methane digester and per day. Therefore, various advanced processes are experimented in laboratories or at the pilot-scale.

5.1. Upflow processes with active biomass trapping

A partially-packed anaerobic filter, combining the traditional anaerobic filter with the upflow anaerobic sludge bed reactor, was developed. The packed bacteria-covered rocks are organized in 2 or 3 layers, leaving empty spaces between them. This way avoids clogging. Much higher efficiencies were obtained. The process is presently being extended to the pilot-scale and even the full-scale.

A modified design was developed by installing tubes vertically in the spaces between the layers. Furthermore, a horizontal design was tested. All upflow or tubular processes are aiming at two goals: first, effective separation of biogas bubbles and microbial flocs, and secondly, increase of settling velocity of the microbial flocs.

5.2. Two-phase processes

By splitting acid-forming and methane-generating stages into separate facilities for individual processing, gas yield and gas production rates can be enhanced up to two-fold over those obtained by conventional processes. The process is more stable and the pH control less difficult. In a trial addition of hydrogen-producing bacteria in the methane digester, the CH_4 content in the produced biogas increased from the normal average of 65 per cent to over 95 per cent (Chen 1983).

5.3. Other processes

Plug-flow processes have been tried with a 50 per cent increase in gas production rate.

Thermophilic digestion was adopted in 1967 and 1978, chiefly to treat alcohol fermentation spent liquors. Also, a full-scale night soil treatment plant was established adopting the process.

Dry fermentation was adopted to 'compost' plant material containing a 15–30 per cent solid content for 2–5 days. No chemicals are needed and less labour is required to handle the resulting solids making the process acceptable to farmers.

6. Economy of Chinese methane digesters

The rural domestic methane digesters are chiefly used to supply energy for farmers' cooking and lighting needs, while community methane digesters are usually running as power generation sets for modified diesel or gasoline driven engines, ranging around 15 kW or with electricity generators of 6–12 kW capacity. Waste heat from the engine is utilized for heating and drying.

The digested slurries and sludges have been proven to be even better fertilizer than their respective raw organic wastes. The major nutrients are retained and humates are formed. Thus, they are utilized in fish and earthworm raising, mushroom and tree cultivation, and cattle feed rations, besides their major use as farm fertilizer and soil conditioner.

In the chemical industry sector, biogas and the residues have been used as raw materials for the production of chloromethane compounds, tetrafluoro-ethylene, formaldehyde, methanol, and formic acid, etc., and also of vitamin B12. Utilization of methane-digestion residues further shortens the payback period for the methane digesters.

The Chinese biogas digesters are mostly built with locally available materials. The materials and resulting investment cost for a typical 6 m^3 domestic methane digester are detailed in Table 8.1. Annual savings from the utilization of a family-size methane digester are detailed in Table 8.2. They are estimated as follows: (i) fuel was formerly purchased from the market; (ii) labour was used in firewood collecting; (iii) fertilizer obtained from digested slurry overflow or sludge cleaned out which substitutes and/or supplements the chemical fertilizer which formerly had to be purchased. An average family usually raises three pigs. When using a biogas digester, two more pigs have to be raised in order to obtain a larger amount of manure to feed the digester. This 80 per cent increase is considered as a 'saving'.

Assuming a 50 per cent cut in the monthly coal consumption by using biogas, each month RMB¥ 2.86 is saved. As far as labour saved from collecting firewood formerly is concerned, the women and grown-up children had to spend 2.5 working days each month. Each working day, the labourer can earn ¥ 3.50 (¥ 4.00 for adults), but since using biogas these working days can be spent on farm work instead.

From Tables 8.1 and 8.2, one can see that the period required for recovery of the outlay for a rural domestic biogas digester can be estimated at 14–15 months. Considering that such methane digesters have a lifespan of 10–15 years, it is clear that biogas utilization is obviously very attractive in China.

For rural community biogas plants, it is calculated that the period required for recovery of the outlay lies around 4 years, by comparing the investment cost with the cost saved after setting up the biogas plant and its annual benefits (Li 1982).

When evaluating the benefits originating from biogas utilization, one

Table 8.1

Investment costs for a 6 m^3 domestic methane digester

Material	Quantity	Cost (RMB¥)*
Cement	366 kg	29.28
Gravel	0.725 m^3	17.55
Bricks	795	79.50
Steel bars	4 kg	2.40
Ceramic tiles	3	2.40
Aluminium tube	1	2.00
Plastic pipe	40 m	4.00
Gas valve (plastic)	1	0.50
U-barometer (glass)	1	0.80
Burner	1	3.50

Sub-total ¥ 141.93

Labour	Working days
Material preparation	3
Excavation	4
Brick-laying	4
Concrete casting, sealing, and earth refilling	6
Total	17

at RMB¥ 4.00/working day,   Sub-total: ¥ 68.00

Total investment   RMB¥ 209.93

* RMB¥: Renminbi (people's currency) Yuan; 1 RMB¥ is equal to 0.346 US $.

Table 8.2

Annual saving of a Chinese family using a methane digester

Item	Amount saved	Annual saving (RMB¥)*
Fuel	65 kg of coal: ¥ 2.86/month (assuming 50% saving rate)	34.32
Labour	2.5 working days: ¥ 8.75/month	105.00
Fertilizer	1.2 tons of pig manure plus 1.15 tons of additional weed (equivalent to the cost of 600 kg of pig waste) or ¥ 28.80 assuming an increase of pig raising to feed the digester, or the equivalent amount of chemical fertilizer costing ¥ 36.00	28.80 or 36.00
	Total annual saving RMB¥	168.12 or ¥ 175.32

* RMB¥: see table 8.1.

Table 8.3

Lethal temperature and the duration for killing pathogens and germs

Nomenclature	Lethal temp. (°C)	Duration (min.)
Bacillus paratyphoidus B.	60	60
Dysentery bacilli	60	60
Anthrax bacilli (growing)	50–55	60
Tuberculosis bacilli	60	15–20
Bacterium mallei	50–60	60
Bacterium brucella	65	15
Streptococcus equi	70–75	60
Lymphnoditis pathogens	65	60
Swine erysipelas bacilli	50	15
Hydrophobia viruses	50	60
Infectious horse encephalo- myelitis viruses	50	60
Hog cholera viruses	60	30
Leptospira	50	10
Wheat smut germs	54	10
Rice blast germs	51–52	10
Suppurate germs	54	10
Rust germs	54	10

should not forget the convenience it contributes particularly to women and children who do household chores. Also worth considering is the improved sanitation conditions it brings with it, to human beings, fish and cattle, and the environment. The lethal temperature for bacteria, etc., are shown in Table 8.3 (Xu 1981). The higher the temperature, the shorter the time required for killing. For example, tuberculosis bacilli are killed in 10 minutes at 70°C, but it requires 15–20 minutes at 60°C. At 70°C, 5 minutes suffices to kill *Bacterium brucella*, but at 55°C, the bacteria can withstand 120 minutes. Nevertheless, practice has proven that anaerobic digestion at ambient temperature kills all intestine pathogens, as well as *Leptospira*, in 30–31 hours.

7. Chinese policy for the implementation of biomethanation

The Chinese government has formulated a comprehensive policy for the development of biogas production based on the experiences gained (Li 1984). The major factors are summarized below.

7.1. Clustered development in a programmed way

The utilization of biogas in rural areas influences not only its energy, but also its fertilizer, sanitation, and household management. Benefits can only be demonstrated when clustered development within a general area is accomplished. In these cases, the biogas programme is extensively accepted. Clustered development also favours extension of services, financial support, and collective operation.

7.2. Progressive dissemination guided by appropriate reactor designs

The failure of a massive methane digester construction in the late 1950s discouraged farmers from building new ones. The expansion in the mid-1970s came from successful demonstrations which led to a rapid yet gradual dissemination resulting from the confident building of methane digesters by the farmers themselves, with technical assistance.

7.3. Adaptation of global programming to local conditions

China has a territory which covers subtropical, temperate, and cold zones, with the associated different resources and energy requirements. Any national programme should match local conditions, so as to bring it to a success.

7.4. Focus on implementation in areas of fuel shortage and where epidemic diseases are prevalent

Biogas production supplies energy while improving sanitary conditions; thus, it is most welcome in areas in need for both. Farmers accept biogas technology when they realize that they are not only freed from firewood collection and coal purchasing, but also that schistosomiasis and formerly common skin diseases caused by bacteria in fresh pig manure spread on fields and in fish ponds are eliminated. They then tell others of the advantages and this is an even more convincing form of propaganda than the government's.

7.5. Supply of both domestic and community methane digesters

Once domestic methane digesters which meet the requirements of individual family units have been introduced, community methane digesters can then be constructed to meet the massive agricultural needs which cannot be satisfied by the decentralized small methane digesters. In a given village, the *per capita* income for 1977 was US $ 130 before biogas plants were constructed. This figure increased to US $ 240 in 1979, after biogas was introduced. Though this increase in *per capita* income was not entirely due to biogas technology, the farmers still saw the latter as the main reason.

7.6. Self-support with proper government back-up

Since both resources and consumption are small-scale and scattered, constructions for rural energy should rely mainly on self-support. A policy has been applied that those who built the digesters should own them, manage them, and benefit from the system. At present, however, not all Chinese farmers are rich enough to afford their own methane digester, so that government subsidies are still necessary in some cases.

7.7. Institution and educational courses

A state 'Leading Office' for biogas production has been established in the central government, and corresponding offices have been set up in province, municipalities, autonomous regions, counties, and districts. These offices supervise technology transfer and information dissemination, as well as allocation of funds and subsidies. The institutions organize 2–4-week training courses and annual meetings on biogas, exchanging experiences and training technicians.

8. Conclusion

Biogas is produced as a result of the metabolic activity of micro-organisms. Regular biogas production could only be mastered when the taxonomy, living conditions, and mechanisms of metabolism of the fermenting micro-organisms were recognized. Biogas fermentation is the concentrated expression of metabolic activities of various coenocia of micro-organisms, and the result of interaction of these micro-organisms and their metabolic products. Appropriate living conditions for the growth of micro-organisms are the foundations of biogas fermentation technology. Advances in microbiology and biochemistry result in new and improved technologies, and in turn in demands for new biogas plants.

Yet, the biogas fermentation technology and the fermentation equipment are only ways and means, while the utmost goal is for improved gas production, stable fermenting processes and more economic operation. For the development of biogas fermenting equipment, the principles of adapting to local conditions, drawing on local resources, keeping up quality demands, and striving for actual effects, namely real benefits to the users, are of major concern.

Besides biogas, the digested residue produced is a high-quality fertilizer and a chemical feedstock which serves as fuel for energy. During application, pathogens are destroyed and environmental sanitation improved, contributing to a social and ecological equilibrium.

In the future, further study of biogas fermenting micro-organisms and the interaction of their metabolism, fermentation mechanisms, and practical processes for biogas digestion, are needed to explore and approach increasing gas production performances. Biogas utilization, as a form of biomass energy supply, should be broadened day by day, in order to contribute more and more to energy and environment for mankind.

References

Chen, R. (English version: Li D. N.) (1982). Building rural digesters. In *Anaerobic Digestion 1981* (eds Hughes, D. E. *et al.*) pp. 293–314. Elsevier Biomedical Press.
—— (1983). Up-to-date status of anaerobic digestion technology in China. (Paper presented at the 3rd Int. Symp. on Anaerobic Digestion, Boston, USA. 1983.)
Li, D. N. (1982). The economic feasibility of the Xinbu system. *Abstracts of Selected Solar Energy Technology*, 4/5, pp. 22–4. The United Nations University, Tokyo.

—— (1984). Biogas in China. *Trends Biotechnol.* Vol. 2, No. 3, pp. 77–9. Elsevier Publications (Amsterdam).

—— (1985). Brief review on China's renewable energy. *China Business Rev.* Vol. No. 4, (ed. Ross, M.) pp. 32–5. National Council for US-China Trade.

McCarty, P. L. (1981). One hundred years of anaerobic treatment. *Anaerobic Digestion 1981*, pp. 3–22. Elsevier Biomedical Press (Amsterdam).

Wu, W. and Chen, E. (English translation: Li, D. N.) (1982). Our views to the resolution of China's rural energy requirements. Paper presented for the Joint CAS-NAS Science Policy Conference. Nov. 1982, Chengdu. Also published in *Biomass*. Vol. 3, No. 4 (1983). pp. 207–312. Applied Science Publishers (UK).

Xu, Z. (ed.) (1981). *Biogas Technology*. Agricultural Publ., Beijing. China.

9

Fermentation technology and its impact on culture and society

H. W. Doelle, Eugenia J. Olguin, and Poonsuk Prasertsan

1. Introduction

Any technological development is aimed at improving the quality of life of a community of people. It may lead to longer life expectancy and higher survival rates through better health conditions.

Since his appearance, man has always lived in an uncertain, and sometimes precarious, symbiosis with nature (King and Cleveland 1980), obtaining his nourishment and the small amount of energy needed from plants and animals personally accessible to him. Fermentation technology was accidentally introduced very soon in the form of wine, followed by the brewing of met (honey beer), beer, the making of bread, the development of milk products such as yoghurt and cheese, the development of a variety of meat products, and the fermentation of the sweet juice (aguamiel) extracted from the heart of a Mexican Agave to produce pulque, an alcoholic beverage, in Aztec country of Latin America. After the conquest of Mexico by the Spanish in 1521, a further improvement of the pulque fermentation led to the now well-known tequila, a spirit produced from Agave tequilana. Further developments in fermentation technology were mainly concerned in the improvement and expansion of the alcoholic beverage industry, which had a profound impact, often detrimental, on the quality of life and also the society thousands of years ago.

2. Impact of fermentation technology, and the industrial revolution on culture and society in the now developed countries

The impetus of the industrial revolution during the 18th and 19th centuries transformed the very nature of society in many parts of the world, which are now referred to as the developed countries. Society was now not only using renewable resources, but also consuming vast amounts of non-renewable resources. The industrial society developed by the accumulation of scientific knowledge, the spread of technological innovations, and the exploitation of enormous natural resources. Traditional vegetable and animal fibres were increasingly replaced, or extended, by synthetics manufactured by an ever increasing chemical industry from coal and petrochemicals. This development in the 20th century fundamentally altered the pattern of consumption, land use, international trade, and the distribution of wealth (King and Cleveland 1980). Longer life expectancy and higher survival rates followed through better housing and sanitation, and the production of antibiotics and vaccines. The quality of life was improved by the introduction of petrol and the motor industry among others. The impact on society was dramatic and on culture devastating as a large proportion of the traditional way of life was lost through this development owing to an ever increasing urbanization.

The reasons for such an impact on culture and society are manifested in the principles of the industrial systems' organization (Fernandez and Ocampo 1980), which dominates today's society in the developed countries. This organization is based on short-term profit, with a production to sell attitude, with preference given to the production of luxury consumer goods over goods required for basic needs, particularly at the level of the large energy systems, such as coal, hydrocarbons, and nuclear energy. Such a production framework is an obstacle to the total realization of individuals and society.

The results of these chemical and manufacturing industries are accompanied by ever increasing amounts of effluents of both heat and toxic substances, many of which are non-biodegradable. Modern agriculture is now strongly based on the application of chemical fertilizers and ever-increasing amounts of organic pesticides, mainly as a consequence of an enormous and rapid expansion in world population demanding an ever greater quantity of food and goods of all kinds. This, in turn, encourages the use of still further quantities of non-renewable resources and energy. This development led in the 1970s to a turning point in the perception of man's relation to his natural environment, the biosphere, as well as a shift in man's relationship to the man-made environment, the technosphere (King and Cleveland 1980). The question was raised whether the earth and its atmosphere can provide an infinite sink and absorb the waste products of industry, agriculture, and urban living as they become more and more prevalent. The processes of physical planning are now challenged and well-established procedures are under severe scrutiny. Whereas a successful community has been judged by the amount of resources it would usefully consume, in the future it will succeed only if it manages to conserve resources without loss in quality of life (Meier 1980).

The less-developed or developing countries have, in general, been by-passed by the industrial revolution and chemical industry development. Starvation on some continents, or at best malnutrition, together with a rising population and rising prices for non-renewable or traditional energy sources, coupled with a more or less complete dependency on the importation of goods have led to neglect of agriculture, but a build-up of a large urban population has brought disaster to the economy and society of many developing countries. The majority of these developing countries lie in two extreme climatic zones, the tropical wet zone and the tropical arid zone. Can scientific knowledge and technology improve their quality of life, life expectancy, and increase survival rates without repeating the disastrous effect of the industrial revolution on culture and society?

3. New trends in fermentation technology

The fast-developing new trend in fermentation technology has emerged from a social evaluation of technology. Simmonds (1980) explained this new trend by saying that efficiency or generation of wealth, in its technical and economic sense, will clearly continue to be desirable, but that it will be placed in balance with equity or distribution of wealth, and survival or continuation of wealth. 'We have lived through a period characterized by one major goal, economic growth, fueled by one source of energy, petroleum hydrocarbon, with materialism as king and consumers as his loyal subjects'. The bioresource development will be different. The availability of renewable substrates is much more restricted (Ivory and Siregar 1984) and should therefore be exploited much more carefully. Such a careful exploitation requires a marked rethinking of the scientist and technologist, as the newly developing fermentation technology must use the vast potential of the microbe to provide fuel, food, fertilizer, and feed supplements from the renewable resource (DaSilva 1979; DaSilva and Doelle 1980). The social evaluation thus turns into a socio-economic bioresource development with multiple goals in place of one goal, the utilization of several sources of energy rather than one, acceptance of heterogeneity as normal rather than homogeneity, a goal of greater overall well-being rather than just more money or possessions (Simmonds 1980). Such a development means that the environmental and the social elements must be approached giving due consideration to technical, economic, and financial aspects of investment, projects, or policies. Bioresource development therefore can only succeed if it is actively integrated into the culture of the developing countries. This is one of the major reasons why new technological advances in developed countries should not be transferred without adjustments to the local conditions. Such a biotechnological transfer should therefore be regarded as an entirely different concept from the development of an 'appropriate biotechnology', which is a concept developed uniquely for the particular nation or region, and which may not be applicable or appropriate to another region (Doelle 1982).

In order to develop an appropriate biotechnology, resources available together with the social structure of the population are of vital importance. It is often the need and not the economics of the process, which is of importance (Sorensen 1979; Rolz 1980). Here seems to lie one of the cardinal differences between the thoughts of appropriate technology in developed, from those in developing, countries. It is evident that microbial or fermentation technology can help to provide solutions to both the industrialized and the developing countries with respect to their own characteristic problems. With the technically advanced nations, the emphasis will be on continued economic growth, energy, and environmental control problems, whereas in the non-industrialized countries the concern will be the problems of hunger, population control, health, and full utilization of natural and human resources (DaSilva 1981).

4. Impact of old and new fermentation technology on culture and society in developing countries

The development of fermentation technology can be divided into three categories (DaSilva 1981).

1. High-capital technology, such as pharmaceuticals and single cell protein production from substrates other than waste materials.

2. Intermediate-capital technology, used in the production of fermented food, biological nitrogen fixation, and single cell protein production from wastes.

3. Low-capital technology, such as biogas and the integrated rural systems.

In developing countries, high and intermediate capital technologies are possible in urban areas, whereas intermediate and low-capital technologies are more practical for the village or rural communities.

The majority of developing countries are tropical countries with vast bioresources, but starvation, with presently immeasurable consequences, exists in extreme arid and desert zones. In both areas, bioresource development could have its greatest impact on society and could help to preserve the culture of that society. In both cases, the introduction of new fermentation or microbial technologies depends very much on government policy and priorities.

In the tropical areas, the extensive availability of renewable resources (Gomez-Pompa 1980) together with the ever-increasing population in urban areas can lead to the development of new fermentation technologies or a revival of old technologies.

4.1. Food

In South-East Asia and in Africa, the indigenous foods play a vital role. Every country within the regions has its own fermented foods, which may have originated thousands of years ago, and some of these are similar throughout an entire region. Examples include soy sauce, fermented fish products, and others in SE Asia (Wang and Hesseltine 1982; Hesseltine 1983), or foo-foo or chikwangue, garri, and others in Africa (Ayanabi 1983; Ekundayo 1978). The use of fermented food can be;

(i) as the main course, e.g. tempe, garri, etc.

(ii) as a flavouring agent, e.g. soy sauce, miso, etc.

(iii) as a colouring agent, e.g. red pigment from strains of *Monascus purpurius* on rice

(iv) to change the physical state and acceptability of the substrate, e.g. fermented vegetables, cassava roots, cereals, etc.

Fermented foods therefore can be produced both in the villages and the cities.

In the urban communities, food production becomes industrialized with mostly non-traditional fermented foods. The reason for the difference is the fact that modern technology has been introduced from developed countries. High-capital cost industry in the developing countries is usually invested by the co-operation between local people and foreigners. The type of products are inevitably alien to the local people in the first instance. Gradually, these alien products are accepted and become part of a mixed diet. Cities are the big market places where people seem to be influenced fastest by Western civilization. It is certainly not surprising, for example, to see Thai people in Bangkok having bread and coffee for breakfast, and beer or wine with their luncheon or dinner, and yakult, a fermented milk, as a drink for children. Such technology transfer certainly leads to the destruction of an ancient culture and does not necessarily have a positive influence on society.

In the village communities, only traditional fermentation, with low and inexpensive cost, is used as the indigenous fermented foods have shown socio-cultural ties and the vital techniques are passed on from generation to generation. The main aim is to preserve the perishable commodities where no refrigeration is available, but other advantages include more income and off-season consumption.

Fermentation technology in the food fermentation area could significantly help. preserve culture and positively influence society in at least three ways.

1. It may help by solving malnutrition problems. Protein and caloric malnutrition is one of the national problems in most of the developing countries, particularly amongst the low income families and in rural communities. It is ironical that while malnutrition is a problem in many developing countries, overweight becomes a serious problem in developed countries. Malnutrition is mainly caused either by lack of a minimal protein content in fermented foods, protein sources which are too expensive such as meat, or non-availability of other protein substitutes such as soybean in some areas, or lack of nutritional knowledge. In South-East Asia, fermented foods have a high nutritive value, whereas in Africa, the cassava- or cereal-bound fermented foods are low in protein. Whereas Saisith (1981) suggests for

South-East Asia the use of fermented foods in a mixed diet, in African countries alternative protein sources or a cheap supply of meat must be sought to reduce malnutrition problems. In each case, however, the additives must be acceptable to local people, should not interfere with the traditional processes and must be of low cost.

2. Fermentation technology can help by improving the quality of life, which means safe and nutritious products together with a clean environment. Safety factors have to be taken into account in considering the production of traditional fermented foods using natural mixed cultures as starters. To overcome this obstruction, pure cultures should be obtainable with the co-operation of scientists and local factories. Steinkraus (1983) pointed out that mycotoxin poisoning can be avoided by using only selected and certified koji strains. The first step is therefore to isolate and identify the micro-organisms involved in fermentation activities, followed by the production of food using pure cultures. The finished product must maintain their original qualities including flavours, texture, and colour. This is a vital requirement, since the social factor, particularly the cultural taste, plays an essential role. It is useless to produce good proteinaceous food, which the local people will not eat. Since the majority of fermented food production involves solid substrate fermentation technology with mixed microbial cultures, the research into the biological constraints are problematic and very difficult (Knapp and Howell 1980; Doelle 1985). In Thailand, where soy sauce is one of the major flavouring agents, the application of the isolated organism *Aspergillus flavus* var. *columnaris* was investigated. The very high protease and no detectable aflatoxin production is used now as controlled koji inoculum (Bhumiratana *et al.* 1980; Flegel *et al.* 1981). After the examination of the enzymatic properties of the strain (Bhumiratana *et al.* 1983), Tochikura and coworkers (1983) stated the potential of the enzyme application.

This example of application in fermentation technology in fermented food production could lead to a significant increase in public health and foster an increased fermented food usage to defeat malnutrition.

3. Fermentation technology may help in the creation of new jobs. The establishment of new industries or the expansion of present industries with the help of appropriate technology will have a great impact on the labour market. New fermentation industries can be developed progressively, if they follow certain criteria. One of these is undoubtedly a Government policy, as is the Government in the developing countries (Ginjaar 1980) which has to locate the so-called demand or social need for a particular technology.

The impact of new fermentation technologies into the fermented food industries therefore can be minimized by improving the presently time-consuming processes and removal of the danger of toxin production. It could lead to the destruction of this old culture by introducing alien processes from the developed countries. In both cases society itself improves provided that such an alien introduction makes food available to the low income families and prevents malnutrition. There is, of course, the danger that this may not be the case and the consequence on society could be similar to the destruction of the old culture.

4.2. Fuel

The Brazilian National Alcohol Fuel Program (Kosaric *et al.* 1980) is a typical example of a Government policy directive to demonstrate to the public the practicality of ethanol as an automotive fuel. This directive has its origin in Brazil spending many hundreds of millions of dollars on oil importation and thus it was a true governmental economical consideration, which led to the directive being issued.

In order to fulfil this directive, an old fermentation technology, originating in developed countries, was introduced with little or no local social consideration. Although such a scheme is of great significance to the country's economy (Rolz 1984), and undoubtedly created thousands of jobs, the production process has not been adapted to fit local conditions. It is uneconomical and requires heavy subsidies by the Government. The danger exists that this National Program could parallel the development of the industrial revolution with a one-substrate-one-goal aim, which has proven to be so destructive in the past to culture and society in developed countries. It is therefore of more than just passing interest for the rest of the world to observe the impact of this particular technology transfer over the next decade.

An entirely different approach has been taken by the Government of Sri Lanka (SRI 1984). This country also suffers from severe economic constraints owing to the structure and economy of the world tea market. The Government therefore announced an agrarian policy which has as its first National Priorities self-sufficiencies in food, feed, and a series of other products. The granting of land to people and incentives in the form of subsidies shows its first success in the almost complete self-sufficiency in their main stable food, rice (Perera 1984). The additional priorities on self-sufficiency in sugar and ethanol, however, calls on the introduction of fermentation technology into the country. Since sugar is a most valuable product, the ethanol demand for pharmaceuticals and hospital requirements must be obtained from molasses fermentation (Doelle 1984). A very close social-economic investigation with the help of foreign aid is under way to find a solution for the most appropriate system, which could also reutilize the stillage waste for either biogas or animal feed production. Such a system would not only meet the Government's priorities in saving import foreign exchange, but would make the National Priorities economical to the industry, help create numerous jobs, attract people from urban areas back into country areas, avoid public health risks owing to excessive wastes, and at the same time increase the production of meat. The impact on society could therefore be quite significant in their life style without affecting the culture.

Both examples represent high-capital technology in the first, and a mixture of high- and intermediate-capital technology in the second instance.

4.3. Energy

A number of low capital technology systems, characterized by low financial investments, small-scale operating procedures, and by being non-waste producing and non-polluting have been introduced in many Asian countries, exhibiting a positive impact on society without affecting seriously the traditional cultures. The biogas (methane) systems in particular have attracted considerable attention as a promising component of decentralized rural development. These systems are of interest from the viewpoints of waste recycling, rural development, public health and hygiene, pollution control, environmental management, and appropriate technology (DaSilva 1980). Since biogas systems constitute a renewable source of energy and also provide fertilizer as a byproduct, they can be used world wide to fulfil these basic needs. The most spectacular impact of biogas systems lies not only in the provision of energy (= electricity and power), but also in the preservation of forests, thus stopping the irreversible conversion of fertile areas into arid zones owing to soil erosion.

Biogas technology involving the anaerobic bacterial conversion of human and animal wastes, organic agricultural, and industrial waste to methane and fertilizer has shown its impact in improving the quality of life and life

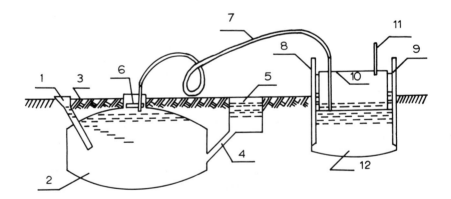

Fig. 9.1. Model for digester with collecting cover. 1, inlet; 2, fermentation compartment; 3, ground level; 4, connecting pipes for overflow; 5, overflow chamber; 6, movable lid; 7, connecting pipes; 8, frame for gas holder; 9, guide trough; 10, floating gas holder; 11, gas outlet; 12, water tank (from Chiao 1986).

expectancy by removing the public health risks of wastes, in preserving the productivity of rural soils, and preventing an ever-increasing urbanization. Biogas itself is constantly and conveniently renewable and thus of significance in the energy economics of the country. The greatest expansion of this system has occurred in the Republic of China (Fig. 9.1; Chiao 1986). On the other hand, social and traditional structures have hindered the full expansion of this technology in some areas of India. The problems encountered in India emphasize again the resistance of the rural people against new technologies if a thorough social study and social evaluation of a technology has not been carried out in advance.

It has to be realized that the incentives for energy recovery from wastes are strongest in rural areas of developing countries, where the level of energy consumption is low (LaRiviere 1979) and the population is deprived of adequate heating and electricity.

4.4. Feed

The high rate of increase of the world population has led to increasingly gloomy forecasts of widespread famines in the future. Starvation and extreme food shortages in Ethiopia, and protein malnutrition in Central Africa already show the truth of these predictions. These problems could be allayed through a tremendous effort in fermentation technology together with other programmes such as birth control to save a number of cultures and societies.

The concept of producing single-cell protein (SCP) from fermentation of agricultural and industrial wastes has been considered to be one of the promising ways of additional and/or complementary sources of protein, either directly as a food component or indirectly as animal feed (Berk 1979; Yousri 1982). SCP includes the protein from bacteria, fungi, yeast, and algae. Micro-organisms which are considered as potential sources of food, and which can be used to produce SCP, have to meet a set of requirements and criteria, and desirable characteristics concerned with economic, nutritional, and health aspects. Furthermore, the production costs of cultivation and harvesting must be able to compete with that of other sources of food proteins available in the region, such as soybean (Reed 1982). SCP plants range from large scale in developed countries (Ringpfeil and Heinritz 1986) to small-scale processes (Rolz et al. 1979) in developing countries, depending upon the demand, substrate availability, and the social-economic situation. Although the main emphasis is given to animal feed to improve the protein supply and keep the animals off precious agricultural land areas, serious consideration is given towards supplementing low protein fermented foods in some of the Central African Regions. The gross failure of SCP as a human food supplementation was due to the complete lack of understanding of the culture and the society of the developing countries by companies from the developed countries.

Increased animal feed production could have a considerable impact on society, if wastes from domestic, animal, and agricultural sources were to be used as raw materials in the tropical areas of developing countries. Such systems would not only provide the required SCP, but also fertilizer from effluent to increase soil fertility and remove a very acute public health hazard. A society suffering under malnutrition would be able to increase its meat production, increase agricultural output at low fertilizer costs and the removal of wastes would give a cleaner environment thus increasing life expectancy and the quality of life.

Most of the fermentation technologies mentioned above are concerned with extra crisis situations, which are either due to a down-swing in the national economy or owing to the lack, shortage, or too expensive basic commodities together with the increasing population. Provided these fermentation technologies have been properly adjusted, or developed according to the socio-economic structure of the population, they will have undoubtedly a great impact on society without affecting the traditional culture of the population. The chances are that the culture of a nation will be best preserved in rural communities.

5. Impact of integrated rural fermentation technology on culture and society in developing countries

Concurrently with the above fermentation technology programmes, a new concept has been developed specifically for rural communities: the so-called integrated rural fermentation technology. This technology, which is discussed elsewhere in this volume by Lewis (see Chapter 3), aims at rural progress, conservation of the rural environment and rural self-reliance in relatively primitive economics (DaSilva 1981). Biological waste conversion into food, food products, fuel, and fertilizer could be the basis of a long-term strategy to alleviate not only the 'food crisis', but also the 'energy crisis'. Long-term strategies require not only a proper technological and economical assessment, but also a deep evaluation of the social and environmental impact produced by the strategy (Olguin 1982). Conventional and non-conventional food and feed production, together with a considerable production of energy and fertilizer would have a positive economic and environmental impact (Olguin 1978; Olguin and Vigueras 1981). Caution should, however, be incorporated when choosing the best and most adequate technological alternatives for waste processing to ensure the social relevance of the final product.

The development of an integrated rural fermentation technology with a certain flexibility in product formation according to social demand and relevance depends primarily on the climatic conditions of the regions.

Developing countries are concentrated in either tropical or tropical arid areas.

5.1. Tropical wet zones

Pilot plant schemes for the introduction of integrated systems have been developed in Mexico and the Philippines. The first integrated system approach is the connection of animal waste with aquaculture and, more specifically, with algae and fish production, and not just provision of biogas and fertilizer. There is no doubt that algae are among the most efficient converters of radiant energy (DaSilva *et al.* 1980; Olguin 1982; 1984), with a solar energy conversion efficiency of approximately 7 per cent. The cultivation of algae has also the advantage of utilizing semi-arid land. Such land utilization is the most efficient by using algae to produce protein in comparison to any conventional source. The production of one tonne of *Spirulina* (Ciferri 1983) requires only 0.03 ha/year compared to 452.5 ha/year to produce one tonne of beef or 1.55 ha/year to produce one tonne of soybeans (Leesley et al. 1980). The integrated system consists of anaerobic digesters, similar to those used in the biogas system, from which the aqueous effluent is connected to algae cultivation ponds. Whereas biogas and fertilizer are constantly produced, the type of algae used depends on the demand of the social community. *Spirulina platensis*, for example, can be used directly as a protein supplement for cattle, pigs, or poultry. *Spirulina* has also found entry into health food shops in the developed countries owing to its high vitamin content and represents the main staple food for the people of Chad (Africa). *Spirulina* cultivation with a multi-purpose approach has been reported by Tel-Or and coworkers (1980), whereby besides protein, chlorophyll-*a*, xantho-phylls, and ß-carotene can be produced. This versatility of the algae *Spirulina* means that the village or society now has the advantage of not only satisfying their energy and food requirements, but also obtaining income from sales of the various products.

A change from *Spirulina* to other microalgae leads to an excellent feed for fish production. Fish, of course, is one of the most important human protein resources.

In the Philippines, an integrated livestock-meat processing and canning operation system has been established at the Maya Farms in the Antipolo Hills outside Manila (Judan 1980). In January 1983, this system had over 4000 sow units, 1000 duck or hen units, and 25 cow units. From the animal waste, some 3510 m^3 biogas are produced per day, which supplies all the energy requirements of the livestock farm, and 90 per cent of the requirements of an associated meat processing and canning plant. A few months ago, the farm increased its biogas output to become completely self-sufficient with regard to its energy requirements. In addition to this large integrated livestock-meat processing and canning operation system, Maya Farms (Liberty Flour Mills Inc.) have developed a crop–livestock–fish-farming system, which brings about intensive use of land, full utilization of farm wastes, and near energy-sufficiency of the farm. These model farms have a size of about 1.2–1.5 hectares.

Such self-sufficient farm units are very important for the Philippines as many rural communities are spread over hundreds of individual islands. They provide a very good example for rural communities being able to maintain their culture, but at the same time lifting their quality of life by removing poverty, malnutrition, and public health risks.

5.2. Tropical arid zones

In the arid zones of Mexico the situation is quite different from the tropical wet zones of this country. Following the concept of maximum use of resources, saline water in these regions could become an essential element for biomass production and an integrated rural fermentation technology of a different kind (Olguin 1986). This system takes advantage of those resources which are most available to arid lands: solar energy, saline water, harvested rain water, organic wastes, non-conventional crops, halophyte plants or

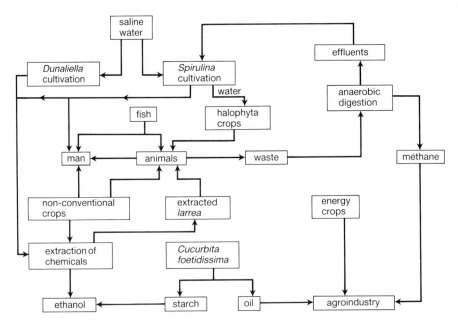

Fig. 9.2. Integrated system for the production of food, feed, fuel, and chemicals in arid zones (from Olguin 1986).

micro-organisms, and energy crops (Fig. 9.2). The result of such an integrated system can lead to a wide diversity of products: food (fish, cattle, pigs, chicken), feed (*Spirulina*, halophyte crops), fuel biogas, ethanol), and chemicals. The alga *Dunaliella* is able to accumulate large amounts of glycerol in response to the externally imposed osmotic pressure caused by high salt concentrations. This production of glycerol has been reported to be favourably compatible with the present petrochemical process (Chew and Chi 1981). The proper application of fermentation technology in arid zones can therefore provide communities or societies for the first time with energy, food, feed, and fertilizer from either animal waste or sea/saline water. The final pollution free effluent can be used for irrigation of halotolerant plants which in turn provide feed for animals. The overall impact on society is immeasurable.

The greatest challenge for integrated rural fermentation technology, however, still lies ahead of us. Starvation in its greatest dimensions occurs in the arid and semi-arid lands which have a high annual solar energy input, e.g. Ethiopia and surrounding countries in Africa. Dry scrub, dry desert, and chaparral lands constitute 19 per cent of the continental area, but only 2.3 per cent of the primary photosynthetic productivity is found here (Bassham 1977). Without irrigation, the photosynthetic productivity is naturally extremely low, and the availability of water becomes the limiting factor. The extremely high evaporation rate in these zones does not allow the Mexican system to be applied. Whereas the land areas are a challenge to the agronomist, the coastal areas could be used for the development of marine algae and aquatic plants as food components (Mitsui 1977). Since 71 per cent of the Earth's surface is covered by ocean, marine systems should be investigated for new food resources and technology.

It is of great importance for the future of these countries to find an agriculture-fermentation technology system, whereby sea water can be used in a manner as shown in the Mexican example. Many thoughts have been given for combating the high evaporation in these areas. It is a great challenge to agriculturists and engineers, as well as to fermentation technologists and microbiologists; a truly biotechnologist challenge to save a number of very important cultures and societies.

6. Future developments and conclusions

It was the aim of this chapter to stress the impact which fermentation technology has had, and may have, in the future development of developing countries on culture and society. There is no doubt that the impact in urban areas and in rural communities will be different, since cultural heritage is much more vulnerable in urban areas. It very much depends therefore, on the developed countries whether culture can be preserved or not in the urban society. The question has to be asked whether science and technology, which is generated in the developed countries is transferable to the developing world for their benefit (Jarrett 1979; Doelle 1982). Where such transfer has occurred, the society may have benefited, but not necessarily its traditional culture. It is therefore important to recognize the impact, not only on the individual and society (Millington 1979), but also on the traditional culture. Priorities in the field of biotechnology/fermentation technology are different in developed countries and developing countries (Baltimore 1982). Whereas in developed countries, the high value added products, especially those for use in medical fields, may dominate the aspirations of biotechnologists (Bull *et al.* 1982) in order to maintain the present living standards, it is the basic needs of the society (ICAITI 1981), without interference in the traditional culture, which will dominate in the developing countries. These basic needs are in food, feed, fuel, and fertilizers; thus, integrated systems of agriculture-waste-fermentation technology will have the greatest appeal. Bioresource utilization programmes (Hermosillo and Gonzales 1981) are the best system to combat malnutrition and starvation, and if coupled with waste treatments could lead to high public health standards. Multiple-goal or integrated-system development can also lead to employment in the villages and higher living standards, avoiding the trend of increasing urbanization. Great challenges are still ahead of us if we want to ascertain the fulfillment of basic need requirements in the tropical arid zones, where starvation and basic needs are greatest.

It is therefore of great importance that scientists, technologists, and biotechnologists/fermentation technologists come to the realization that the social, political, and economic environments in which science and technology is to be embedded in the developing countries are enormously complex, and different from those which pertain in the developed world. If we can achieve such a realization, train people from developing countries accordingly and familiarize ourselves with the basic needs and resource availability of the particular country before designing and establishing a fermentation technology, we may succeed in not repeating the destruction of culture on cost of society improvements, but rather achieving a positive impact on society and preserving their culture, particularly in the rural communities of the developing world.

References

Ayanabe, A. (1983). The state of biotechnology in Africa, with special emphasis on biological nitrogen fixation. In *Biotechnology in developing countries* (eds. P. A. van Hamert, H. L. M. Lelieveld, and J. W. M. laRivere). Delft University Press.

Baltimore, D. (1982). Priorities in biotechnology. In *Priorities in biotechnology. Research for international development*, pp. 30–7. National Academic Press, Washington, USA.

Bassham, J. A. (1977). Potential of arid zone vegetation as a source of substrate. In *Microbial conversion systems for food and fodder production and waste management* (ed. T. G. Overmire), pp. 33–53. KSIR, Kuwait.

Berk, Z. (1979). Production of food as an objective for bioconversion systems. In *Bioconversion of organic residues for rural communities*. United Nations Univ., Tokyo.

Bhumiratana, A., Flegel, T. W., Impoolsap, A., Kaenjak, A., Chancharoensin, S., Shinmoyo, A., and Okada, H. (1983). Research and development in soy sauce manufacturing technique in Thailand. In *Microbial utilization of renewable resources* (ed. H. Taguchi) Vol. 3, pp. 99–107. ICCRD in Microbial Engin., Osaka.

Bhumiratana, A., Flegel, T. W., Impoolsap, A., Okada, H., and Oshima, Y. (1980). Isolation and analysis of molds from soy sauce koji in Thailand. In *Microbial utilization of renewable resources* (ed. H. Taguchi) Vol. 1, pp. 92–100. ICCRD in Microbial Engin., Osaka.

Bull, A. T., Holt, G., and Lilly, M. D. (1982). *Biotechnology. International trends and perspectives*. OECD Publication, Paris.

Chew, B. J. and Chi, C. H. (1981). Process development and evaluation for algal glycerol production. *Biotech. Bioeng.* 23, 1267.

Chiao, J. S. (1986). Biogas production in China. In *Applied Microbiology* (eds. H. W. Doelle and C. G. Heden). pp. 97–110. Unesco, Paris.

Ciferri, O. (1983). Spirulina, the edible microorganisms. *Microbial. Revs.* 47, 551–78.

DaSilva, E. J. (1979). Microorganisms as tools for biomass conversion and energy generation. *Impact Sci. Soc.* 29, 361–74.

——, (1980). Trends in microbial technology for developing countries. In *Renewable resources. A systematic approach* (ed. E. Campz-Lopez), pp. 329–68. Academic Press Inc., London.

——, (1981). The renaissance of biotechnology: man, microbe, biomass and industry. *Acta Biotechnol.* 1, 207–46.

——, and Doelle, H. W. (1980). Microbial technology and its potential for developing countries. *Process Biochem.* 15, 2–6.

——, Shearer, W., and Chatel, B. (1980). Renewable bio-solar and microbial systems in 'eco-rural' development. *Impact Sci. Soc.* 30, 225–33.

Doelle, H. W. (1982). Appropriate biotechnology in less developed countries. *Conserv. Recycling* 5, 75–7.

——, (1984). Biotechnology of molasses fermentation. *Second Res.* Planning Workship Sugarcane Res. Inst. Kantale, Colombo, Sri Lanka.

——, (1985). Biotechnology of solid substrate fermentation in the production of food. *ASEAN Food J.* 1, 10–4.

Ekundayo, J. A. (1978). An appraisal of advances in microbial biotechnology. In *Central Africa. Symp. Fungal Biotechnol.* Glasgow, U.K.

Fernandez, I. R. and Ocampo, A. T. (1980). Considerations on the social impact of technologies. In *Renewable resources. A systematic approach* (ed. E. Campoz-Lopez), pp. 223–32. Academic Press Inc., London.

Flegel, T. W., Bhumiratana, A., Impoolsap, A., Takada, N., Oshima, Y., and Okada, H. (1981). Studies on Aspergillus flavus var. columnaris in soy sauce koji. In *Microbial utilization of renewable resources* (ed. H. Taguchi) Vol. 2, pp. 59–63. ICCRD in Microbial Engin., Osaka.

Ginjaar, L. (1980). The role of government in the development of biotechnology. *Biotech. Lett.* 2, 95–100.

Gomez-Pompa, A. (1980). Renewable resources from the tropics. In *Renewable resources. A systematic approach* (ed. E. Campoz-Lopez) pp. 391–406. Academic Press, New York.

Hermosillo, O. and Gonzalez, G. (1981). *Biotecnologia para el aprovachamiento de los desperdicios organicos*. A. G. T. Editor, S. A., Mexico.

Hesseltine, C. W. (1983). The future of fermented food. *Nutrition Rev.* 41, 293–301.

ICAITI (1981). Concepts in bioengineering. *ICRO/UNEP/Unesco/OAS Training Course*. ICAITI, Guatemala.

Ivory, D. A. and Siregar, M. E. (1984). Forage research in Indonesia: past and present. In *Asian Pastures: Recent advances in pasture research and development in Southeast Asia*, pp. 12–29. FFTC Book Series 25. Taiwan.

Jarrett, F. G. (1979). Technological change—creator or destroyer? In *Science and technology for what purpose?* (ed. A. T. A. Healey), pp. 109–24, Australian Academy of Science, Canberra.

Judan, A. A. (1980). *A report on the Maya Farming System in a 1.2 hectare pilot project on its first year of operation*. Liberty Flour Mills Inc., Angono, Philippines.

King, A. and Cleveland, H. (1980). The renewable way of life. In *Bioresources for development* (eds. A. King and H. Cleveland), pp. xii–xxii. Pergamon Press, Oxford.

Knapp, J. S. and Howell, J. A. (1980). Solid substrate fermentation. *Topics Enz. Ferment. Technol.* 4, 85–144.

Kosaric, N., Ng. D. C. M., Russell, I. and Stewart, G. S. (1980). Ethanol production by fermentation: an alternative liquid fuel. *Adv. Appl. Microbiol.* 26, 147–227.

LaRivere, J. W. M. (1979). Environmental goals for microbial conversions in rural communities. In *Bioconversion of organic residues for rural communities*, pp. 26–35. United Nations University, Tokyo.

Leesley, M. E., Newson, T. M., and Burleson, J. D. (1980). A low energy method of manufacturing high-grade protein using blue-green algae of the genus Spirulina. *ASAE Nat. Energy Symp.* 3,619–23.

Meier, R. L. (1980). Creating resources-conserving communities for the 1980s and beyond. In *Bioresources for development* (eds. A. King and H. Cleveland), pp. 65–86. Pergamon Press, Oxford.

Millington, R. J. (1979). The ecological impact of science and technology on the individual and on society. In *Science and technology for what purpose?* (ed. A. T. A. Healey), pp. 151–69. Australian Academy of Science, Canberra.

Mitsui, A. (1977). Marine algae and aquatic plants as food components. In *Microbial conversion systems for food and fodder production and waste management* (ed. T. G. Overmire), pp. 3–31. KISR Kuwait.

Olguin, E. J. (1978). Appropriate technology: The case of a single cell protein (SCP) and biological upgrading of wastes. *Research Fellowship Report. Technol. Policy Unit* University of Aston, Birmingham, U.K.

——, (1982). Conversion of animal waste into algae protein within an integral agriculture system. *Proc. Seminar on Microbiological conversion of raw materials and by-products of agriculture into proteins, alcohol, and other products*. University of Novi-Sad, Yugoslavia.

—— (1984). Microalgae biomass as source of chemicals, fuel and protein. *Proc. VIth Austral. Biotech. Conf.*, Brisbane, Australia.

—— (1986). Appropriate biotechnological systems in the arid environment. In *Applied Microbiology* (eds. H. W. Doelle and C. G. Heden) Unesco, Paris.

—— and Vigueras, J. M. (1981). Unconventional food production at the village level in a desert area of Mexico. *Proc. Second World Congr. Chem. Engineering,* Montreal, Canada.

Perera, K. S. (1984). Funding of sugar research, and the case for establishment of a sugar authority. *Second Res. Planning Workshop* Sugarcane Res. Institute Kantale, Colombo, Sri Lanka.

Reed, G. (1982). Microbial biomass, single cell protein and other microbial products. In *Industrial Microbiology* (ed. G. Reed), pp. 541–92. 4th edn. Avi Publ. Co., Westport, USA.

Ringpfeil, M. and Heinritz, B. (1986). Single cell protein technology. In *Applied Microbiology* (eds. H. W. Doelle and C. G. Heden) pp. 64–96. Unesco, Paris.

Rolz, C. E. (1980). Biotechnology and bioengineering research in Central America. *Proc. VIth Intern. Ferment. Symp.*, p. 190. National Res. Council, Ottawa, Canada.

—— (1984). Perspectives presentes y futuras en la produccion de etanol. In *Advances en la produccion de etanol* (eds. C. Rolz, S. de Cabrera, R. Garcia, J. F. Calzada, and R. de Leon), pp. 1–10. Tercer Simposio Panamericano de Combustibles y Productos Quimicos via Fermentacion, ICAITI, Guatemala.

——, Menchu, J. F., de Cabrera, S., DeLeon, R., and Calzada, F. (1979). Strategies for developing small-scale fermentation processes in developing countries. In *Bioconversion of organic residues for rural communities*, pp. 36–40. United Nations University, Tokyo.

Saisith, P. (1981). Roles of fermented foods in solving malnutrition problems. In *Microbial utilization of renewable resources* (ed. H. Taguchi) Vol. 2, 200–10. ICCRD Microbial Engin., Osaka.

Simmonds, W. H. C. (1980). New directions for economic and social growth: the ecology of change. In *Bioresources for development* (eds. A. King and H. Cleveland), pp. 87–96. Pergamon Press, Oxford.

Sorensen, B. (1979). Energy technology and social structure. In *Appropriate technology for underdeveloped countries* (eds. Univ. Centralamericana, UCA) pp. 38–55. Proc. Second Intern. Symp. Engin., San Salvadore.

SRI (1984). *The second research planning workshop of the sugarcane research institute at Kantale*. Colombo, Sri Lanka.

Steinkraus, K. H. (1983). Traditional food fermentation as industrial resources. *Acta Biotechnol.* 3, 3–12.

Tel-Or, E., Boussiba, S., and Richmond, A. E. (1980). Products and chemicals from Spirulina platensis. In *Algae Biomass* (eds. G. Shelev and J. F. Soeder). Elsevier, Amsterdam.

Tochikura, T., Ito, M., Yano, T., Yamamoto, K., Techiki, T., and Kumagai, H. (1983). Enzymes of soy sauce koji application to food processes. In *Microbial utilization of renewable resources* (ed. H. Taguchi) Vol. 3, 117–21. ICCRD in Microbial Engin., Osaka.

Wang, H. C. and Hesseltine, C. W. (1982). Oriental fermented foods. In *Industrial Microbiology* (ed. G. Reed) 4th edn., pp. 492–537. Avi Publ. Comp., Westport, USA.

Yousri, R. M. (1982). Single cell protein: its potential use for animal and human nutrition. *World Rev. Anim. Prod.* 18, 49–67.

10

Alcohol production: anaerobic treatment of process waste water and social considerations

F. Siñeriz

This chapter focuses on the problems and solutions associated wtih the production of ethanol by fermentation, but not on the fermentation procedures themselves, this particular aspect being covered by other contributors to this book.

1. Ethanol production in Argentina in June 1984 and projections

Argentina at the present time belongs to the countries that, according to a World Bank Report (Kohli 1980) have a surplus in agriculture and a deficit in energy, although the latter factor is not critical since the country is about 90 per cent self-sufficient due to important reserves of natural gas. Most of the gas associated to the oil wells is unfortunately now vented away for lack of pipelines.

Basically, the deficit is more acute in light liquid fuels necessary for transportation (van Broock 1984). Correcting this situation by traditional means would require considerable investments in gas pipelines and in new cracking facilities in the oil refineries. Though investments are also required to build a new alcohol distillery, the cost of between four and six million dollars for a 100 m^3/day unit is fractional compared with the cost of a new cracking facility. Moreover, the investments on new distilleries can be spread over several years, contrary to the oil refinery where the facility has to be completed before going into production. The size of alcohol distilleries is based on the amount of raw material available in close proximity to the distillery to keep transportation costs to a minimum.

As a way to fill the gap at a lower cost and also to give new possibilities to the old sugar cane industry, the alconafta (gasohol) programme was introduced in 1981 in the province of Tucuman, the main sugar producing area of the country (50 per cent of the total production). Alconaftas in the two versions, normal and super, are mixtures of 10–15 per cent anhydrous ethanol in nafta. They were introduced taking into account the fact that these mixtures can be employed without modifications by existing cars and that the mixtures are compatible with gasoline in any proportion. In this way, cars coming from other provinces can use them and cars normally running with alconaftas can use any other gasoline when alconaftas are not available. The use of 96° ethanol requires comparatively major modifications of the motor and the water content of the alcohol makes it incompatible with gasoline.

The introduction of the programme is leading to an increase in the fermentation capacity of existing distilleries and in the establishment of new dehydration facilities. Indeed, ethanol is exclusively produced in Argentina by fermentation. Though there are some distilleries that ferment grains (especially for liquor), the bulk is produced by yeast fermentation of sugar cane molasses, a by-product of the sugar factories. The 96° alcohol so produced is of beverage quality. Much of it is exported, though exports are dwindling as more alcohol is dehydrated for fuel.

Molasses are, in general terms, concentrated sugar cane juice treated with some chemicals from which all the crystallizable sucrose has been extracted. They contain about 50 per cent sugars, mainly sucrose and inverted sucrose (glucose + fructose), and can be stored for periods of several months. In the distillery, that often is part of a sugar refinery, the molasses are diluted to about 18–20 per cent of the total sugars and fermented by yeast. Some distilleries collect the carbon dioxide evolved during fermentation, which is then sold for soft drinks manufacture. Once the fermentation is over, yeast cells are separated in most distilleries by centrifugation to be recycled to new fermentation batches and the liquid is distilled to obtain the alcohol. About 250 000 m^3 ethanol were produced in 1984. This amount is increasing since the alconafta programme, started for Tucuman on March 15th 1981, has been recently extended to all the northern provinces.

2. Impact of increasing ethanol production on the environment

The residue left after the distillation of the alcohol is known as 'vinaza', (stillage). It contains organics from the sugar-cane sap and from the metabolic processes of yeast, pentoses (not used by *Saccharomyces* spp.), and salts. The ratio of carbon compounds to salts is here several times lower than in sugar cane juice since most of the carbohydrates have been removed by yeast. For each litre of 96° ethanol produced, about 10 litres of stillage are discharged into rivers. Hence, as distilleries are concentrated in the sugar cane-growing region of the country, which conforms a closed basin, the impact is particularly severe. The average composition of the stillage produced in Tucuman is shown in Table 10.1.

As can be seen, the COD is 200–300 times higher than in urban waste water (about 0.25–0.35 g/l). The 2.5 million m^3 stillage that are presently dumped into the rivers will climb to about 10 million in the next few years (van Broock 1984).

Table 10.1
Composition range of stillage in Argentina

Chemical oxygen demand (COD)	77–100 g/l
Total solids	80–110 g/l
Volatile solids	60–90 g/l
Ashes	20–30 g/l
pH	4.6–5.1
N_{TK}	1.0–2.5 g/l
P	0.2–0.3 g/l
SO_4^{2-}	6.0–7.0 g/l
K^+	11.7–17.5 g/l
Ca^{2+}	0.9–1.5 g/l
Mg^{2+}	0.6–1.2 g/l
Na^+	1.1–2.5 g/l

One interesting aspect of the use of anhydrous ethanol mixtures is that ethanol enhances the octane number of the gasoline, avoiding then the use of tetraethyl lead with evident benefits to the environment.

3. Solutions to the pollution problem

Most of the sugar factories that survived closures in the late sixties have financial problems of one type or the other. To force these factories to put in operation stillage treatment plants could lead to more closures in times of increasing unemployment. With this principle in mind, we have been working for some years now on different anaerobic systems that could be used for the treatment of stillage with the production of methane, as a first step to full treatment, recovering the investments needed with the gas produced to make the whole idea economically acceptable.

We established previously (Sanchez-Riera *et al.*, 1982), the feasibility of the anaerobic treatment of stillage at high loading rates using a packed-bed reactor. Studies were extended to other types of reactors: an anaerobic filter and an upflow anaerobic sludge blanket reactor (UASB) (Sineriz *et al.* 1983). These systems (Fig. 10.1), have the ability to retain the micro-organisms needed to carry out the fermentation of the waste water (Hobson 1981), allowing shorter hydraulic retention times, that is the time needed for one volume of load to go through one volume of reactor. To our knowledge, it was the first time that reactors of this type were used for the treatment of these highly concentrated stillages by a two-step process. Treatment of this type of waste did not seem attractive because it is best applied to waste waters with high carbohydrate contents (Verstraete *et al.* 1981).

These preliminary studies yielded the following conclusions. First, the

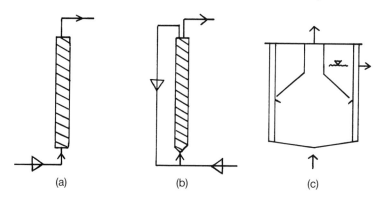

Fig. 10.1. Flowsheet of different reactors used for stillage treatment. (a) Anaerobic filter; (b) packed bed reactor with recirculation; (c) UASB reactor

efficiency for the conversion of COD to methane, 70–80 per cent, was more or less the same in all the designs tested. There was no evident superiority of one design over another. Observed differences were due to differences in the composition of the stillage used since, indeed it was shown (Sanchez Riera *et al.* 1985) that changing batches influenced the performances. Secondly, the systems proved reliable with time and the treatment could proceed over extended periods of time with minimal attention.

The main drawbacks of the anaerobic filter and the packed bed reactor when treating waste waters with suspended solids are associated with the accumulation of undigested material that can lead to plugging of the reactor (Young and Dahab 1982). For the interested reader, there are several recent reviews in which methanogenesis and the performance of different systems tested for the treatment of industrial waste waters is discussed (Nyns *et al.* 1980; Speece 1983; Kirsop 1984; Rolz *et al.* 1984; van den Berg 1984; Kirsop 1984).

As the UASB concept had been applied successfully to other organic wastes (Lettinga *et al.* 1980) and as it is relatively simple to operate, having in mind the large amounts of stillage produced, the impracticability of storage, and the presence of suspended solids, it was the design of choice studied in the laboratory with the aim of obtaining suitable information to operate larger-scale plants.

Quite simultaneously, studies with a pilot-scale reactor were conducted in Cuba (Valdez Gimenez *et al.* 1983) though with lower strength stillage. Recently, results were presented at a pilot-plant level, also using a lower strength stillage by the IPT in São Paulo, Brazil (Villen *et al.* 1985).

The methods used and the results obtained with the semi-pilot scale UASB reactor are detailed by Sanchez-Riera *et al.* (1985). The main conclusions drawn from the experiments with the 100 l reactors are summarized here.

(i) A reduction of about 75 per cent of the COD can be achieved without pretreatment, though substance(s) inhibitory to the fermentation in the raw stillage makes necessary some type of dilution, to be established in each case. The observed accumulation of propionate points to the inhibition of the obligate syntrophic propionate degraders, presumably due to high levels of sulphur compounds (SH^-, $S^=$) produced by reduction of the sulphate present in stillage as reported in consortia by Boone and Bryant (1980).

(ii) When dangerous levels of acids are reached inside the digester (10–12 g/l expressed as acetic acid), and decreasing or stopping the loading rate does not correct the values, though a complete stop of the methane production is not observed, tap water can be used to dilute the contents with good results, and the operation of the digester returns very quickly to normal.

(iii) A good pellet formations of the sludge is obtained with this substrate, allowing extended operation of this type of reactor with a high efficiency and avoiding the need of obtaining pellets from other reactors as seems to be the case with some types of waste waters (Lettinga et al. 1983)

(iv) As stillage contains quite a large amount of salts, these tend to accumulate in the pellets. When the ratio of ashes to total solids becomes high (over 50 per cent of the TS in sludge), some sludge (30–50 per cent) must be removed from the reactor.

(v) Stoppages of up to 60 days are not deleterious if the temperature is allowed to decrease. They are harmful and cause degradation of pellets if the operation temperature of 40°C is maintained.

(vi) The resulting effluent, though still with a high COD, can be applied directly to land since the pH is 7.2 (the original stillage has a pH of about 4.5) and since the rapidly-biodegradable material has been removed. The pH inside the reactor remains practically constant because of the high alkalinity and the buffering action of bicarbonate so that no prior neutralization of the stillage is required.

Table 10.2
Composition range in stillages from sugar cane juice

Chemical oxygen demand	15–33 g/l
Total solids	14–35 g/l
Volatile solids	12–30 g/l
Ashes	4–6 g/l
pH	3.7–4.6
N_{TK}	0.5–0.8 g/l
P	0.01–0.2 g/l
K^+	1.2–2.1 g/l
Ca^{2+}	0.1–1.5 g/l
Mg^{2+}	0.2–0.4 g/l
SO_4^{2-}	0.6–0.8 g/l

Though these results were obtained with stillage from molasses fermentations, it does not seem risky to believe that the system can also be applied to
stillages produced by the autonomous distilleries since the COD of these
stillages is considerably lower (Table 10.2).

With the results of the 100 l UASB reactor, it was considered that more
information was required at pilot-plant level before being able to design a
full-size plant. So, a 30 m³ plant was designed, its components built, and it is
in the process of being assembled and instrumented. A general flow-sheet of
the pilot plant is presented in Fig. 10.2.

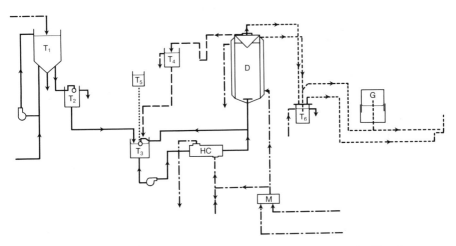

Fig. 10.2. Flowsheet of a pilot plant for the anaerobic treatment of stillage.
D, digester; G, gasometer; HC, heat exchanger; M, water temperature regulator;
$T_{(1-6)}$, mixing and holding tanks. (——) stillage; (—·—) water; (—···—) gas;
(— — —) effluent; (······) alkali.

The fermenter is jacketed in order to avoid the use of serpentines. It has a
capacity of 30 m³, 6 m height and 2.5 m in diameter. The loading rate can be
varied, and dilution and recirculation possibilities are incorporated. The
temperature is controlled by hot water circulation. As the stillage from the
distillation column has a temperature of 104°C, it will be cooled in a heat
exchanger. The biogas will be vented from a gasometer that will act as a
pressure regulator and will avoid the reflux of air into the digester whenever
the latter has to be partially drained. The main instrumentation will include a
control loop for the influx of stillage from the distillery, a control loop in the
mixing tank pumping system, temperature control of the incoming stillage
and in the jacket, a pH probe in digester, and a gas flow-rate measuring
service.

The construction of the pilot plant has been made possible through a
contract between the Secretary of Science and Technology, and a leading

distillery. The instrumentation is provided for by the Secretary (about 25 000 US $). The construction, general services and in-plant staff is provided by the company for a similar amount of money. The cost does not include development, scientific personnel, and the previous or simultaneous work at laboratory level. Under this agreement, the distillery will receive engineering support for a full-scale plant and will not pay royalties for other plants if belonging to the company.

A full-scale plant should be modular in design since the distillery outputs can vary. In this way, some modules can be shut down if necessary without impairing the performance of the plant as a whole.

4. Economic advantages derived from the anaerobic treatment

There have been many discussions going on whether the energy balance for the production of ethanol by fermentation is positive or negative (Costa Ribeiro and Castello Branco 1981; Beba 1983). These balances, taking into account managing of the crop, transport costs, transport of the finished product, performance as a fuel, etc., are difficult to extrapolate from location to location and have to be done in each particular case.

We calculated the energy benefit ratio (EBR), that is the ratio between the energy contained in the product and the energy required to produce it, for several distilleries in the Tucuman area. The results are shown in Table 10.3. It can be seen that there is a wide range of values, depending upon the obsolescence of the distillery material and management factors. The EBR values are displayed in the bottom line of Table 10.3, considering a 70 per

Table 10.3

Energy benefit ratio (EBR) in five distilleries of Tucuman. Current and projected values

	Dist. #1	Dist. #2	Dist. #3	Dist. #4	Dist. #5
Current EBR	2.5	1.9	2.1	1.3	2.1
Expected EBR*	9.7	3.8	5.3	2.1	3.8

* If stillage is fermented to biogas.

cent efficiency for the anaerobic digestion, if the methane produced is used to fuel the boilers. The values shown refer to the operation of the distillery as a whole, that is from the moment the molasses are diluted till the pure alcohol is obtained. There is a decrease of EBR valued when the dehydration process is also included, but no reliable data could be obtained so far from existing distilleries. Only in one case, in which a procedure involving water adsorbtion is used, we were able to assess that only 10 per cent of the energy was lost during the dehydration process.

As there are no autonomous distilleries in Argentina, we were not able to assess the energy balance for this type of distillery. The data obtained in Brazil, however, indicate that there is a surplus of bagasse. Hence, from the strict energy point of view, alcohol production has a positive energy balance since sugar cane bagasse is used for fuel.

If methane was also recovered by treatment of the waste water (stillage), the factory could be left with considerable amounts of bagasse suitable for paper production. The use of bagasse as a source of cellulose for the production of newsprint paper requires the prior separation of the long fibre bagasse from the short fibre bagasse or bagasse pith. This procedure is already employed to furnish a paper mill in the area about 30 km south of San Miguel de Tucuman. Bagasse surplus, would create a very beneficial situation for the paper factory, as the bagasse has presently to be paid with its fuel replacement value since the sugar factories, selling the bagasse, have themselves to buy natural gas to be used as fuel for the boilers.

The calculations made for two of the most efficient distilleries of the province of Tucuman, indicate that 65 and 75 per cent of the energy needs of the distillery could be supplied by the gas produced by the anaerobic treatment of the stillage. This value decreases when the industries are older or mismanaged (Cordoba *et al.* 1982).

In the case of the distillery, we work in co-operation with, the gas produced would allow an extra economy by cutting the financial and insurance costs of keeping reserves of fuel-oil, since this plant is not hooked to the natural gas pipelines and the supply of fuel-oil is at times not reliable.

5. Social implications associated with the ethanol fermentation industry

The social impact associated with an increasing production of ethanol has two main aspects: a potentially negative if not corrected in terms of environmental degradation, and a positive one in terms of job creation.

The disposal into rivers and streams of the high-strength stillage poses particularly serious problems since the sugar-growing region is the rural area with the highest population density in the whole country. This is due to the fact that sugar cane cultivation is labour intensive, especially during the harvest season, which extends from May until October. Many of these seasonal workers, coming from neighbouring countries, other provinces or from the higher, drier zones settled in the area with time.

Many people settled down in villages and cities close to the rivers, and as these foul up, water-quality related illness, like diarrhoea, viral hepatitis, become a recurrent problem. Also, a valuable source of protein, the abundant fish, is being wiped out by the discharges of stillage. Foul smells are common-place near a distillery and downstream.

In the neighbouring province of Santiago del Estero, a huge dam, providing electricity and water for irrigation for about 100 000 hectares of the Corporacion del Rio Dulce, a semi-public enterprise, collects all the discharges form the sugar processing plants.

All these considerations make mandatory the treatment of the soluble wastes. However, although legislation has been passed to this effect, very little has been done to treat the efluents. (Until 1983, San Miguel de Tucuman, the main city in the area, with about 400 000 inhabitants, did not have a waste water treatment plant.)

It would now be possible, with the expertise gained in the area, to provide plant design and operation for the anaerobic treatment of waste water at a fraction of the cost of similar systems available on the international market. Courses for the training of plant personnel could also be given.

The undoubtedly positive impact would be on the regional economy. Currently, sugar production is being subsidized in Argentina, but only the amount of sugar needed for the internal consumption is covered by this. Extra sugar has to be exported, and although there is also a support price, many times at a loss. The internal price is higher than the market price.

This mechanism led to the establishment of 'quotas' among the planters. The planter is paid the subsidized price only for the quota sugar cane. The rest remains unharvested or has to be diverted to other uses. There was, until 1983, no economic incentive to increase the area planted with sugar cane because the profitable operation (for sugar) is restricted to the 'quota' plantations that are alloted among all the existing planters.

The use of sugar cane for alcohol production would allow the increment of the planted area and a steady use of the labour force and the capital equipment. However, shifting from the profitable production of alcohol directly from sugar cane without the production of sugar needs also a price structure. This is the political ingredient of the alconafta programme. Sugar cane for ethanol should not be priced as sugar cane for sugar, but a minimum price for the ethanol, as a percentage of the retail gasoline price, should be established. Agreement has been reached along these lines and though it actually implies a loss of revenue since alcohol for fuel is not taxed, these losses are considered to be offset by the number of jobs created and the savings in oil imports.

The direct production of ethanol from sugar cane requires a new type of distillery, called 'autonomous' in Brazil, in which the steps required especially for sugar production are avoided. In these autonomous distilleries, sugar cane is milled and the juice fermented directly. This however produces only about 6–7 per cent alcohol in the fermentation medium. Alternatively, the juice is concentrated about two-fold (total reducing sugars, TRS, about 20 per cent).

About ten of these distilleries are in different stages of planning and construction, directly in the middle of the sugar cane fields to minimize transportation costs.

The overall socio-economic impact of such a programme is seen primarily in the permanent settling of farm workers and the great number of jobs created in connection with the new distilleries. This is especially true when the sugar producing area has a high degree of unemployment due to the fact that economically inefficient sugar factories were closed down in the second part of the sixties.

Each new autonomous distillery (with a capacity of about 100 m³ day) would create directly about 250 jobs and indirectly the jobs associated with sugar cane cultivation and harvest. If we consider that the unemployment figures for the last years in the Tucuman area are more than double the national average (Table 10.4), the socio-economical impact can be easily appreciated.

Table 10.4

Percentage unemployment in the sugar cane area compared with the national average

	1979	1980	1981	1982	1983	1984
Tucuman area	4.9	8.3	10.6	8.7	7.5	10.5
National average	2.4	2.5	5.3	4.6	3.9	4.5

Source: INDEC (1984).

Other crops, especially sweet sorghum, could be used for the fermentation of ethanol. However, as sugar cane is the best possible crop for this end, especially because it provides the fuel for the whole operation (Silva 1976), these other crops would be utilized to be able to operate the distillery beyond the sugar cane harvest season or 'zafra' that extends itself for about 180 days from May to October (EEAOC 1981).

6. Concluding remarks

Though it is difficult to extrapolate from the conditions of one particular country, it can be concluded that the social and economical implications associated with the production of ethanol by fermentation for fuel use are positive, provided certain steps are taken simultaneously to preserve the environment. The solution provided by the anaerobic treatment of stillage is not only environmentally sound, but also economically sound by the savings in fuel needs for the distillery. The technology involved is a relatively simple, soft technology, and countries with the capability to build alcohol distilleries can very well build this type of treatment plant if some bioengineering expertise is available.

Though anaerobic treatment alone cannot eliminate completely the COD of the stillage, it can be the first, necessary step, in this direction. Once the industries have recovered the investments needed for the treatment plant (about 3.5 years), they could be induced to reduce the COD levels of the effluent by further treatments. Especially in developing countries, where the greed for quick returns is generally established, only an approach involving direct economic benefits can have chances of success.

Acknowledgements

The authors with to express their gratitude to the International Foundation for Science (Sweden), the International Atomic Energy Agency, UNESCO, and the Secretary of Sciences and Technology (Argentina) for continued support.

References

Beba, A. (1983). Energy efficiency in the process of ethanol production from molasses. In *Biomass utilization* (ed. W. A. Cote) Nato Series A: Life Sciences, Vol. 67, pp. 659–68. Plenum Press, New York.

Boone, D. R. and Bryant, M. P. (1980). Propionate degrading bacterium, *Syntrophobacter wolinii* sp. nov. gen. nov. fom methanogenic ecosystems. *Appl. Environ. Microbiol.* 40, 626–32.

Cordoba, P. R., Sanchez Riera, F., and Sineriz, F. (1982). Possibilities of energy recovery in alcohol distilleries of Argentina. In *Energy conservation and use of renewable energies in the bio-industries*, International Seminar Proceedings, Trinity College, Oxford, UK, 6–10 Sept 1982 (Ed. F. Vogt) pp. 196–202. Pergamon Press, London.

Costa Ribeiro, C. and Castello Branco, J. F. (1981). Stillage: a resource disguised as a nuisance. *Process Biochem.* 16, 8–15.

EEAOC (1981). El alcohol combustible, recurso energetico nacional. *Avance Agroindustrial* (Revista de la Estacion Experimental Agroindustrial Obispo Colombres, Tucuman), 2(7).

Hobson, P. N. (1981). Microbial pathways and interactions in the anaerobic treatment process. In *Mixed culture fermentation* (eds. M. E. Bushell and J. H. Slater), pp. 54–9. Academic Press, London.

INDEC (1984). *Instituto Nacional de Estadistica y Censo, Octubre*. Ministerio de Economia, Buenos Aires.

Kirsop, B. H. (1984). Methanogenesis. *CRC Crit. Rev. Biotechnol.* 1, 109–59.

Kohli, H. S. (1980). Renewable energy: alcohol from biomass. Finance and development. *World Bank Report*, Washington, DC, USA.

Lettinga, G., Roersma, R., and Grin, P. (1983). Anaerobic treatment of raw domestic sewage at ambient temperature using a granular bed UASB reactor. *Biotechnol. Bioeng.* 25, 1701–23.

——, van Velsen, A. F. M., Hobma, S. W., de, Zeeuw., W., and Klapwijk, A. (1980). Use of the upflow sludge banket reactor concept for biological wastewater treatment, especially for anaerobic treatment. *Biotechnol. Bioeng.* 22, 699–734.

Nyns, E. J., Naveau, H. P., Chrome, R., and Bertrand, Y. (1980). Digesters—a worldwide review in anaerobic digestion. in *Anaerobic digestion* (ed D. A. Stafford, B. I. Wheatley, and D. E. Hughes) p. 37. Applied Science, London.

Rolz, C., Calzada, J. F., and De Leon, R. (1984). Biogas—Basic principles and application for Latin American. *Interciencia* 9, 8–20.

Sanchez-Riera, F., Cordoba, P. R., and Sineriz, F. (1985). Use of the UASB reactor for the anaerobic treatment of stillage from sugar cane molasses. *Biotechnol. Bioeng.* (in press).

——, Valz-Gianinet, S., Callieri, D. A. S. and Sineriz, F. (1982). Use of a packed-bed reactor for anaerobic treatment of stillage of sugar cane molasses. *Biotechnol. Lett.* 4, 127–32.

Silva, J. G. (1976). Balanço energetico cultural da produçao de alcool étilico de cana-de-açucar, mandioca e sorgo sacarino. Fase agricola e industrial. Brasil Açucareiro.

Sineriz, F., Diaz, H. F., Cordoba, P. R., and Sanchez-Riera, F. (1983). Continuous production of methane for stillage. In *Avances en digestion anaerobica*. Mircen-Biotecnologia, ICAITI, Guatemala.

Speece, R. E. (1983). Anaerobic biotechnology for industrial waste water treatment. *Environ. Sci. Technol.* 17, 416–26.

Valdez Gimenez, E., Obaya-Abreu, M. C., and Garcia Pena, A. (1983). Treatment of waste from the alcohol industry. In *Advances on digestion anerobica*. Mircen-Biotecnologia, ICAITI, Guatemala.

van Broock, M. R. G., de (1984). Industrial fermentation ethanol production in South America. *CRC Crit. Rev. Biotechnol.* 1, 209–28.

Vann den Berg, L. (1984). Developments in methanogenesis from industrial waste water. *Can. J. Microbiol.* 30, 975–90.

Verstraete, W., de Baere, L., and Rozzi, A. (1981). Phase separation in anaerobic digestion: motives and methods. *Trib. CEBEDEAU*, 453–4, 367–75.

Villen, R. A., Craveiro, A. M., Gonçalves, A. C. R., and Iglesia M. R. D. L. (1985). Anaerobic treatment of organic residues. Vinasse treatment. *Energy from Biomass*, Third EC Conf. March 25–29, Venezia, Italia.

Young, J. C. and Dahab, M. F. (1982). Effect of media design on the performance of fixed bed anaerobic reactors. *Proc. AWPR Semin. anaerobic treatment of waste water in fixed film reactors*, 16–18 June 1982, Copenhagen.

11

Single-cell protein: past and present developments

J. C. Senez

1. Introduction

The term 'single-cell protein (SCP) was coined in 1966 to describe the dried cells of micro-organisms grown in large industrial systems for use as a protein source for human and animal alimentation. This term is now universally accepted. However, one should point out that it refers to a whole microbial biomass, i.e. to a complex mixture of proteins, nucleic acids, carbohydrates, lipids, and other cell constituents. On the other hand, the term SCP applies to biomass not only of single-cell organisms, such as yeasts, bacteria, and unicellular algae, but also to coenocytic multicellular moulds.

The utilization of microbial biomass for food and feed purposes dates backs to the origins of modern microbiology. A small and semi-artisanal production of food and fodder yeast from molasses has existed in Western Europe since the very beginning of the century (Rose 1979a). However, these first developments did not contribute significantly to overall protein consumption, except during World Wars I and II, where up to 15 000 tons per year were produced in Germany and was substituted for about 60 per cent of pre-war protein importation.

In the 1950's, a new situation resulted from the menace of a world protein shortage, and from scientific advances in the field of microbial physiology and biochemistry. Following the pioneer development in France (1957) of SCP production from petroleum hydrocarbons, many major international firms in the petroleum and chemical industry engaged themselves activity in this new field of biotechnology.

The technology, sanitary control, and economical evaluation of SCP have been the subject of much literature and of several comprehensive books (Pontanel 1972; Champagnat and Adrian 1974; Tannenbaum and Wang 1975; Rockwell, 1976; Rose, 1979b; Ferranti and Fiechter 1983; Senez 1983; Shenan 1984; Goldberg 1985). In this chapter, special emphasis will be put on the prospects of SCP and closely related biotechnologies in the developing countries.

2. The world protein problem

The human consumption of protein and, still more strikingly, the ratio of animal versus total protein in the diet vary considerably from one part of the world to the other. According to recent estimation of FAO, 1.1 thousand million people are presently suffering from protein malnutrition in the Third World.

Owing to demographic growth, this situation is expected to rapidly worsen. The world population will reach 6.5 thousand million at the turn of the century and most of the increase will take place in the poorest countries, which are already starving. Therefore, merely to maintain the world food production standards at the present level, the production of protein should be multiplied by 1.5 (Table 11.1). Moreover, during this period of time, the Gross Domestic Product will increase substantially in a number of developing countries (United Nations 1977).

These developments will be reflected in the demand for protein, both quantitatively and qualitatively, economical progress being always correlated with a larger consumption of total protein and a higher ratio of animal versus total protein (Hoshiai 1978). Consequently, it is estimated that from now to the year 2000, the demand for feed protein will be multiplied by 1.8 (Table 11.1) (Hoshiai 1983; Senez 1979).

Table 11.1
Prospects of protein demand

	1980	2000	Increase factor
World demographic and economical trends*			
Population†	4400	6405	1.46
G.D.P.‡	1165	2071	1.78
Protein demands§			
Human consumption	48.6	78.4	1.61
Animal feeding	43.1	106.3	2.47

* Data from UN (1977).
† Millions;
‡ G.D.P.: Gross Domestic Product in constant US dollars (1977);
§ Million tons crude protein (data from Hoshiai 1978, 1983; Senez 1979).

The present situation of feed protein is quantitatively and economically dominated by soybeans of which the production (90 million t) corresponds to 30 million t crude protein, i.e. 27 per cent of total protein consumed for animal feeding in the world. The international market of crude soya protein (14.7 million t) is in the hands of three exporting countries, namely the USA

(62 per cent), Brazil (26 per cent), and Argentina (12 per cent). On the other hand, the quasi-totality of this international market is absorbed by Western Europe and Japan which are importing 80 per cent of their feed protein at a cost of 5 thousand million dollars. However, for the reasons mentioned above, some parts of the developing world are beginning to import rapidly increasing quantities of feed proteins, mainly in the case of Mexico, Venezuela, Saudi Arabia, and other oil-producing countries, as well as South-East Asia and China which in 1980 started to import 1.5 million t of soybeans per year.

Probably, agriculture alone will not be able to cover the supplementary demand of feed protein in the coming decades (Hoshai 1983; Senez 1979). From 1969 to 1979, world production of soybeans increased by 80 per cent, but this increase has slowed down and increased only by 2.4 per cent in the course of the last 6 years (Anon 1984). The main limitation to the production of soybeans is the shortage of suitable farmland in Northern and Southern America and this situation is affected by the large programmes for the production of ethanol from sugar cane in Brazil and fructose from corn in the USA.

On the other hand, the amelioration of productivity by selection of new varieties and by improved agricultural practice seems near to have been optimized, as suggested by the fact that, since 1969, the average yield of soybeans per hectare in the USA (FAO 1985) remained quite stable (1.8–2.0 t/ha). Moreover, efforts made in Europe, Africa, and Asia to produce soybeans at a competitive price have met with little success.

According to FAO (1978), the world production of crude feed protein is expected to reach 63 million in the year 2000. This forecast would correspond to a deficit of about 50 million with regard to the prospective demand (Table 11.1). The potential of the SCP market is thus of the same order of magnitude.

3. Production of SCP from petroleum, methanol, and natural gas

3.1. Petroleum processes

The industrial production of SCP from petroleum was initiated in the 1950s. Fundamental studies initiated in France by the CNRS established that *Candida* yeasts can be cultivated from either pure *n*-paraffins (alkanes) or gas–oil with a growth rate comparable to that obtained from glucose and with a conversion factor of 100 g dry yeast per 100 g of paraffins consumed. Following these findings, the French Branch of British-Petroleum (SF-BP), in collaboration with CNRS, developed two industrial processes (Shenann 1984), one utilizing gas–oil as a raw material, and the other using highly purified alkanes preliminarily extracted from petroleum by filtration on a molecular sieve. Both processes were based on the principle of continuous culture developed shortly after World War II by Monod and Szilad, but not yet utilized on an industrial scale.

The SCP produced in large pilot plants were submitted to extensive animal testing which conclusively demonstrated the nutritional value and safety of the products (Pontanel, 1972), and in 1970 a first generation of production plants was built. One of these, using the *n*-paraffin process and having a capacity of 4000 tons per year, was located at the Grangemouth refinery, Scotland. The other one, at Cap-Lavéra near Marseilles, was based on the gas–oil process and had a capacity of 16 000 t/year.

In the *n*-paraffin process, the yeast *Candida (Saccharomycopsis) lipolytica* is grown aseptically from a purified mixture of saturated linear hydrocarbons with a predominant chain length of C_{11}–C_{13}. The organism utilized in the gas–oil process, *C. tropicalis*, is grown non-aseptically in a air lift fermenter. The raw material is constituted by a fraction with a boiling temperature of 300–380°C, containing 10–25 per cent *n*-paraffins, predominantly in the range C_{14}–C_{16}. Both processes are run in the continuous mode. Bacterial contamination

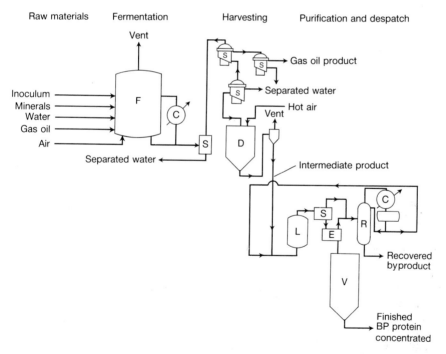

Fig. 11.1. The BP–Lavéra process of single-cell protein production from gas–oil. S: separator; C: cooler or condenser; F: fermenter; E: evaporator; D: dryer; V: storage vessel; L: leaching; R: recovery column.

is prevented by growing the organisms at an acidic pH (3.5–4.0), monitored by the addition of the nitrogen and phosphorus sources (gazeous NH_3 and liquid $H_3 PO_4$. The biomass is harvested by static decantation followed by centrifugation, and the resulting yeast cream, containing 15 per cent dry weight, is spray-dried. The final humidity of the product is below 5 per cent.

In the gas–oil process (Fig. 11.1) the residual fuel, of which only the alkane fraction is utilized by the culture, is recycled to the refinery and the dried biomass is completely freed of contaminating hydrocarbons by extraction with isopropanol. In the case of the n-paraffin process, the raw material is completely consumed and no final purification is necessary. The production costs of the two processes, requiring either a final purification of the product or a preliminary extraction of alkanes from petroleum, are practically equal. However, the gas–oil process has the important advantage in that it does not require costly sterile conditions. Moreover, the final extraction of residual hydrocarbons results in a delipidation which is beneficial to the nutritional quality of the product. In addition, the separated lipids provide a commercially valuable by-product.

In both processes, biological deparaffinization associated with SCP production results in lowering the freezing point of residual petroleum, the commercial value of which is thus appreciably increased.

From 1972 to 1976, some 40 000 tons of SCP were produced and successfully commercialized in the European Community under the trade names of Toprina-L (Lavéra) and Toprina-G (Grangemouth). On the other hand, a second generation plant of much larger capacity (100 000 t/year) and based on the n-paraffin process was built from 1974 to 1976 in Sardinia by the Italproteine Society, instituted jointly by BP and Ente Nationale Idrocarburi (ENI) with a financial contribution from the Italian government. The yeasts were produced continuously under aseptic conditions in three vertical fermenters each of 1500 m^3, with mechanical stirring and under pressurized air in order to optimize oxygen transfer.

Following these first industrial applications, similar R & D programmes for the production of SCP not only from petroleum, but also from methanol and natural gas were actively undertaken all over the world (Table 11.2).

In Italy, another 100 000 t plant was built by the Liquimica (Liquigas) company. This plant, located at Salinas, Calabria, was based on a n-paraffin process independently developed by the Japanese firm, Kanegafuchi Chemical Industry. The yeast used, Candida maltosa, was cultivated continuously and aseptically in a series of eight air-lift fermenters of 600 m^3. The product was to be commercially produced under the trade name of Liquipron. However, for reasons to be mentioned later, the two Italian plants of Italproteine and Liquichimica did not come into operation.

A considerable and rapidly increasing production of SCP from petroleum is taking place in the USSR, where a number of plants up to 300 000 t/year have been reported to be in operation at Krasnodar, Gorky, Kirishi, and Mozyr.

Table 11.2
Industrial production of SCP from petroleum, methanol, and methane

Substrate	Organism	Process		Production (t/year)
Gas–oil	*Candida tropicalis*	SF–BP Lavera*	Toprina-L	16 000
	Lodderomyces elongisporus	DDR–Schwedt	Fermosin	50 000
n-Paraffins	*C. lipolytica*	BP–Grangemouth*	Toprina-G	4 000
	C. lipolytica	Italproteine*	Toprina–G	100 000
	C. maltosa	Liquichimica–Kanegafuchi*	Liquipron	100 000
	Candida spp.	USSR–Mozyr		3 000 000
	C, paraffinica	Roniprot–Dainippon Ink	Roniprot	120 000
	C. tropicalis	IFP/Technip		Pilot
	C. tropicalis	Assam (India) IFP		Pilot
	C. tropicalis	Shanghai (Academa Sinica)		Pilot
Methanol	*Methylophilus methylotrophus*	ICI–Billingham	Pruteen	55 000
	Methylomonas clara	Üdde–Hoechst	Probion	Pilot
	Yeast	KISR–Kuwait		Pilot
	Yeast	Phillips Petroleum	Provesteen	Pilot
	Pichia pastoris	IFP/Technip		Pilot
	Pichia aganagobii	Mitsubichi		Pilot
Methane	*Pseudomonas methyltropha*	Shell		Pilot
	Methylococcus	BP		Pilot

* Not in operation (see text).

According to recent information, a total of 3 million tons SCP are presently produced in the USSR from *n*-paraffins, gas–oil, methanol, and other raw materials, but the respective quantities from these different substrates are not known and no information on biotechnological processes are available. It seems, however, that over 1 million tons are produced from *n*-paraffins.

Important developments have also been achieved in Eastern Europe (Senez 1978). In Romania, the Roniprot Company, with the technical assistance of these Japanese firm Dainippon Ink and Chemicals is producing 120 000 t/year of *Candida paraffinica* from alkanes. The petrochemical

Kombinat of Schwedt (DDR) produces 50 000 t of SCP by a gas–oil process (Ringpfeil 1983) similar to that of Lavéra but with a different yeast species (*Lodderomyces elongisporus*) under the name of Fermosin. The gas–oil fractions used have a boiling point of 240–360°C and contain up to 45 per cent alkanes in the length carbon range of C_{14}–C_{23}.

3.2. Methanol and methane processes

Methanol synthesized chemically from methane or coal is metabolized by a large variety of methylotrophic micro-organisms including yeasts and bacteria. The first process of SCP production from this raw material has been developed in the UK by the Agricultural Division of Imperial Chemical Industries (ICI). In operation since 1980, the Billingham plant, having a capacity of 50 000 t/year, is producing a bacterial biomass from methanol, under the trade name of Pruteen. The organism, *Methylophilus methylotrophus*, is grown in a continuous process and under sterile conditions in a huge vertical fermenter of 1500 m^3 based on a new air lift system (Senior and Windas 1980). In order to prevent inhibition of the culture by methanol, the substrate is supplied through an elaborate injection system, maintaining a homogenous concentration not exceeding 2 ppm for the entire height of the fermenter. Cells are separated by a flocculation/flotation procedure, followed by centrifugation. A treatment by oxygen peroxide was recently introduced to improve the nutritional quality of the product and the thick biomass (25–30 per cent dry weight) is finally dessicated in a pneumatic air drier.

A most remarkable scientific achievement of the ICI research team was the improvement of the conversion rate of methanol into protein by genetic manipulation (Windass *et al.* 1980; Haber *et al.* 1983). Ammonia is normally assimilated by *M. methylotrophus* via the glutamate and glutamine synthetase pathway, requiring ATP. The ICI group succeeded in substituting this energy-consuming process with the energy-conserving pathway of the enzyme glutamic dehydrogenase cloned via plasmids from *Escherichia coli*. The resulting *M. methylotrophus* strain showed a 3–5 per cent increase in growth yield. Industrial utilization of the modified organism was prevented by its genetic instability in large-scale continuous culture. However, this first application of genetic engineering to an organism of industrial interest will most likely be followed by others in a near future.

Several other processes for SCP production from methanol have been developed at the pilot plant scale. One of the most advanced is the Udde–Hoechst process (FRG) using the bacterial species *Methylomonas clara* (Faust and Sittig 1980). A 1000 t/year pilot plant is in operation in Frankfurt and a large production unit is planned. The product is intended to be commercially produced or animal and/or human consumption under the trade-mark of Probion.

Processes utilizing methylotrophic yeasts have been reported (Table 11.2) by the Institut Français du Pétrole (IFP), Philips Petroleum (Provesteen, USA), Mitsubichi (Japan), and Kuwait Institute of Scientific Research (KISR) (Faust and Sittig 1980; Ballerini and Thonon 1980). Yeasts have usually a growth yield on methanol slightly lower than that of bacteria. However, they have the advantage of growing optimally at acidic pH (3.5–4.0), thus reducing the risk of bacterial contamination. The *Torulopsis candida* strain of IFP can be utilized for SCP production in non-sterile conditions. A further advantage of this strain is to be relatively more thermo-tolerant ($+ 35°C$) than most other yeasts, and therefore well adapted to the production of SCP under tropical climates.

The production of SCP from gaseous methane, which constitutes 44–94 per cent of the North Sea natural gas, has been studied at the laboratory and pilot level (Shenann 1984). Methane is exclusively utilized as growth substrate by a specific group of obligate methanotrophic bacteria. The SCP processes from methane are based on the stable mixed cultures of a methanotrophic organism, such as *Pseudomonas methylotropha* (Shell) or *Methylococcus capsulatus* (BP), associated with heretotrophic bacteria, the essential role of which is to prevent feed-back inhibition of the methanotroph by the intermediary accumulation of methanol.

The cultures are grown at 44–50°C, thus minimizing the cooling expenses. However, there are several technical constraints, namely the explosion risk of the methane/air mixture in the fermenter off-gases, and the low solubility of methane requiring to grow the cells under high-pressure conditions (4–9 bars).

3.3 Nutritional value and safety control of SCP

SCP produced from either petroleum hydrocarbons or methanol are characterized by a high and well-balanced content of essential amino-acids, particularly L-lysine, L-tryptophane, and L-isoleucine which are the main limiting amino-acids in cereals. They are rather low in thio-amino-acids (methionine and cysteine), but only to the same extent as soybeans. For poultry farming, this relative deficiency of SCP and soya has to be corrected by addition of 0.2–0.3 per cent DL-methionine commercially available at low cost (3.5 US $/kg).

The biological efficiency of food and feed proteins is expressed in term of net protein value (NPV). From the ratio of the respective NPV (Table 11.3), 1 ton of alkane yeast or of methylotrophic bacteria is nutritionally and commercially equivalent to 2.26 or 2.15 tons of soybean meal, respectively.

Guide lines for testing the nutritional value and safety of SCP in human and

Table 11.3

Nutritional value of protein sources

	Dried skim-milk	Soybean meal	Alkane grown yeast*	Methylo-trophic bacteria[†]
Analytical composition				
Total nitrogen	5.7	7.0	11.2	11.5
Crude protein				
(N × 6.5)	35.9	44	70	72
Nucleic acids	–	–	6.3	13.8
Biological coefficients				
NPU	87	64	91	84
Biological value	93	65	96	88.4
Net protein value (NPV)	31.2	28.2	63.7	60.5
Metabolizable energy	2510	2240	2540	3468
	(10 507)	(9 377)	(10 632)	(14 517)

* Toprina-L, BP. † Pruteen, ICI, NPU: net utilization of protein. Biological value of soybean mean, Toprina or Pruteen experimentally determined with addition to the diet of 0.2 or 0.3 % DL-methionine. Net protein value: crude protein (N × 6.25) × NPU. Metabolizable energy for chicken: kcal/g (kj/g).

animal food have been edicted by the Protein Advisory Group of the United Nations (PAG) in 1974 and revised in 1973 (PAG/UNU 1983). Similar guidelines for animal feed have been set up in 1974 by the International Union of Pure and Applied Chemistry (IUPAC 1974).

The testing of SCP is specific for a given microbial strain, raw material and processing, and the constancy of each industrial product has to be controlled. However, in this last regard, SCPs are in an especially favourable situation (Senez 1983). Provided that stable culture conditions are maintained, the selection pressure tends to optimize the biomass production and stabilize the genetic characteristics of the micro-organism utilized. These considerations have been confirmed by the remarkable stability of industrial-produced SCP for several successive years.

SCPs from petroleum hydrocarbons (Shacklady and Gatummel 1972; Engel 1972; Shennan 1973) and from methanol (Stringer 1983; Lloyd 1983) have been submitted to nutritional and toxicological testing of an unprecedented level, on a large variety and multiple generations of laboratory and target animals. In 1976, the results from these studies were considered by the PAG (Scrimshaw 1976) as having conclusively demonstrated the high nutritional value and complete safety of the SCP tested (Toprina, BP and Pruteen, ICI) for animals and for the human consumers of animal products.

The possibility of utilizing SCP for direct human consumption is of obvious interest. In this regard, a specific problem is the relatively high content of nucleic acids in yeast (5–6 per cent) and bacteria (10–13 per cent). Prolonged ingestion of nucleic acids increases uricaemia and, in some predisposed individuals, may provoke intestinal problems. From extensive studies performed at the Massachusetts Institute of Technology (MIT) (Garattini *et al.* 1979), it was concluded that a daily intake of nucleic acids up to 2.0 g is completely safe. This limit, corresponding to 25 g of dry yeast, would be sufficient to provide a most significant supply of protein in the diet. On the other hand, a number of SCP treatments are available in order to effectively reduce the nucleic acid content at a normal level (Garattini *et al.* 1979). Mild cutaneous and/or gastro-intestinal troubles have been reported in consumers of some SCP products, but can be prevented by appropriate modification of the industrial processing (Scrimshaw and Udall 1983).

3.4. Present situation and prospects of the SCP industry

The energy crisis of 1973 and 1980 thwarted the expected development of the SCP industry in the Western world. For several years, the production of feed protein from petroleum and methanol was economically non-competitive, owing to both the considerable rise in cost of the raw materials and the low price of reference proteins, namely soybeans. A number of projects for large scale SCP production in Japan, Mexico, and Venezuela were thus abandoned.

Moreover, in several countries, SCP were faced with political obstacles. In Italy, the Ministry of Public Health presented a series of potential health dangers, amongst which carcinogenic risks for the consumers of animal products, occasional pathogenicity of some yeasts belonging to the genus *Candida*, and accumulation of uneven fatty acids in the adipose tissues of the animals. These unsubstantiated theories were scientifically disproved (D'Agnolo 1979). However, under the pressure of a hostile press campaign, the government authorization for commercial production in Italy of alkane-grown yeast was withdrawn.

A strikingly similar situation developed in Japan in 1975, where public opinion had been rightly sensitized by the Minamata disease and radio-active contamination of fish. The concerns of consumer groups regarding carcinogenic risks, were completely annuled by extensive experimentation under the guidance of the Ministries of Agriculture, and of Health and Welfare (Arima 1983). Nevertheless, the production of SCP was temporarily prohibited by the Japanese government.

Because of these economical and socio-political vicissitudes, the two large Italian plants of Italproteine and Liquichimica did not come into operation, and the production of the BP plants at Lavera and Grangemouth was discontinued in 1977. At the present time, the only production of SCP in Western Europe is that of the ICI-Pruteen plant, the activity of which was reduced to about half capacity. However, as already mentioned, the production of SCP from petroleum is still actively developed in the USSR and East European countries.

The economic prospects of SCP in Western Europe are rapidly improving. Since 1981, the price of crude oil decreased from 270 to 132 US $ per ton, i.e. by 50 per cent and this favourable evolution is generally expected to continue still further in the coming years. The yield of SCP from petroleum (132 US $/t) and from methanol (179 US $/t) being of 1:1 and 1:2, respectively, the corresponding costs of raw material are presently US $ 132 and 358 $/t of product. However, the cost of methanol can be expected to be substantially reduced in the near future, owing to the falling price of natural gas from which methanol is produced.

The conjuncture is still more favourable in the oil-producing countries. For these, the actual production cost of petroleum is considerably under the international market price. On the other hand, the quantity of crude oil required for a large industrial production of SCP would be practically negligible compared to their petroleum resources and income. From 0.46 per cent of the petroleum annually extracted in Latin America, 0.41 per cent in Africa or 0.18 per cent in Middle East, 1 million ton SCP could be obtained. Thus, the production of SCP may provide the oil producing countries with financial compensation for the slumping price and export of oil, as well as a major contribution to their own protein sources.

The Arab countries of North Africa and the Middle East, a region with a fastly growing population of 150 million and limited agricultural resources, are presently importing 90 per cent of their food and feed consumption. The potential SCP market in these countries was recently estimated at 800 000 t and anticipated to reach 1.1 million tons in the 1990s (Senez 1984; Hamer and Hamdam 1985). From these considerations, a joint project for construction of a 100 000 ton plant was set-up by the Organization of the Petroleum Arab Exporting Countries (OPAEC) and is now in progress. Similar projects are in existence in several heavily populated countries disposing of petroleum, such as Mexico, Nigeria, China, Indonesia, and others.

Methanol processes are especially attractive for countries having unexploited supplies of natural gas presently burnt-off and wasted. However, the large investments required for conversion of methane into methanol cannot be justified by SCP production alone, but only by a large scale market for energy and other industrial uses.

On the other hand, the processes based on purified alkanes or gas–oil can be implanted in any oil refinery of a suitable size, i.e. of about 3 million tons capacity per year. Moreover, the biological deparaffinization associated with the production of SCP is of special interest for those countries such as China,

India, Libya and others, producing heavy crude oil with a high paraffin content.

4. SCP from renewable raw materials

4.1. Agricultural products and residues

The production of protein for food and feed from agricultural raw material and wastes is the subject of a considerable and rapidly growing literature, and of intensive research and development carried all over the world. However, amongst the proposed substrates and processes, only a few are presently industrialized (Table 11.4) or have a chance of being so in the near future.

Table 11.4
Industrial production of SCP from renewable raw materials

Substrate	Organism	Process		tons/year
Molasses	*Candida utilis*	Cuba	Speichim (France)	80 000
Molasses and starch	*Corynebacterium melassicola*	France	Orsari*	10 000
	Brevibacterium lactofermentum	France	Eurolysine*	8 000
Starch hydrolysate	*Fusarium graminarium*	UK	Rank Hovis– McDougall	50–100
Fecula plant wastes	*Endomycopsis fibuliger + Candida utilis*	Sweden	Symba	10 000
Corn wastes	*Trichoderma viride*	USA	Denver	Pilot
Confectionary wastes	*Candida utilis*	UK	Tate & Lyle	500
Sulphite liquors	*Paecilomyces variotti*	Finland	Pekilo	10 000
		USA	Rhinelander	5000
	Candida utilis	USA	Boise Cascade Co.	5000
		Finland	Metsaluton Co.	10 000
Whey	*Kluyveromyces fragilis*	France	Bel (Protibel)	8000
	Kluyveromyces fragilis	USA	Amber Laboratories	5000
Cellulosic wastes	*Chaetomium cellulolyticum*	Canada	Waterloo Univ.	Pilot

* By-products of amino-acid production.

One of the main reasons is that, in order to be economically feasible, the production of SCP must be performed continuously and at a rate of at least 10 000 tons per year. Such prerequisites exclude those agricultural raw materials which are locally available in small quantity and the geographical dispersion of which would require over-costly transportation. They also exclude some raw materials, such as vegetable and fruit surpluses, which are available seasonally and cannot be stored. Another obstacle is that the production of SCP from some agricultural products or wastes is economically in competition with other uses, such as direct animal feeding, composting, or biogas production.

Amongst the potential substrates for SCP production, much attention is commonly attributed to ligno-cellulosic materials, which are available at low cost and in enormous quantities. According to UNEP estimations (Sasson 1980, Ward 1982), the world production of cereals is annually providing about 1.7 thousand million tons of straw, the major part of which is wasted. On the other hand, 50 million t of stalks and 67 million t of bagasses from sugar cane are available in the developing world.

The chemical composition of ligno-cellulosic materials is complex and variable from one source to the other. Basically, it consists of highly polymerized phenolic compounds (lignin) and sugars in the form of cellulose and hemi-celluloses. These last components, made of glucose and pentoses, constitute 30 per cent of straw, 25 per cent of hardwood and 10 per cent conifer wood.

One approach for the production of SCP from these raw materials is a preliminary chemical hydrolysis of cellulose and hemi-cellulose in strongly acid conditions and at high temperature, followed by cultivation of a glycolytic yeast. This type of processing was industrialized for several years in the Scandinavian countries and USSR, but is now practically abandoned as being not economically viable. A further difficulty inherent to SCP production from ligno-cellulosic materials is the toxicity of some phenolic by-products of lignin degradation which prevents the utilization of these proteins for feed for mono-gastric animals.

Another possible approach is the direct production of ligno-cellulolytic bacteria and moulds in either mono-specific or mixed cultures (Crawford 1981). At the present time, the growth rate and yield of these organisms are low and insufficient for the industrial production of SCP. However, fundamental studies on the chemical structure and biological degradation of lignin are rapidly progressing, and some promising micro-organisms (Reid *et al*. 1985) have been isolated recently. On the other hand, new developments can be expected in the field of genetic engineering and interspecific transfer of DNA from ligno-cellulolytic moulds to other micro-organisms.

The main objectives and industrial prospects of these fundamental advances are the delignification of cellulose for paper mills and the production from lignin of by-products of industrial interest. However, delignified cellulose may in the future provide a suitable raw material for direct protein enrichment by the techniques of solid state fermentation to be mentioned in a subsequent section of this chapter. As it will be pointed out, these techniques are also of interest for protein production from other agricultural raw materials.

4.2. Industrial by-products and wastes

Molasses from sugar cane and sugar beets, the world supply of which is of about 10 million tons per year, are the first raw material to have been utilized for SCP production. Currently, owing to economic considerations, molasses are chiefly utilized for direct feeding of cattle and for production by the fermentation industry of ethanol, chemicals, amino-acids, antibiotics, and other pharmaceuticals. However, in Cuba, seven plants built by a French engineering firm (Revuz 1981) are producing 80 000 t/year of fodder yeast from sugar-cane molasses. This utilization certainly has a future in other developing countries such as India, disposing of large and presently under-exploited supplies of molasses.

Other quantitatively important raw materials are the sulphite liquors of paper-mills, containing 18–26 per cent of fermentescible sugars with a large proportion of pentoses originating from the hydrolysis of wood hemi-cellulose. Some 25 000 t/year are produced in Europe and 10 000 t in the USA from this substrate (Table 11.4). However, the product is of mediocre nutritional value owing to contamination of the raw material by large quantities of lignine-sulphonate (50 per cent) and SO_2. Moreover, since the introduction of the kraft process in the paper industry, the supply of sulphate liquors suitable for SCP production is rapidly declining (Forage and Righelato 1979).

Whey has been used as a substrate for yeast protein since 1940 (Forage and Righelato 1979; Moebus and Teuber 1983). On a dry basis, it contains approximately 70 per cent lactose and 9–15 per cent protein. About 13.6 million m^3 of liquid whey are annually produced in the USA, corresponding to some 880 000 t solids. SCP produced from whey by the lactose fermenting yeast *Kluyveromyces fragilis* is well accepted as a food source for both animals and humans.

Processes have been proposed for SCP production from a number of wastes from the food industry locally available in large quantities, such as fruit-pulp and waste waters from the confectionery and canning industry. For example, in the process of pineapple canning the liquid residues contain 7 per cent fermentescible sugars which can be used as a growth medium yielding up to 15 g per litre of *Candida utilis*.

The residual lipids from the industrial processing of vegetable oil may become another important source of protein, as they constitute about 2.5 per cent of the processed oil, the world production of which now amounts to 4.2 million tons. It has been recently shown (Montet *et al.* 1983) that the fatty acids liberated by preliminary saponification of the residual lipids contained in soapstocks and/or liquid effluents are actively metabolized by *Candida* yeasts, and can thus be utilized for SCP production. These new prospects are of obvious interest for tropical countries producing groundnut or palm-oil, such as Malaysia and Senegal.

In the industrialized countries, the main economic incentive for SCP production from agro-industrial wastes is to comply with anti-pollution measures imposed by public authorities. A similar situation will probably result from industrialization and increasing awareness of conservation of the natural environment in a number of developing countries.

An indirect source of SCP, which is rapidly developing, is the microbial biomass from industrial fermentations for purposes other than protein production. The world production of amino-acids (2 million tons) yields approximately 600 000 t SCP, of which 18 000 t are presently produced commercially in France, as by-products of glutamic acid and lysine (Table 11.4). Likewise, the production of ethanol from sugar cane in Brazil (Rothman *et al.* 1983), which is expected to reach 2 million tons, will correspond to about the same amount of SCP. Similarly, in France (Byé and Mounier 1984) the projected production of 60 000 t ethanol from wheat and sugar-beet will yield some 56 000 t of biomass. The commercial production of biomass for animal feeding is economically an essential part (30 per cent) of the return from amino-acids and ethanol production.

The utilization of unicellular algae and photosynthetic cyano-bacteria (blue-green algae) for SCP production has been and still is the subject of much studies and speculation (Becker 1982; Mitsui *et al.* 1982). The interest of these organisms is that they can potentially be produced in purely mineral media with free solar light and atmospheric CO_2 as the energy and carbon sources. Some cyanobacteria have the further advantage of fixing atmospheric nitrogen. Since time immemorial, photosynthetic bacteria (Cyanobacteria) proliferating spontaneously in Lake Tchad (*Spirulina platensis*), and in Lake Texcoco near Mexico City (*S. maxima*) were consumed by local populations, and those of Lake Texcoco are still nowadays harvested industrially in small quantities (320 t/year) and produced commercially as human food.

Under artificial conditions, the algae are grown in batch or semi-continuous cultures, and the productivity is optimized by water circulation and the monitored addition of CO_2 or bicarbonate. The organisms utilized include green unicellular algae (*Chlorella*) and freshwater or marine cyanobacteria (*Spirulina*, *Nostoc*). Under optimal conditions, the reported rates of biomass production vary between 5 and 50 g/day/m^2. In Japan, a relatively important quantity of *Chlorella* and *Spirulina* is presently produced commercially as high-priced dietetic food. Another prospect actively developed in Japan is the use of unicellular algae as feed in aquaculture (Mitsui *et al.* 1982).

However the future of large-scale production of algal SCP is limited by several constraints, namely climatic variations and the cost of CO_2 supply and of biomass harvesting. Moreover, the high pigment content of the cells (chlorophyll and carotenoids) reduces the digestibility and nutritional value, and requires costly bleaching.

5. Protein enrichment by fermentation

A great variety of fermented staple foods traditionally produced in Asia, Africa, and elsewhere have been described (Hasseltine 1965), and are the subject of another chapter of this book. On a similar line, modern biotechnology can provide both the industrialized countries and the Third world with new and valuable protein sources for human and animal feeding.

Properly speaking, the products of direct protein enrichment by fermentation, although usually referred to as SCP, are fundamentally different, and deserve specific denomination and attention. This type of biotechnological processing does not involve a separation step, and the product is thus constituted by the whole culture, i.e. a mixture of microbial biomass and residual raw material.

Such simplified technology curtailing the investment and production costs has important advantages and prospects, in particular for the developing countries. It provides the possibility of operation in relatively small and inexpensive production units, well adapted to the utilization of raw materials which would be too costly or in quantities too small for economic SCP production. A further advantage of direct protein enrichment is that it can be performed with filamentous fungi or yeasts cultivated at acidic pH (3.5–4.0), thus preventing bacterial contamination and permitting operation in non sterile conditions.

Many substrates have been proposed and a number of processes have been developed at the laboratory or pilot level. (Rose 1979*d*, *b*; Goldberg 1985; Ward 1982). Among those utilizing ligno-cellulosic material, the Waterloo process (Moo-Young *et al.* 1979) is one of the most advanced and promising. In this three-step process, wood residues or straw are pre-treated by steaming or acid hydrolysis, fermented by the filamentous fungus *Chaetomium cellulolyticum*, and the solid residues discarded. With bovine manure, a preliminary anaerobic fermentation yielding biogas provides an energy supply for the operation of the whole system.

In the USA, the General Electric Company developed a process for conversion of feedlot wastes to feed proteins using a thermophilic actinomycete (Bellamy 1974). In the recent Institut National de la Recherche Agronomique, France (INRA) process of solid fermentation (Durand *et al.* 1983), sugar-beet pulp pre-treated by mild acid hydrolysis is fermented by the yeast *Candida utilis*. A similar process using the same organism and partially hydrolysed straw is presently industrialized at a small scale in Czechoslovakia (Volfova 1984).

For reasons already mentioned, (Volfova 1984) protein enrichment of ligno-cellulosic materials is a slow process and the utilization of the products is restricted to the feeding of ruminants.

Protein enrichment of starchy materials such as cereals and potato in the temperate climates, and cassava, banana, and other crops in the tropical regions, is particularly attractive. Agriculturally-produced with a high productivity per hectare, these substrates are available in large quantities. On the other hand, starch can be fermented with a high rate of bioconversion ($\geq$ 50 per cent) by a great variety of fast-growing yeasts, moulds, and bacteria. Several processes of protein enrichment in liquid medium have been experimented with at the pilot-plant scale. The one developed at Guelf University (Canada) utilizes the fermentation of crude cassava flour with a thermophilic fungus, *Aspergillus fumigatus* (Aidoo et al., 1982). In another process developed in France (Adour), the same raw material is fermented continuously in an air-lift fermenter by a amylolytic strain of the yeast *Candida tropicalis* (Revuz 1981). The product containing 20 per cent protein is destined for human consumption.

The possibility of directly obtaining a concentrated product not requiring expensive dehydration by solid-state fermentation (Brook et al. 1969) has received much attention in recent years. The first application of solid-state fermentation to starchy raw materials, such as cassava flour (Senez et al. 1980), were of low efficiency, providing only 3–4 per cent protein enrichment. Technically, the main problem is to maintain in a concentrated mash of substrate aerobic conditions and a rate of oxygen transfer excluding an anaerobic contamination of the culture.

In the INRA-ORSTOM process (Senez et al. 1980), the raw material is preliminarily steamed for 10 minutes at 70°C for glutinization of starch and inoculated with spores of an amylolytic fungus (*Aspergillus hennerbergii*). Under appropriate conditions, i.e. an initial water content of 55 per cent, the inoculated substrate spontaneously takes a granular structure freely permeable to air. After 24 hours incubation under controlled aeration, temperature (35°C), pH (3.5), and humidity, the resulting product contains up to 20 per cent protein of good nutritional quality and 25–30 per cent residual carboyhydrates. This process has been successfully operated at the pilot plant level with a variety of substrates, including cassava, banana, and potato wastes from fecula plants, and with a number of other amylolytic fungi (*Asp. oryzae, Penicillium* sp.) isolated from tempe, koji, and other fermented foods of South-East Asia. In Mexico, similar technique and results were reported using cassava flour and the mould *Rhizopus oligosporus* (Ramos-Valdivia et al. 1983).

Concerning the general agro-economical prospects of protein enrichment, it is important to note that, as shown in Table 11.5, via fermentation starchy

Table 11.5

Productivity of protein-rich feeds and of protein enriched feeds

	Yield (t/ha)	Protein content (%)	Protein (t/ha)
Soybeans	1.8	34	0.6
Rapeseed	3.0	20	0.6
Sunflower	1.8	16	0.3
Horse-bean	3.2	28	0.9
Protein enriched feeds*			
Cereals	8.5	20	1.7
Potato or cassava†	15.2	20	1.7

* Fermented product with 20% protein.
† On the basis of 40 t/ha fresh weight, 38% dry material and 55% conversion rate (protein/carbohydrate).

raw materials can provide two to three times more protein per hectare than the cultivation of soybeans or other protein-rich crops. This fact should retain the attention of every country willing to develop its own protein resources.

Large supplies of potential raw materials are available, e.g. the European surplus production of cereals, sugar-beets, and potatoes. In the tropical regions, cassava is potentially the most abundant and appropriate substrate. In spite of a world production amounting to 120 million tons, cassava is usually cultivated on small parcels of land, with a low productivity averaging only 8.9 t/ha. However, with modern agricultural practice, this yield could reach 40–50 tons per hectare or even more. Mass production of cassava has been developed in Thailand, and this country is presently exporting 5.5 million tons dried cassava at a low price (45 US $/t) to the European Community. By solid-state fermentation, 1 ton of dried cassava containing initially 80 per cent starch and less than 1 per cent protein is equivalent to 0.5 ton of soybean meal.

Banana, with a world production of 35 million tons, is another potentially important tropical raw material. In the exporting countries of Central America and Africa, 20–30 per cent of the harvested fruits are discarded and wasted, owing to insufficient protein content for animal feeding. These wastes could provide some countries such as Honduras, which is exporting annually 1.4 million tons of bananas, with a valuable source of feed protein.

One may expect from protein enrichment of foods and feeds important developments not only at the industrial, but also at the rural level. Existing processes, particularly those based on solid-state fermentation (Senez *et al.* 1980), can be operated in integrated agro-systems combining the production of raw material with protein enrichment and utilization of the product for human or animal feeding. Such prospects are of obvious interest for most developing countries which can thus increase efficiently and at low cost their own protein resources.

Conclusions

Single-cell proteins can be produced industrially from a large variety of raw materials including agricultural and industrial wastes. However, petroleum, methanol, and natural gas are the only substrates from which protein can be obtained in quantities corresponding to the present and future needs of the world.

In the industrialized countries, the development of the SCP industry was curbed for several years owing to economical circumstances, i.e. the petroleum crisis and the increasing prices of substrates. A more favourable situation is presently arising especially in the oil-producing countries.

New biotechnological methods for direct protein enrichment of foods and feeds by fermentation have recently been developed. Such relatively simple and inexpensive processes are of special interest for the developing world. They provide the possibility of producing protein for food and feed from a number of raw materials which would be too costly or in insufficient supplies for production of SCP by more sophisticated technologies. A further important prospect of protein enrichment is its application in rural communities.

The possible contributions of microbiology to the protein problem should receive from the developing world the priority they deserve.

References

Aidoo, K. E., Hendry, R., and Wood, B. J. B. (1982). Solid state fermentations. *Adv. Appl. Microbiol.* 28, 201–37.

Anon. *Tourteaux et autres matières riches en protéines* Ann. Report (1984), INRA, Robert S. A. and Sido, Y., Paris, France.

Arima, K. 1983). The problems of public acceptance of SCP-S. In *International Symposium on SCP* (ed J. C. Senez) pp. 145–62. Lavoisier, Paris.

Ballerini, D. and Thonon, C. (1980). Single cell protein: IFP processes. In *Proc. OAPEC Symp. on Petroprotein*. Kuwait, pp. 153–79.

Becker, E. W. (1982) The production of microalgae as source of biomass. In *Biomass Utilization* (ed. W. A. Coté) pp. 205–27. Plenum Press, New York.

Bellamy, W. D. (1974). SCP from cellulosic wastes. *Biotechnol. Bioeng.* 16, 869–80.

Brook, E. J., Stanton, W. R., and Wallbridge, A. (1969). Fermentation methods for protein enrichment of cassava. *Biotechnol. Bioeng.* 11, 1271–84.

Byé, P. and Mounier, A. (1984). *Les futurs alimentaires et énergétiques des biotechnologies*. Presses Univ. Grenoble.

Champagnat, A. and Adrian, J. (1974). *Pétrole et Protéines*. Doin, Paris.

Crawford, R. L. (1981). *Lignin biodegradation and transformation*. Wiley Interscience, New York.

D'Agnolo, G., Donelli, G., Macri, A., and Silano, V. (eds). (1979) Livieti cultivati su *n*-alcani (Biproteine). *Ann. Ist. Sup. Sanita* 15, 348–689.

Durand, A., Arnoux, P., Teilhard de Chardin, O., Chereau, D., Boquien, C. Y., and Larios de Anda, G. (1983). Protein enrichment of sugar beet pulp by solid state fermentation. In *Production and feeding of single cell protein* (eds H. P. Ferranti and A. Fiethter) pp. 120–2. Applied Science Publishers, London.

Engel, C. (1972). Safety evaluation of yeast grown on hydrocarbons. In *Protein from hydrocarbons* (ed. H. G. de Pontanel) pp. 53–81.) Academic Press, New York.

FAO (1978). *Oilseeds, fats and oils, oilcakes and meals.* Supply demand and trade projection 1985.

—— (1985). *Mon. Bull. Stat. B.* 8(1), 18.

Faust, V. and Sittig, W. (1980). Methanol as carbon source for biomass production in a loop reactor. *Adv. Biochem. Eng.* 17, 63–99.

Ferranti, M. P. and Fiechter, A. (eds) (1983). *Production and feeding of single cell protein.* Applied Science Publishers, London.

Forage, A. J. and Righelato, R. C. (1979). Biomass from carbohydrates. In *Economic microbiology* (ed. A. H. Rose) vol. 4, pp. 289–313. Academic Press, New York.

Garattini, S., Paglialunga, S. and Scrimshaw, N. S. (eds) (1979). *Single cell proteins. Safety for animal and human feeding.* Pergamon Press, New York.

Goldberg, I. (1985). *Single cell protein.* Springer Verlag, New York.

Haber, C. L., Allen, L. H., Zhao, S., and Hanson, R. S. (1983). Methylotrophic bacteria: Biochemical diversity and genetics, *Science* 221, 1147–53.

Hamer, G. and Hamdan, I. Y. (1985). The transfer of SCP technology to the petroleum exporting Arab Countries. *MIRCEN J.* (Unesco), 25–32.

Hasseltine, C. W. (1965). A millenium of fungi, food and fermentation. *Mycologia* 2, 149–97.

Hoshiai, K. (1978). The world protein demand in future. *Chem. Econom. Eng. Rev.* 10, 1–9.

—— (1983). Present and future of protein demand for animal feeding. In *International Symposium on SCP* (ed. J. C. Senez) pp. 34–63. Lavoisier, Paris.

IUPAC (1974). Proposed guidelines for testing of single cell protein destined as major protein source for animal feed. Techn. Bull. no. 912, IUPAC Secretariat.

Lloyd, D. R. (1983). The nutritional evaluation of Pruteen. In *Proc. Int. Symp. on SCP from Hydrocarbons for Animal Feeding* (ed. I. Y. Hamdam) pp. 159–75. Fed. Arab. Scient. Res. Coun.

Mitsui, A., Kumazawa, S., Philips, E. J., Reddy, K. J., Matsunaya, T., Gill, K., Renuka, B. R., Kusumi, T., Reyes-Vasquez, G., Miyazawa, K., Hayes, L., Duerr, E., Leon, C. B., Rosner, D., Ikemoto, H. *et al.* (1982). Mass cultivationn of aglae and photosynthetic bacteria: concepts and application. In *Biotechnology and bioprocess engineering* (ed. T. K. Ghose). Indian Inst. Technol. Delhi. pp. 119–55.

Moeubs, O. and Teuber, M. (1983). General aspects of production of biomass by yeast fermentation from whey and permeate. In *Production and feeding of single cell protein* (eds H. P. Ferraite and A. Fiechter) pp. 124–35. Applied Science Publishers, London.

Montet, D., Ratomahenina, R., Ba., A., Pina, P., Graille, J. and Galzy, P. (1983). Production of SCP from vegetable oils. *J. Ferment. Technol.* 61, 417–20.

Moo-Young, M., Dangulis, A., Charal, D., and Macdonald, D. G. (1979). The Waterloo process for SCP production from waste biomass. *Process. Biochem.* 14, 38–45.

PAG/UNU (1983). Guidelines no. 6, 7, 12 *Food Nutr.* (UNU) 5, 59–70.

Pontanel, H. G. de, (ed.) (1972). *Proteins from hydrocarbons.* Academic Press, New York.

Ramos-Valdivia, A., de la Torre, M., and Casas-Campillo, C. (1983). Solid state fermentation of cassava with *Rhizopus oligosporus.* In *Production and feeding of single cell protein* (eds M. P. Ferranti and A. Fiechter) pp. 104–11. Applied Science Publishers, London.

Reid, I. D., Chkao, E. E., and Dawson, P. S. S. (1985). Lignin degradation by *Phanerochaete chrysosporium* in agitated cultures. *Can. J. Microbiol.* 13, 222–5.

Revuz, B. (1981) Production industrielle de protéines a partir d'hydrates de carbone. In *Single cell protein II* (eds S. R. Tannenbaum and D. I. C. Wang) pp. 126–39. MIT Press, Cambridge, Mass.

Ringpfeil, M. (1983). Technology of SCP production from fossil raw materials. In *SCP from hydrocarbons for animal feeding* (ed. I. Y. Hamdan) pp. 85–95. Fed. Arab. Scient. Res. Coun., Kuwait.

Rockwell, P. J. (1976). *SCP from renewable and non renewable resources.* Noyes Date Corp. Park Ridge, New Jersey.

Rose, A. H. (1979*a*). History and scientific basis of large-scale production of microbial biomass. In *Economic microbiology* (ed. A. H. Rose) Vol. 4, 1–29. Academic Press, London.

——, (1979*b*). Microbial biomass. In *Economic microbiology* (ed. A. H. Rose). Vol. 4. Academic Press, New York.

Rothman, H., Greenshields, R., and Calle, F. R. (1983). *The alcool economy: fuel ethanol and the Brazilian experience.* Frances Pinter Ltd, London.

Sasson, A. (1983). *Les biotechnologies: défis et promesses.* Unesco Paris, France.

Scrimshaw, N. S. (1976), *PAG Symp on SCP in Animal Feeding, Brussels*, 29–30 March 1976. Press Release, PAG/UN, New York.

——, and Udall, J. (1983). The nutritional value and safety of SCP for human consumption. In *International Symposium on ECP* (ed. J. C. Senez) pp. 102–13. Lavoisier, Paris.

Senez, J. C. (1979). Pour une politique nationale en matière de protéines alimentaires *Prog. Scient.* 203, 5–32.

—— (ed.) (1983). *International Symposium on SCP.* Lavoisier, Paris, France.

—— (1986). The economical aspects of single-cell protein production from petroleum derivatives. In *Perspectives in biotechnology and applied microbiology* (ed D. I. Alani and M. Moo-Young) pp. 33–48. Elsevier, London.

——, Raimbault, M., and Deschamps, F. (1980). Protein enrichment of starchy substrates for animal feeds by solid state fermentation. *Anim. Rev. FAO* 35, 36–9.

Senior, P. J. and Windas, J. (1980). The ICI single cell protein process. *Biotechnol. Lett.* 2, 205–10.

Shacklady, C. A. and Gatumel, E. (1972). The nutritional value of yeast grown on alkanes. In *Proteins from hydrocarbons* (ed. N. G. de Pontanel) pp. 21–52. Academic Press, New York.

Shennan, J. L. (1983). Microbiological aspects of SCP production. In *Economic microbiology* (ed. A. H. Rose) Vol. 4, pp. 103–217. Academic Press, London.

——, (1984). Hydrocarbons as substrates in industrial fermentation. In *Petroleum microbiology* (ed. A. I. Atlas). Macmillan, New York.

Stringer, D. A. (1983). Toxicological evalation of Pruteen. In *Proc. Int. Symp. on SCP from hydrocarbons for animal feeding* (ed. I. Y. Hamdan) pp. 147–58. Fed. Arab. Scient. Res. Coun.

Tannenbaum, S. R. and Wang, D. I. C. (eds) (1975). *Single cell protein. II.* MIT Press, Cambridge, Mass.

UN (1977). *The future of world economy*. Oxford University Press, Oxford.

Volfova, O. (1984). Microbial biomass. In *Modern biotechnology* (eds V. Krumphanzl and Z. Rehacek) pp. 525–54. Unesco, Paris.

Ward, R. F. (1982). Food, chemicals and energy from biomass. In *Biomass utilization* (ed. W. A. Coté) pp. 23–49. Plenum Press, New York.

Windass, J. D., Worsey, M. J., Piopi, E. M., Pioli, D., Barth, P. T., Atherton, K. T., Dart, E. C., Byrom, D., Powell, K., and Senior, P. G.(1980). Improved conversion of methanol to single-cell protein by *Methylophilus methylotrophus. Nature Lond*. 287, 396–401.

12

Algae associated with sewage treatment

H. W. Pearson

1. Introduction

Modern methods of biological sewage treatment are derived from the basic self purification process which occurs with time in all natural water bodies that become polluted with organic matter. Man has exploited this ability for many centuries disposing of his wastes into convenient rivers and lakes, and in coastal regions by discharging them directly into the sea. However, rapid increases in population and the development of large densely populated communities frequently led to the overloading of many receiving water bodies with sewage, turning them into festering anoxic and odorous environments. Such polluted rivers and lakes were aesthetically displeasing, but more importantly were a public health hazard, since they harboured human pathogens and increased the risk of spreading excreta-related diseases through the water-borne route. It was to prevent such problems occurring that sewage treatment systems were designed.

Studies of the natural purification processes identified chemoorganotrophic micro-organisms as being the major active components responsible for the destruction of the organic matter and also showed that both aerobic and anaerobic processes were operating. Modern sewage treatment systems still rely on the same types of micro-organisms to carry out purification as those at work in the natural environment. The difference is that they are contained within installations designed to optimize their activities and thus speed up the rate of treatment. To this end, aerobic and anaerobic processes are often separated into individual reactors, e.g. anaerobic sludge digestors and aerated, activated sludge tanks or trickling filters.

In these so-called conventional sewage treatment systems aerobic conditions are maintained by forced aeration, and thus require plant and energy. In contrast, in lakes and ponds the photosynthetic activity of the phytoplankton population provides much of the oxygen to the water column. The full potential of microalgae as oxygen generators for use in sewage treatment was, however, only recognized with the development of waste stabilization ponds for treating waste waters in the USA during the 1940s and 1950s (Gray 1940; Caldwell 1946; Oswald and Gotaas 1955; Gloyna and Herman 1956). It should be mentioned in passing that the moats around medieval castles could be considered to be the forerunners of modern waste stabilization ponds linked to aquaculture re-use, since they received the wastes from the castle inhabitants and supported a carp population which fed on the algae growing profusely in the water.

Involvement of microalgae in the aerobic component of the sewage treatment process is frequently explained, somewhat simplistically, in terms of a mutual association between the algae and bacteria. The algae release photosynthetically produced oxygen into the water where it is utilized by the heterotrophic bacteria during the aerobic degradation of the organic matter in the sewage. This produces carbon dioxide and releases inorganic nutrients which in turn are utilized by the algae to grow and photosynthesize (Fig. 12.1). The assimilation of these nutrients by the algae, particularly ammonia and phosphate, which is taken up to a luxury extent (Kuhl 1974), improve effluent quality and reduce the risk of these nutrients triggering eutrophication in the receiving water body.

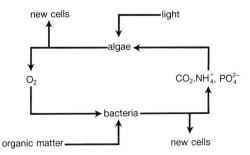

Fig. 12.1. A schematic representation of the mutual association existing between the microalgae and organotropic bacteria in sewage treatment ponds (from Mara 1976).

Efficient operation of algal treatment systems depends on the establishment of an equilibrium between the overall oxygen production by the algae and its consumption by the bacteria, although it is obvious that the absolute concentration of dissolved oxygen at any time of the day will vary and be lowest at night. To this end it is generally considered that an optimum biomass ratio for bacteria to algae to be in the range 1–2: 3.

One of the major attractions of using algal sewage treatment systems which is particularly pertinent to the poorer developing countries of the world is their low-cost construction, maintenance, and operation. They require little or no plant since, for example, the aeration system is built-in and most if not all of their energy requirements come from the sun. Algal sewage treatment systems also present a range of potential re-use options which include the production of high protein food and safe water for irrigation. These are important advantages in their own right, but they also help to defray the costs of sewage treatment.

2. Types of algal systems and waste stabilization ponds

2.1. Introduction

Waste stabilization ponds are also referred to as sewage lagoons and algal ponds or hybrids of these terms. They warrant detailed discussion here since an understanding of the way they are designed and how they perform is important to our understanding of all algal sewage treatment systems.

Waste stabilization ponds are a low cost, low technology, but efficient method of sewage treatment, and are particularly suited to hot or warm temperate climates where the ambient temperatures support vigorous microbial activity. Even so, ponds are used successfully in Alaska, where they become covered with a layer of ice in winter, and in New Zealand, where the City of Auckland has some 5 km^2 of ponds. In fact, waste stabilization ponds are used anywhere where there is sufficient land available.

Ponds are used in Africa, Asia, Central and South America, and in parts of the Middle East. They are widely used in the USA and increasingly so in

Fig. 12.2. A secondary facultative waste stabilization pond at San Juan near Lima, Peru, receiving effluent from an anaerobic pond and emptying into the first maturation pond (in the background) of a series of two. Note the simple inlet structure and earthen embankments which have been stabilized by the establishment of natural vegetation. This vegetation needs cutting on occasions to prevent mosquito breeding in the shaded zone produced at the pond edge.

Europe, notably in Germany, France, and Portugal. Waste stabilization ponds are essentially shallow impoundments with earth embankments (or occasionally retaining walls), into which waste water continuously flows and from which treated effluent is discharged. They are thus very simple to construct in engineering terms although some attention to the design and siting of the inlet and outlet structures can greatly improve pond performance, particularly in small ponds, by minimizing hydraulic short circuiting and the loss of any surface scum or floating solids into the effluent (Fig. 12.2).

Compared to many other sewage treatment processes, pond systems have long hydraulic retention times (weeks rather than hours) and hence a large land area requirement. However, this does not pose a serious problem in most developing countries where sufficient low-cost land is frequently available for the construction of ponds close to large towns and cities.

Even in Europe and the United States, where land costs are often relatively high, careful cost analysis may show that when the cost of construction is considered together with projected operation and maintenance costs ponds may be cheaper than conventional treatment systems incorporating trickling filters or activated sludge plants (Arthur 1983).

Ponds have several advantages over other forms of sewage treatment; they are simple to construct, maintain, and operate, and therefore do not require the use of skilled or highly trained personnel. Furthermore, they do not require any input of external energy (except solar energy). Their long hydraulic retention times also make them less susceptible to shock hydraulic and organic loads, or inhibition by toxic substances (they can tolerate up to 60 mg/l of heavy metals). Above all, however, they combine a high efficiency of organic matter removal (BOD) with the production of the best micro-biological quality effluent. The ability of waste stabilization ponds to destroy pathogenic bacteria, viruses, and the ova of intestinal parasites more efficiently than other forms of sewage treatment requires special emphasis since in tropical developing countries water-related diseases are a prime cause of death and illness. It is doubly important when the effluent is to be re-used in agriculture and aquaculture.

There are three main types of waste stabilization ponds in common use, these are facultative and maturation ponds, and anaerobic pretreatment ponds. Different combinations and numbers of these pond types can be put together to form a series capable of treating any biodegradable waste water to any required effluent quality standards.

2.2. Facultative ponds

A facultative pond followed by one or more maturation ponds in series is the most common format used in the design of existing waste stabilization pond systems. There are two types of facultative ponds, primary facultative ponds receiving raw sewage and secondary facultative ponds which receive settled or pretreated sewage, usually from anaerobic ponds. This difference influences the design equations used for the two types of facultative ponds (see later).

The main role of facultative ponds is to remove BOD and removal efficiencies in the range of 60–80 per cent are usual. Some die-off of pathogens is also achieved (e.g. a log reduction in faecal coliforms, the faecal indicator bacteria), but pathogen removal occurs predominantly in the subsequent series of maturation ponds. Facultative ponds are usually between 1 and 2 m deep. During daylight they are a biphasic treatment process comprising an oxygenated surface layer some 30–50 cm deep which is sustained by the photosynthetic activity of a dense bloom of microalgae (phytoplankton) which stratify within the surface layer and a lower anerobic zone, which in primary facultative ponds can be sub-divided into an aqueous layer and a bottom sludge (solids) layer. Rapid oxidation of soluble organics is brought about by facultative aerobic bacteria in the surface layer using the oxygen generated by the microalgae and with which they form the mutualistic association described previously. The aerobic zone is replenished with soluble organics present in the influent waste water, and also from the activities of predominantly fermentative organotrophic bacteria utilizing colloidal and settleable solids in the lower anaerobic zones, in which methanogenesis also occurs.

Photosynthetic activity by the microalgae varies with the incident light intensity, and this causes diurnal changes in the depth of the aerobic zone and in the concentration of oxygen to be found at a particular depth within the zone. Concentrations of oxygen in excess of 20 mg/l are not uncommon in the surface layers during periods of rapid photosynthesis.

In contrast, at night the water column frequently turns anaerobic except in a narrow layer (a few millimetres deep) just below the water–air interface. This is because in the absence of photosynthetically-produced oxygen the respiratory oxygen demand exerted by the microbial population (including the microalgae) and zooplankton present exceeds the rate of oxygen transfer from the air to the pond water column. In response to this situation zooplankton can frequently be seem forming a thin dense surface film during the hours of darkness.

Despite the relatively efficient buffering capacity of the influent sewage the photosynthetic activity of the microalgae causes diurnal and depth variations in pH. Values around pH 9 are frequently recorded in the surface, algal-rich layers of facultative ponds during periods of rapid photosynthesis when dissolved carbon dioxide is being removed from the water by the algae more rapidly that it can be replenished from the atmosphere, or through the combined respiratory activity of the microbes and pond fauna. At night the pH drops to between 7 and 7.5. It is the need for at least part of the microbial pond population to accommodate to these physico-chemical fluctuations in their environment which has given the term 'facultative' to the naming of these ponds.

2.3. Maturation ponds

These ponds receive the effluent from facultative ponds, and the size and number of maturation ponds govern the quality of the final effluent (both chemical and microbiological) of a pond series. Their surface BOD loading is much lower than that applied to facultative ponds. They are designed principally to removed pathogens rather than BOD (although some further destruction of BOD does occur) and their function therefore is quite different from that of facultative ponds.

Maturation ponds are usually designed to be 1–1.5 m deep. They remain aerobic throughout their depth except possibly for a short time before dawn when the bottom may turn anaerobic. The more extensive aerobic conditions prevailing in maturation ponds when compared to facultative ponds can be linked to the reduced organic loading they receive and thus reduced turbidity. This allows greater penetration of light through the water column, which in turn supports net algal photosynthesis to greater depths and affords better distribution of the photosynthetically produced oxygen (Fig. 12.3). The total oxygen demand is also lower in maturation ponds since the reduced organic loading and nutrient status supports smaller heterotrophic microbial and algal populations. The latter are also more vulnerable to predation by zooplankton which are better able to survive in larger numbers in the prevailing aerobic

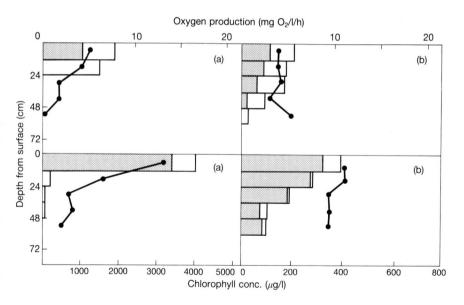

Fig. 12.3. Typical data showing variations in gross and net primary production (full and solid histograms, respectively) with depth and chlorophyll concentration (●) in a typical primary facultative pond (a) and a maturation pond (b) in N.E. Brazil. The water temperature was 26°C.

conditions. The different photosynthetic regime in maturation ponds together with the less efficient buffering capacity of the now partially treated waste water gives rise to extended periods of high pH throughout most of the water column during the day. Values of pH 9.5 and higher are common and even at night the pH stays around 8.5.

The number of maturation ponds to be used in a particular system depends on the desired microbiological quality of the effluent i.e. the number of faecal coliforms per 100 ml. Marais (1974) showed that each maturation pond of 5–7 days retention time was able to reduce faecal coliform numbers by 90–95 per cent. However, this should be seen within the context of a need for overall reductions in faecal coliforms of at least 99.999 per cent (5 log reductions) to ensure that the final effluent is pathogenically safe.

The mechanisms of pathogen removal in waste stabilization ponds have received considerable attention, but nevertheless in many instances they can only be considered tentative. In the case of helminth eggs and protozoan cysts sedimentation during the relatively long retention periods in the ponds is the key removal mechanism (Mara 1976). Although this provides an effluent low in these pathogens their numbers inevitably build up in the sludge layers where they remain viable for several months or even years, but since ponds require only very occasional desludging, if ever, this does not constitute a problem. Feachem et al. (1983) have suggested a minimum of two ponds in series with a total retention time of 20 days to ensure an effluent free of protozoan cysts and helminth eggs.

The die-off of bacterial pathogens is assumed to follow the die-off pattern of the indicator organisms, the faecal coliforms (and faecal streptococci). This is supported for example by the work of Yanez (1984) and Bartone et al. (1985) who have reported that estimates of faecal coliform die-off correlated well with salmonellae removal rates. The rate of die-off increases with increasing temperature and this forms the basis of design equations for maturation ponds (see later). However, since the rate of die-off is faster in maturation ponds than in facultative and anaerobic ponds it follows that the relationship between die-off and temperature is an indirect one, linked to the increased rates of microbial metabolism which result from an increase in ambient temperature.

High concentrations of dissolved oxygen resulting from algal photosynthesis (Marais 1974), the release of toxins by the algae (Toms et al. 1975), and increased ultraviolet light penetration through the water column (Moeller and Calkins 1980) have been suggested as direct causes of faecal coliform die-off in maturation ponds. Whilst these parameters may indeed have a synergistic effect, results of field and laboratory studies (to be published), on faecal coliform die-off rates in ponds, recently carried out jointly by my own group and Professor Mara's group at the University of Leeds, collaborate the findings of Parhad and Rao (1974). These showed that periods of high pH (> 9) resulting from algal photosynthesis under the conditions of reduced nutrient availability existing in maturation ponds, are mainly responsible for faecal-coliform die-off and thus presumably for that of bacterial pathogens as well.

Virus removal is also efficient in ponds, but there is little information available on the precise mechanisms involved. It is generally assumed that adsorption on to settleable solids is the principle method of removal from the effluent, but survival times in the bottom sludge may be prolonged (see Feachem *et al.* 1983).

2.4. Anaerobic ponds

These are not algal systems, but are mentioned here for completeness since they are frequently used as a pretreatment stage with their effluent feeding into a secondary facultative pond of a pond series. An anaerobic pond may contain a thin surface film of flagellate algae, usually *Chlamydomonas*, but their presence can be summed up as 'opportunistic' in the sense that they appear to play no obvious role in the treatment process, but just exploit the available ecological niche. In this respect they are not unlike the film of microalgae that forms on the surface layer of rocks in a trickling biological filter.

As their name denotes, anaerobic ponds are designed to operate under anoxic conditions and any small amount of oxygen produced right at the surface by the algal film is inconsequential. Although these flagellate algal species are frequently capable of organotrophic growth (Albeliovich and Weisman 1978) they represent such a small proportion of the microbial biomass that it is unlikely that they make a significant contribution to BOD removal either.

The use of anaerobic ponds can make a significant saving on overall land area requirements since they are deep (2–5 m) and have short retention times (2–3 days) and yet bring about a BOD_5 reduction of some 40–70 per cent or more (Mara *et al.* 1983). They are most advantageously used with strong wastes ($BOD_5 > 300$ mg/l). A single anaerobic pond is usually sufficient when treating raw domestic and municipal wastewaters, but with high stength industrial wastes ($1000 > BOD_5 < 30\,000$ mg/l) several anaerobic ponds in series may be necessary prior to the effluent entering a secondary facultative pond.

The BOD_5 volumetric loading ($100–400$ g/m^3/d) is so high that the ponds remain completely devoid of dissolved oxygen and the suspended solids settle to the bottom where they undergo vigorous anaerobic digestion releasing methane provided the ambient temperature is above 15°C. At temperatures below 15°C the anaerobic ponds just act as settling tanks and accumulate more sludge. In warm temperate and tropical climates anaerobic ponds only need desludging ever 5 years or so, i.e. when they become half filled with sludge. Current experience suggests that facultative and maturation ponds never need desludging, although removal of sludge accumulating near the inlet of primary facultative ponds may be necessary very occasionally.

Contrary to popular belief anaerobic ponds do not cause serious smells (due to hydrogen sulphide release), provided the volumetric loading is less than 400 g/m^3 and the concentration of SO_4^{2-} (the substrate for the anaerobic sulphate reducers) is below 500 mg/l (Gloyna 1971). The misconception that anaerobic ponds will, of necessity, produce foul odours is rooted in the misconception that since overloaded facultative ponds smell then anaerobic ponds (incorrectly considered to be grossly overloaded facultative ponds) must also smell. This argument does not take account of the differences in the microbiological activity of the two systems the details of which are, however, beyond the remit of this chapter.

3. Pond design and design equations

3.1. Anaerobic ponds

It is usual to design anaerobic ponds on the basis of permissible volumetric loading (λ_v) expressed in g BOD$_5$/m^3/d. The design value of λ_v is a function of pond temperature, but currently insufficient field data exist to permit a mathematical relationship to be developed. Table 12.1 lists suitable

Table 12.1

Design loadings for, and BOD$_5$ removals in, anaerobic ponds

Mean monthly temperature (°C)	Volumetric BOD$_5$ loading (g/m³/day)	BOD$_5$ removal (%)
< 10	100	40
10–20	$20T - 100$*	$2T + 20$*
> 20	300	60

* T = temperature, °C.

design values for λ_v at various temperatures, the design temperature being the mean temperature of the coldest month. Ideally, this would be the mid-depth water temperature, but usually has to be based on air-temperatures. However, since the water temperature in the pond is usually 2–4°C above the amblent air temperature this adds a degree of conservatism to the design. The organic loading rate (λ_v) is related to the BOD$_5$ of the raw waste water

(L_i, mg/l), the waste water flow (Q, m^3/day) and the anaerobic pond volume (V_a, m^3) by the equation

$$\lambda_v = L_i Q / V_a,$$

thus allowing the pond volume to be calculated.

3.2. Facultative ponds

Primary facultative ponds are at present best designed according to the empirical method of McGarry and Pescod (1970) as modified by Arthur (1983). The design areal loading rate λd_s (kg/ha/day) is given by

$$\lambda d_s = 20T - 60 \tag{1}$$

where T is the mean ambient air temperature of the coldest month (it would be preferable to use water temperatures, but these are not usually available). The mid-depth pond area (A, m^2) is then calculated from the equation

$$A = 10L_i \, Q / \lambda d_s \tag{2}$$

where L_i is the influent BOD$_5$ concentration in mg/l and Q is the flow (m^3/day). Knowing the area the mean hydraulic retention time in the pond (t, days) is given by

$$t = AD/Q \tag{3}$$

where D = pond depth (m).

Secondary facultative ponds have no or only a minimal sludge layer at the bottom because all the settleable solids have been removed during pretreatment in anaerobic ponds. Since it has been calculated that the sludge layer is responsible for removing 30 per cent of the influent BOD$_5$ via methane production (Marais 1970) a correction to equation (1) is necessary. Thus, for secondary facultative ponds

$$\lambda d_s = 0.7 \, (20T - 60). \tag{4}$$

3.3. Maturation ponds

These are designed to remove pathogens and the equations are based on the studies of faecal coliform removal kinetics in ponds assuming first-order removal in a completely mixed reactor (Marais 1966, 1974). The recommended equation for a series of ponds is

$$N_e = N_i / (1 + K_b\theta_a) \, (1 + K_b\theta_f) \, (1 + K_b\theta_m)^n \tag{5}$$

where N_e and N_i are the faecal coliform numbers in 100 ml of the final effluent and raw waste water, respectively, K_b is the first-order constant for faecal coliform removal per day, θ is the mean hydraulic retention time in days, the subscripts a, f, and m refer to the anaerobic, facultative, and maturation ponds, respectively, and n is the number of maturation ponds. The value of N_i is either known from measurement or can be conservatively estimated as $1 \times 10^8/100$ ml and N_e is usually stipulated by the local regulatory agency. Marais (1974) gives the following equation for the very marked variation of K_b with temperature:

$$K_b = 2.6 \, (1.19)^{\,T-20} \tag{6}$$

where T is again the mean air temperature of the coldest month.

Thus, at the design stage there are two unknowns in eqn (5), θ_m and n, so a trial and error solution is required. The recommended minimum value of θ_m is 3 days to avoid hydraulic shortcircuiting (Marais 1974) and it is not usual for θ_m to be greater than θ_f. Thus, eqn (5) should be solved for $n = 1, 2, \ldots$ etc., and a suitable combination of θ_m and n chosen, bearing in mind that the areal loading on the first maturation pond should be less than that on the facultative pond. For this purpose an overall removal efficiency of 70 per cent in the anaerobic (if any) and facultative pond can be reasonably assumed.

4. Pond layout

Waste stabilization pond systems always comprise a series of ponds which include a facultative pond followed by several maturation ponds with or without an anaerobic pond or ponds ahead of the facultative pond. It is recommended that, in general, all systems should include anaerobic ponds because of the land area they save.

Once the land area required by the individual pond types has been determined and the number of maturation ponds calculated it is imperative, particularly with large installations, to design the pond series to minimize land wastage. It may therefore prove useful to further subdivide the area required, say for the facultative pond, into two or three ponds instead of having one large one. This also has the additional advantage of providing a further margin of safety and flexibility into the system in the event of a pond malfunctioning. In Fig. 12.4 a design is shown in which the facultative pond area is divided up into three ponds and instead of a single six-pond series of maturation ponds two series of smaller ponds were used. This enables all the ponds to fit neatly into a rectangle using the minimum of pipework. This unitized design has the advantage of allowing several units to be put together for use with larger communities or to accommodate expansion of the existing

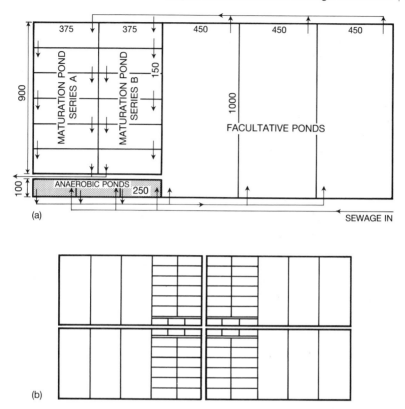

Fig. 12.4. (a) A diagrammatic representation of a Unit Waste Stabilization pond complex (for a population of approximately 250 000 people). The dimensions are in metres. (b) Four units combined. Note how the anaerobic ponds can be arranged so that they are in the centre of the complex.

community. It also minimizes the area of preliminary treated sewage open to direct access by the community, by surrounding the anaerobic ponds with the cleaner ponds.

5. Waste stabilization pond algae

The algal genera and the concentrations of algal biomass in waste stabilization ponds vary with pond type and organic loading, and are good, rapid indicators of pond performance and status.

Many algal genera have been identified in waste stabilization ponds, typical examples are presented in Table 12.2. Palmer (1969) recorded 21 genera in one sewage pond on a single day and a total of 83 genera during a 2-year study

Table 12.2
Examples of some typical algal genera found in waste stabilization ponds

Algae		Facultative ponds	Maturation ponds
Euglenophyta	*Euglena*	+	+
	Phacus	+	+
Chlorophyta	*Chlamydomonas*	+	+
	Chlorogonium	+	+
	Eudorina	+	+
	Pandorina	+	+
	Pyrobotrys	+	+
	Ankistrodesmus	−	+
	Chlorella	+	+
	Micractinium	−	+
	Scenedesmus	−	+
	Selenastrum	−	+
	Carteria	+	+
	Coelastrum	−	+
	Dictyosphaerium	−	+
	Oocystis	−	+
	Rhodomonas	−	+
	Volvox	+	−
Chrysophyta	*Navicula*	+	+
	Cyclotella	−	+
Cyanophyta	*Oscillatoria*	+	+
(Cyanobacteria)	*Arthrospira*	+	+

+ = present, − = absent.

of three ponds in Indiana, USA. The dominant genera are usually members of the Chlorophyta and Euglenophyta, and to a lesser extent the Chrysophyta and Cyanophyta (Cyanobateria). (Palmer 1969; Raschke 1970; Shillinglaw and Pieterse 1977, 1980).

Species diversity in ponds generally decreases as the organic loading increases and consequently fewer species are found in facultative ponds than in maturation ponds. Flagellate genera such as *Chlamydomonas*, *Euglena*, and *Pyrobotrys* tend to dominate in the more turbid conditions of facultative ponds where presumably their motility and thus their ability to move towards surface light gives them a competitive advantage over non-motile forms such as *Scenedesmus*, *Chlorella*, and *Micractinium* which are abundant in the more transparent waters of maturation ponds.

In highly loaded facultative ponds the algal population may become almost a monoculture of *Chlamydomonas* or *Euglena*, and under overloading conditions in which high sulphide concentrations develop, the ponds may change colour from green to purple as purple sulphur bacteria—members of the Chromatiaceae—predominate. These changes in facultative ponds are frequently apparent before a serious reduction in the quality of the final effluent occurs and allow time for adjustments to be made such as reducing the load. Conversely, a so-called facultative pond which is seen to have a broad diversity of algal species with numerous non-flagellate species present is clearly underloaded and could withstand increases in organic loading without unacceptable losses in effluent quality.

Thus, a knowledge of the algal genera and their relative abundance can provide rapid and easily obtainable preliminary information on the state of a stabilization pond which can be useful when trouble-shooting malfunctioning pond systems, particularly when access to a laboratory to perform detailed physico-chemical analyses is either not possible or a considerable distance from the site under study.

The total algal biomass in facultative ponds is usually greater than in maturation ponds, but concentrations also depend to some extent on the location. The algal standing crop in efficiently operating primary facultative ponds, for example those in North-East Brazil, frequently reach values of 2,500 μ g/l chlorophyll-*a* or even higher, while in maturation ponds (BOD$_5$ loading < 50 kg/ha/day) values from 100 to 1000 μg/l chlorophyll-*a* were recorded, the lower values being obtained at the lowest organic loads (Mara *et al*. 1983; Konig 1984).

The concentrations of algal biomass fluctuate with environmental changes associated with the seasons particularly in temperate climates, but other more frequent fluctuations have been observed which may result from zooplankton grazing, transient chemical poisons (Shillinglaw and Pieterse 1977), or attack by pathogenic microbes (Abeliovich and Dikbuck 1977; Dor and Svi 1980).

During daylight and in the absence of wind-induced vertical mixing the algal biomass in facultative ponds frequently stratifies into a narrow dense band some 20 cm thick. This algal band moves up and down through the water column (usually within a zone extending from the surface down to a depth of 40–50 cm), in response to changes in incident light intensity. This concentrated band of algae (which disperses at night) can cause large diurnal fluctuations in effluent quality (i.e. BOD, COD, and suspended solids) as it passes through the effluent removal zone (Mara *et al*. 1983). It can also lead to false estimates of the pond algal standing crop. These are best taken by a total pond water column sampler (Fig. 12.5) since in this way the entire water column is sampled and a representative estimate of the mean algal biomass can be obtained at any time of the day regardless of the distribution of the algae within the water column. Figure 12.6 shows the discrepancy that can occur in estimating algal biomass on the basis of effluent grab samples (these were taken at 0.800 hours when the algal band was usually in the effluent

Fig. 12.5. A total water column sampler made from plastic drainpipe tubing fitted with a simple flap end for taking samples from waste stabilization ponds.

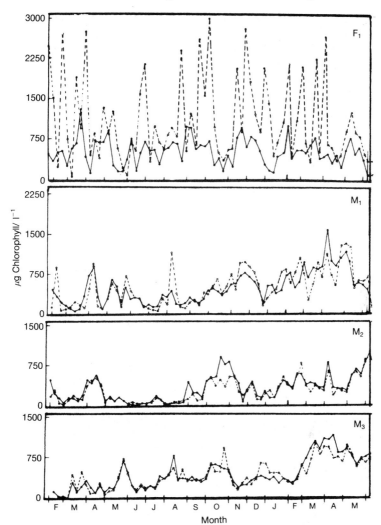

Fig. 12.6. Algal biomass concentration (chlorophyll-*a*) for in-pond samples (solid line) and 08.00 hours grab effluent samples in a series of ponds (F = facultative and M = maturation pond).

removal zone), compared to in-pond estimates with the column sampler. It also shows the sort of frequent, irregular, fluctuations that occur in the algal biomass concentration of a facultative pond.

In contrast, the algal population of maturation ponds tends to be more uniformly distributed with depth (see Fig. 12.3) and thus diurnal fluctuations in final effluent quality due to the algae is less pronounced. However, the algae present in the effluent do significantly affect the final effluent quality. For example, an algal concentration equivalent to 1 mg/l chlorophyll-*a* contributes a COD of 300 mg/l (Fig. 12.7) and algae may be responsible for more than 60 per cent of the total BOD$_5$ and the total suspended solids in the final effluent. In the USA this has sometimes made it necessary to remove the algae from the effluent prior to discharge to a water course to meet the stringent effluent standards in force. However, in many countries including those of the developing world, effluent standards are less stringent particularly in terms of BOD and suspended solids, and the presence of algae may actually be beneficial to the re-use strategy to which the effluent is to be put (see later).

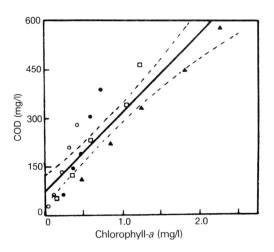

Fig. 12.7. Regression between chlorophyll-*a* and COD based on data from four pond algal types Chlorella (○), Pyrobotrys (●), Oscillatoria (□) and Euglena (▲). The dotted line represents the 95 per cent confidence limit. (For the combined algal data $n = 20$ and $r = 0.904$.)

Although, as has been mentioned previously, ponds are less susceptible to shock organic loads and toxic substances than conventional sewage treatment processes, the algae are probably the most sensitive component of the system, especially their photosynthetic apparatus. In this respect, the two substances most likely to cause problems to the algae of facultative ponds are ammonia and sulphide.

Ammonia toxicity increases with increased pH since it is the unionized form, ammonia (NH$_3$), which most readily passes across the algal cell membranes and which predominates at alkaline pH's. The data of Abeliovich and Azov (1976) suggests that ammonia concentrations above 2 mM are toxic to certain sewage pond algae at pH values above 8.0. However, recent studies in my own laboratory (manuscript in preparation) would suggest that 2 mM may be a very conservative estimate since several waste stabilization pond algae including species of *Chlorella*, were able to withstand 10 mM ammonia (140 mg/l NH_3-N at pH 9.0). It also seems likely that in pond systems with overall retention times of several weeks that the algal population may adapt to the increasing ammonia concentrations. Furthermore, in situations where high ammonia levels are present from the onset, development of the algal population during the initial stabilization period will favour ammonia tolerant strains which will thus predominate. Nevertheless, ammonia and pH require careful and regular monitoring in pond systems receiving wastes suspected to be high in ammonia.

Sulphide is similarly more toxic in the unionized H_2S form, but unlike ammonia this predominates as the pH decreases. In facultative ponds where the pH range is between 7 and 9 sulphide concentrations of the order of 200 mM (~ 8 mg/l) inhibit algal photosynthesis (see Howsley and Pearson 1979; Pearson and Howsley 1980).

The inhibitory effects on photosynthesis of both ammonia and sulphide are reversible at least over the first few hours and in the case of ammonia, reducing the pH to 7.0 to 7.5 may eleviate the inhibitory effect and allow the isolated pond to recover.

Sulphide increases in concentration in facultative ponds at night due to the activity of the sulphate reducing bacteria and the lack of dissolved oxygen to oxidize it away. This may cause algal photosynthesis to be significantly reduced in efficiency just after dawn. However, the anaerobic purple photosynthetic sulphur bacteria which are frequently a significant component of the microbial flora in highly loaded facultative ponds are able to oxidize the sulphide initially to elemental sulphur and so reduce the sulphide concentration to a level where oxygenic algal photosynthesis can be switched on. Therefore, in some ponds there exists an interesting interaction between the microalgae and the anoxygenic photosynthetic bacteria. With the onset of algal photosynthesis and the development of aerobic conditions in the surface waters the motile obligate anaerobic purple photosynthetic bacteria must move down in the pond away from the oxygen. However, such is the pigment complementation of their photosynthetic apparatus that they can use long wavelength light (i.e. in excess of 700 nm) not used by the microalgae to photosynthesize and so can exist below the microalgae in the anaerobic zone of the water column and act as a 'sulphide filter'.

5.1. Polishing ponds

Polishing ponds are equivalent in operation to the maturation ponds described earlier, but this term is frequently used to describe ponds used to produce tertiary quality effluent from the secondary effluents of activated sludge units and trickling filters (Toms *et al.* 1975). Their retention times are longer than those of humus tanks used at many sewage works and they develop a similar algal population to that found in maturation ponds. The use of polishing ponds is frequently essential in several states of North America where final effluent standards are very stringent in terms of physico-chemical and microbiological quality. Even so in some instances the concentration of algae in the final effluent may still make too high a contribution to the BOD and SS levels, and make it necessary to remove them. In general, it is better to use a series of small polishing ponds rather than one large one as this reduces the algal biomass (phytoplankton) in the final pond in the series, and thus the final effluent by encouraging the predominance of larger filamentous algae such as *Cladophora* which remain within the pond.

5.2. Night soil ponds

Night soil ponds can be considered to be primary facultative ponds which instead of receiving a continuous flow of raw waste water receive batch loads of night soil (faeces and urine) transported to them in vehicles ranging from vacuum tankers to hand carts (Shaw 1962). They are often used as treatment systems in areas without sewerage, but where a municipal night soil collection service is provided such as in many countries in South-East Asia.

Two loading strategies seem to operate. A single pond can be fully loaded with night soil and then left to purify for several weeks without further additions, before being emptied and refilled once more. Alternatively, several ponds in parallel can be kept in use at the same time so preventing the overloading of individual units. In either case, just enough additional water is added to maintain the water level which would otherwise decrease as a result of evaporation or seepage.

In night soil ponds which are being continuously loaded, high concentrations of inorganic salts may build up since night soil is inherently more concentrated than sewage, and this might ultimately inhibit algal growth and thus impair treatment quality. However, based on the limited information available, this takes several years to occur unless ammonia concentrations are high and is easily remedied by draining the pond, desludging and refilling again. It is worth noting that since night soil ponds only receive excreta and urine they should not contain any of the domestic and industrial contamination generally associated with municipal waste waters. The effluent and algal biomass should therefore be suitable for a wide range of re-use strategies.

Night soil ponds are also very conservative of water an important fact in arid areas.

5.3. High rate algal ponds

High rate algal ponds (HRAP) also called high rate algal-bacterial ponds or accelerated algal-bacterial ponds, can be considered as highly modified waste stabilization ponds designed to maximize algal yields while still reducing waste water organics and destroying pathogens. The current state of the art stems from the pioneering work of Oswald and his co-workers in California (Oswald 1962, 1963, 1969; Oswald *et al.* 1957) and, more recently, from studies in South-East Asia (McGarry and Tongkasame 1971), South Africa (Buhr and Miller, 1983) and in particular Israel (Azov *et al.* 1980; Azov and Shelef 1982). Most of the information has come from small-scale pilot installations and even after some 20 years of research few full-scale systems are in operation. The operation of HRAP has been most recently reviewed by Fallowfield and Garrett (1985) who also discuss studies in Northern Ireland.

A high rate algal pond usually takes the form of a shallow channel 2–3 m wide with a water depth of between 20 and 60 cm, and arranged in a tight 'zig-zag' or 'race track' configuration (Fig. 12.8). In order to prevent the algae settling out, the pond is mixed by stirring, either continuously or at regular

Fig. 12.8. A HRAP system operating in Portugal. Stirring is achieved by pressurized entry of the influent through a line of nozzles on the T-shaped inlet (see arrow). The pond walls are made of corrugated asbestos cement sheet and the base of the system was sealed with asphalt. The algae in the effluent were not being harvested so effluent quality was poor.

intervals, by paddles located along the pond's length, this stirring is vital to the efficient operation of the pond. Detention times are between 2 and 6 days, and are therefore much shorter than those in conventional ponds, but they can still reduce the influent BOD_5 by between 93–96 per cent (Azov and Shelef 1982). The shallow depths of HRAP and their short retention times make them more sensitive to changes in environmental conditions and shock loads than waste stabilization ponds. Their short retention time may appear to offer a significant reduction in land area requirements, but this is offset by their shallow depth and relatively low removal of excreted pathogens, which in some instances may require the inclusion of maturation ponds in the system to ensure a satisfactory final effluent.

The influent waste water (domestic, agricultural, and industrial) is usually pretreated by primary sedimentation to remove settleable solids. These solids can be digested anaerobically and the biogas so produced used as an energy source for sterilization and drying of the final algal product.

HRAP can be heavily loaded with sewage up to 350 kg BOD/ha/day in the tropics and subtropics, and still, so it is claimed, produce an effluent with less than 20 mg/l of filtered BOD_5. Algal yields can exceed 100 000 kg dry weight/ha/year (Shelef *et al.* 1980). Since HRAP are designed to maximize algal biomass production, it follows that efficient harvesting of the algae is crucial to effluent quality (in terms of BOD_5 and suspended solids) and also to the economic viability of the system. Algal harvesting techniques are mentioned later in this chapter.

Algal species common in HRAP include *Euglena gracilis*, *Scenedesmus dimorphus*, *Chlorella vulgaris*, *Micractnium pussillum* and *Ankistrodesmus falcatus*. *Euglena* and *Scenedesmus* species appear to predominate over *Chlorella* and *Micractinium* below 15°C because of their increased tolerance to lower temperatures (Azov *et al.* 1980). The biology and operation of HRAP are more complex than those of ordinary waste stabilization ponds, but models to explain the theory of their operation based on light and temperature as key controlling factors have been developed (Oron and Shelef 1980; Azov and Shelef 1982; Buhr and Miller 1983). Studies by Abeliovich and Welsman (1978) on the nutrition of *Scenedesmus obliquus* in HRAP suggest that the microalgae exhibit heterotrophic metabolism and play a primary role in reducing the BOD of the waste water by consuming more organic material than the relatively small bacterial population.

In warm temperate climates reduced treatment efficiency during winter months has been countered by increasing the retention time. This has been achieved by maintaining a constant flow (and area), but increasing the depth as ambient temperatures fall; for example, in Israel Azov and Shelef (1982) increased the depth from 0.3 m (retention time 2.5 days) in July to a maximum of 0.60 m (retention time 5 days) in January. Such a strategy can also reduce dilution and washout of the algae from ponds during periods of heavy rainfall frequently experienced in the tropics.

Reduced waste water treatment efficiencies and reduced algal yields have also been associated with predation by zooplankton such as *Daphnia* and *Moinia* and as a consequence of fungal infections. Correction of such problems is possible, but requires a very high degree of microbiological competence. Careful manipulation of HRAP performance may also be necessary in an attempt to control algal speciation in instances where the re-use strategy requires a particular type of algal product; however, this sort of control is extremely difficult. Even when HRAP are being skilfully operated the microbial quality of the effluent is not usually as good or as reliable as that from ordinary pond systems incorporating maturation ponds.

In conclusion it should be emphasized that HRAP are highly-tuned and sensitive biological reactors requiring careful control and maintenance by highly skilled personnel. They are much more complicated to operate than activated sludge systems, and they cannot be considered as a simple alternative to waste stabilization pond systems. Their use in sewage treatment should only be contemplated when the necessarily highly-trained and experienced technical staff are routinely available, and when the algal product whose removal is fundamental to effluent quality can be economically utilized.

5.4. Fish ponds

Fish culture in ponds fertilized by the addition of night soil or raw sewage has been practised in Eastern Europe and Asia (particularly China) for centuries. The processes operating are comparable to those in lightly loaded night soil or facultative ponds with the bacteria mineralizing the organics, and stimulating the growth of the algae on which the fish feed just as in the castle moat. Care must be taken to ensure that the fish are not killed at night by the entire pond water column turning anoxic. Ammonia is another problem since fish are much more sensitive to it than algae (1.0 mg/l NH_3-N is frequently lethal to fish in the pH range of ponds). This sort of integrated single pond treatment system should not be encouraged, however, since fish cultured in this way in water of poor microbiological quality will present real health risks to the community handling and eating them (Feachem *et al.* 1983).

It is far better to integrate herbivorous fish culture into the sewage treatment process by growing them in final maturation ponds or in separate fish ponds receiving maturation pond effluent. In this way the fish ponds constitute a sort of quaternary treatment system. Species of the tilapia and carp families grow well in such systems where problems from ammonia and anoxia are diminished, and by feeding on the microalgae the fish and zooplankton greatly improve the final effluent quality.

6. The utilization and disposal of algal-rich effluents

6.1. Algal removal and harvesting

There are two reasons for removing algae from pond effluents: to improve effluent quality in terms of BOD and SS; and because the algae are required as an exploitable resource. These two reasons are not necessarily compatible when it comes to the choice of removal technology to be applied. The research and development into HRAP systems has included numerous studies into suitable and efficient algal harvesting techniques since both the economic success and the sewage treatment efficiency of these systems depend on them.

Algal removal techniques include:

(i) coagulation and chemical flocculation of the algae using substances such as alum and lime;

(ii) filtration through various sand filters, horizontal rock filters, micro-strainers (fine mesh screens), or continuous-flow centrifuges;

(iii) dissolved-air flotation in which the previously chemically coagulated algae are floated to the surface using pressurized air;

(iv) autoflocculation, a term used to describe the settlement of microalgae under still ambient conditions without the addition of chemicals (this natural settlement process is considered to be a stress response to high light intensity, high pH, or low nutrient or dissolved oxygen levels);

(v) operational techniques such as the use of variable level discharge outlets to avoid the algal layers in the water column when taking off the effluent;

(vi) biological removal strategies like the use of final treatment macrophyte ponds where rooted or floating water plants are used to 'shade-out' the phytoplankton in the water column or the use of controlled predation by herbivorous fish or zooplankton;

(vii) land application where, for example, the effluent is passed over gently sloping grass plots to remove suspended solids including algae and incidently bacteria (improving the pathogen quality).

Techniques 1–4 are suited to algal harvesting prior to further processing of the algal product, while techniques 5–7 are only suited to algal removal for the sake of improving final effluent quality. Details and discussions of the various techniques outlined here have been reviewed in the literature Benemann *et al.* 1980; Moraine *et al.* 1980; Middlebrooks *et al.* 1982; Ellis 1983; Fallowfield and Garrett 1985).

6.2. Re-use

Although waste stabilization ponds and their variants are commonly the cheapest form of waste water treatment option in warm or tropical climates when inexpensive land is available they nevertheless represent a real cost in terms of pond construction and maintenance. It is therefore essential, particularly in the poorer developing countries, to link their use to appropriate re-use strategies such as crop irrigation, fish and algae biomass production, and energy production to maximize the general benefits to the community and to defray pond maintenance costs.

In poor rural communities waste water treatment in ponds integrate with fish ponds or irrigation of agricultural land seem the most appropriate re use options at present.

The use of treated waste water in agriculture is of prime importance in arid and semi-arid regions of the world. In its simplest form sewage effluent represents a supply of valuable water for irrigation, but the mineral nutrients and organic content also make it a valuable fertilizer. Waste stabilization ponds are an ideal treatment system prior to irrigation because they can produce an effluent of excellent microbiological quality and are flexible in operation so that their effluent quality can be matched to irrigation needs (Table 12.3). The mineral salts in pond effluents which are immediately available to plants are also present in a well balanced combination to promote efficient crop growth. The algae in pond effluents have the potential to act as

Table 12.3

Tentative microbiological quality guidelines for treated waste water re-use in agricultural irrigation*

Re-use process	Intestinal nematodes† (geometric mean no. of viable eggs per litre)	Faecal coliforms (geometric mean no. per 100 ml)
Restricted irrigation‡ irrigation of trees, industrial crops, fodder crops, fruit trees§ and pasture¶	1	Not applicable
Unrestricted irrigation irrigation of edible crops sports fields and public parts**	1	1000‡‡

* From IRCWD (1985). In specific cases, local epidemiological, sociocultural, and hydrogeological factors should be taken into account, and these guidelines modified accordingly.

† Ascaris, Trichuris and hookworms.

‡ A minimum degree of treatment equivalent to at least a 1-day anaerobic pond followed by a 5-day facultative pond or its equivalent is required in all cases.

§ Irrigation should cease 2 weeks before fruit is picked, and no fruit should be picked off the ground.

¶ Irrigation should cease 2 weeks before animals are allowed to graze.

** Local epidemiological factors may require a more stringent standard for public lawns, especially hotel lawns in tourist areas.

‡‡ When edible crops are always consumed well-cooked, this recommendation may be less stringent.

'slow release fertilizers' releasing N, P, and K (and important trace elements) as they gradually decompose in the soil. Therefore, algal-rich effluents for agricultural use should be actively encouraged, as decomposing algal cells also improve the water holding capacity of the soil and its humus content.

Where spray and trickle irrigation systems are used in preference to furrow and flood techniques it has been suggested that the sewage algae will clog the equipment. There is in fact little evidence to support this as it is filmentous algae growing on the spray and drip nozzles, and which are not derived from the effluent which cause the trouble and thus constitute quite a different problem. Where filters are to be used they should be of such a pore size as to hold back coarse and medium-sized solids in the effluent, but let the microalgae pass through.

There can be problems with long-term irrigation with waste waters if things like the conductivity, nitrogen content, salinity, and sodium adsorption ratio (SAR), and the presence of industrial toxicants are not considered (Moore 1981). In this respect algal pond effluents are no exception. However, all these problems can be relatively easily solved by applying the right irrigation regime and taking suitable precautions. The benefits of waste water irrigation usually far outweigh the disadvantages (Pettygrove and Asano 1985).

The processed algal product from HRAP systems which is usually in pellet form has been satisfactorily used as a feed supplement for chickens and pigs and as a complete fish food (Lipstein and Hurwitz 1980; Walz and Brune 1980). It should be mentioned, however, that this is frequently not a pure algal product, but comprises several algae plus bacteria and protozoa.

Generally speaking, the HRAP product has a low customer acceptability as a human feed and also requires careful toxicological evaluation. Since the product is also very variable such toxiciological evaluations would need to take the form of a continuous screening process.

7. Current and future developments

Waste stabilization pond systems have been operating successfully for over 40 years, but there is still a need to improve their design. This will only come with an increased understanding of the complex microbial interactions and processes in these algal-bacterial systems. In particular there is a need to reduce the land area requirement. Studies are already underway on the use of deep (> 2 m) facultative and maturation ponds in Spain and Brazil, but as yet designs and details of efficiencies are tentative.

Other sorts of ideas that deserve investigation include, for example, ultra-shallow final maturation ponds or channels with short retention times designed to rapidly elevate the pH in the presence of high daytime light intensities and so increase the rate of bacterial die-off. The phytoplankton could also be replaced with attached (epilithic) film-forming algae so giving the benefits of algal activity without their contribution to the BOD and SS in the effluent. This approach would necessitate increasing the surface area available for attachment in the light zone.

The idea of using natural film-forming algae or artificially stabilized layers of algal cells (in polyacrylamide or carragheen, itself an algal product), may have applications for treating special industrial wastes. The potential use of selected ammonia-tolerant *Chlorella* strains to treat nitrogen fertilizer wastes has already been reported (Matusiak 1976) and the selection for natural tolerance or the genetic manipulation of algal strains to treat specific toxic wastes must be a realistic proposition. The use of algae as biofilters to remove and concentrate substances such as heavy metals for subsequent recovery is another possibility (Mouchet 1986). The growing of algae on waste waters has not only provided us with an efficient sewage treatment process, but also a means of off-setting the costs of algal mass culture. Refinements to systems like HRAP to improve quality control of the algal product holds exciting prospects for the future production of high value biochemicals from algae such as vitamins, amino acids, lipids and possibly, algae-produced antibiotics (see Fallowfield and Garrett 1985).

8. Conclusions

In conclusion, it must be re-emphasized that algal sewage treatment systems have an important role to play in the development of the Third World by providing above all else an inexpensive, simple to operate sewage treatment system which produces a much better effluent in terms of its microbiological (pathogen) quality than any conventional sewage treatment system. Luckily, pond systems are at their most efficient in hot climates where most of the world's poorer countries are found. The flexibility of pond systems and their vital role in the re-use of treated waste waters to conserve precious water resources and increase food production cannot be overestimated. The Developing World is aware of its problems, but every effort must be made tö ensure that those problems are solved by the most appropriate solution. Waste stabilization ponds are the solution to sewage treatment in most instances in these countries.

References

Abeliovich, A. and Azov, Y. (1976). Toxicity of ammonia to algae in sewage oxidation ponds. *Appl. Environ. Microbiol.* 31, 801–6.
—— and Dikbuck, S. (1977). Factors affecting infection of *Scenedesmus obliquus* by a *Chytridium* sp. in sewage oxidation ponds. *Appl. Environ. Microbiol*, 34, 832–6.
—— and Weisman, D. (1978). Role of heterotrophic nutrition in growth of the alga.

Scenedesmus obliquus in high rate oxidation ponds *Appl. Environ. Microbiol.* 35, 32–7.

Arthur, J. P. (1983). Notes on the design and operation of waste stabilization ponds in warm climates of developing countries. Technical Paper No. 7. World Bank, Washington.

Azov, Y. and Shelef, G. (1982). Oxidation of high-rate oxidation ponds: theory and experiments. *Water Res.* 16, 1153–60.

——, ——, Moraine, R., and Levi, A. (1980). Controlling algal genera in high rate waste water oxidation ponds. In *Algae biomass* (eds G. Shelef and C. J. Soeder) pp. 245–53. Elsevier/North-Holland Biomedical Press, Netherlands.

Bartone, C. R., Esparza, M. L., Mayo, C., Rojas, O., and Vitko, K. (1985) *Monitoring and maintenance of treated water quality in the San Juan lagoons supporting aquaculture.* Final Report Phases I & II, CEPIS, Lima, Peru.

Benemann, J., Koopman, B., Weissman, J., Eisenberg, D. and Goebel, R. (1980). Development of microalgae harvesting and high-rate pond technologies in California. In *Algae biomass* (eds G. Shelef and C. J. Soeder) pp. 457–95. Elsevier/North-Holland Biomedical Press, Netherlands.

Buhr, H. O. and Miller, S. B. (1983). A dynamic model of the high-rate algal-bacterial waste water treatment pond. *Water Res.* 17, 29–37.

Caldwell, D. H. (1946). Sewage oxidation ponds—performance, operation and design. *Sew. Works J.* 18, 433–58.

Dor, I. and Svi, B. (1980). Effect of heterotrophic bacteria on the green algae growing in waste water. In *Algae Biomass* (eds G. Shelef and C. J. Soeder) pp. 421–30. Elsevier/North-Holland Biomedical Press, Netherlands.

Ellis, K. V. (1983). Stabilization ponds: design and operation. *Crit. Rev. Environ. Control* 13, 60–102.

Fallowfield, H. J, and Garrett, M. K. (1985). The treatment of wastes by algal culture. *J. Appl. Bacteriol.* (Symposium Supplement) 59, 187S–205S.

Feachem, R. G., Bradley, D. J., Garelick, H., and Mara, D. D. (1983). *Sanitation and disease: Health aspects of excreta and wastewater management.* Wiley, Chichester, England.

Gloyna, E. F. (1971) *Waste stabilization ponds.* World Health Organization, Geneva.

—— and Herman, E. R. (1956). Some design considerations for oxidation ponds *J. Sanit. Eng. Div. Am. Soc. Civ. Eng.* 88, 1047–1059.

Gray, H. F. (1940). Sewerage in ancient and medieval times. *Sew. Works. J.* 12, 939–46.

IRCWD News 1985. Health aspects of wastewater and excreta use in agriculture and aquaculture: *The Engelberg Report*, pp. 11–18.

Konig, A. (1984). 11–18. PhD Thesis, University of Liverpool.

Kuhl, A. (1974). Phosphorus. In *Algal physiology and biochemistry* (ed. W. D. P. Stewart) pp. 636–54. Blackwell Scientific Publications, Oxford.

Lipstein, B. and Hurwitz, S. (1980). The nutritional and economic value of algae for poultry. In *Algae biomass* (eds G. Shelef and C. J. Soeder) pp. 667–85. Elsevier/North-Holland Biomedical Press, Amsterdam.

Mara, D. D. (1976). *Sewage treatment in hot climates.* Wiley, Chichester, England.

——, Pearson, H. W., and Silva, S. A. (1983). Brazilian stabilisation pond research suggests low-cost urban applications. New factors in the design, operation and performance of waste stabilisation ponds. *World Water* 6, 20–4.

Marais, G. V. R. (1966). *Bull. WHO* 34, 737–63.

—— (1970). Dynamic behaviour of oxidation ponds. In *Proc. 2nd Intl. Symp. Waste Treat. Lagoons.* (ed. R. E. McKinney) pp. 15–46. University of Kansas, Laurence, USA.

—— (1974). Faecal bacterial kinetics in stabilization ponds. *J. Environ. Eng. Div. Am. Soc. Civ. Eng.* 100, 119–39.

Matusiak, K. (1976). Growth of Chlorella vulgaris in wastes. *Act. Microbiol. Polon.* 25, 8233–42.

McGarry, M. G. and Pescod, M. B. (1970). Stabilization pond design criteria for tropical Asia. (ed. R. E. McKinney) pp. 114–32. University of Kansas, Laurence.

—— and Tongasame, C. (1971). Water reclamation and harvesting. *J. Wat. Pollut. Cont. Fed.* 43m 824–35.

Middlebrooks, E. J., Middlebrooks, C. H., Reynolds, J. H., Watters, G. Z., Reed, S. C., and George, D. B. (1982). *Wastewater stabilization lagoon design, performance and upgrading.* Macmillan, New York.

Moeller, J. R. and Calkins, J. (1980). Bacterial agents in wastewater lagoons and lagoon design. *J. Wat. Pollut. Cont. Fed.* 52, 2442.

Moore, C. V. (1981). Economic evaluation of irrigation with saline water within the framework of a farm, Methodology and empirical findings: A case study of Imperial Valley, California. In *Salinity in irrigation and water resources* (ed. D. Yaron). Marcel Dekker, New York.

Moraine, R., Shelef, G., Sandbank, E., Bar-Moshe, Z., and Shvartzburd, L. (1980). Recovery of sewage-borne algae: fiocaviation, flotation and centrifugation. In *Algae biomass* (eds G. Shelef and C. J. Soeder) pp. 531–46. Elsevier/North-Holland Biomedical Press, Amsterdam.

Mouchet, P. (1986). Algal reactions to mineral and organic micropollutants, ecological consequences and possibilities for industrial scale application: a review. *Water Res.* 20, 399–412.

Oron, G. and Shelef, G. (1980). An optimization model for high rate algal ponds. In *Algae biomass* (eds G. Shelef and C. H. H. Soeder). pp. 497–504. Elsevier/North-Holland Biomedical Press, Amsterdam.

Oswald, W. H. J. (1962). Water reclamation, algal production, and methane fermentation in waste ponds. *Adv. Water Pollution Res.* 2, 119–27.

——, (1963). Fundamental factors in waste stabilization pond design. In *Advances in biological waste treatment*, (eds W. W. Eckenfelder and B. J. McCabe) pp. 357–72. Pergamon, Oxford.

—— (1969). Current status of microalgae from wastes. *Chem. Eng. Prog. Symp. Ser.* 65, 87–92.

—— and Gotaas, H. B. (1955). Photosynthesis in sewage treatment. *Proc. Am. Soc. Civ. Eng.* 81, 1–34.

——, ——, Golueke, C. G., and Kellen, W. R. (1957). Algae in waste treatment. *Sewage Ind. Wastes* 29, 437–57.

Palmer, C. M. (1969). A composite rating of algae tolerating organic pollution. *J. Phycol.* 5, 78–82.

Parhad, N. M. and Rao, N. U. (1974). Effect of pH on survival of *E. coli. J. Wat. Pollut. Cont. Fed.* 46, 980–6.

Pearson, H. W. and Howlsey, R. (1980). Concomitant photoautrophic growth and nitrogenase activity by cyanobacterium *Plectonema torucinum* in continuous culture. *Nature* 288, 263–5.

Pettygrove, G. S. and Asano, T. (1985). *Irrigation with reclaimed municipal wastewaters.* Lewis Publishers Inc., Michigan.

Raschke, R. L. (1970). Algal periodicity and waste reclamation in a stabilization pond ecosystem. *J. Wat. Pollut. Cont. Fed.* 42, 598–630.

Shaw, V. A. (1962). A system for the treatment of night-soil and conserving tank effluent in stabilization ponds. CSIR Reprint RW, 166.

Shelef, G., Azov, Y., Moraine, R., and Oron, G. (1980). Algal mass production as an integral part of a wastewater treatment and reclamation system. In *Algae biomass* (eds. G. Shelef and C. J. Soeder) pp. 163–89. Elsevier/North-Holland Biomedical Press, Amsterdam.

Shillinglaw, S. N. and Pieterse, A. J. H. (1977). Observations on algal populations in an experimental maturation pond system. *Water SA* 3, 183–92.

—— and —— (1980). Algal concentration and species composition in experimental maturation ponds with effects of aeration and recirculation. *Water SA* 6, 186–95.

Toms, J. P., Owen, M., Hall, J. A., and Mindenhall, M. J. (1975). Observations on the performances of polishing lagoons at a large regional works. *J. Wat. Pollut. Cont. Fed.* 74, 383–401.

Walz, O. P. and Brune, H. (1980). Studies on some nutritive effects of the green algae *Scenedesmus actus* with pigs and broilers. In *Algae biomass* (eds. Shelef and C. J. Soeder. pp. 733–44. Elsevier/North-Holland Biomedical Press, Amsterdam.

Yanez, F. (1984) *Reduccion de organismo patogenos y diseno de lagunas de estabilizacion en paises desarrollo.* XIX Congresso interamericano de ingenieria sanitaria e ambiental, Santiago, Chile.

13

Biodeterioration biotechnology

K. J. Seal and L. H. G. Morton

Introduction

Biodeterioration may be described as negative biotechnology in the pragmatic sense of its definition. Implicit in the study of biodeterioration
processes, however, is the positive aspect directed at controlling the problem.
Biodeterioration has been defined as 'the study of the deterioration of
materials of economic importance by organisms' (Hueck 1965). Eggins and
Oxley (1980) widened the definition to include structures and processes,
whilst an account of the meaning of biodeterioration and its relationship with
biodegradation have been described by Eggins (1983). Eggins reminds us that
the word 'deterioration' means 'to make worse' whilst 'degrade' means 'to
step down' or 'to break down'. Thus, biodeterioration processes will include
biodegradation activity, but will also include a number of other perhaps less
direct effects of organisms on materials. We often consider biodegradative
process as those which, by their action, result in a product of improved
quality. The treatment of agricultural wastes to yield biomass or the
detoxification of waste pesticides are examples which do not come into the
remit of this chapter.

Hueck (1965) has classified biodeterioration processes into three types
which reflect the wide meaning of the definition.

1. *Mechanical*: insect and rodent attack on non-nutrient materials such as
lead pipe and plastic cable.

2. *Chemical*: (a) assimilatory where the material is a food source;
(b) dissimilatory where waste products or secondary metabolites are able to
degrade the material to no vital benefit of the organism.

3. *Fouling and soiling*: where the organism causes a worsening of the
material, structure, or process by its mere presence or the secretion of toxic
metabolites. It may cause a stain, a blockage or foul the hull of a ship.

This classification helps us to decide upon strategies for the study and
control of biodeterioration problems, but does not necessarily reflect the
metabolic activities or ecological preferences of the organism. To a cellulose-
utilizing fungus there may be no distinction between a natural cellulose fibre
and a cotton fabric carefully processed by man. The distinction is drawn with
pathology which is concerned with living organisms. Although there may be
some overlap biodeterioration does not concern itself with disease, and this

affects the control measures which can be successfully used (see later section).

The factors which encourage biodeterioration are legion, and a combination of the inherent susceptibility of the material by virtue of its chemical composition, surface characteristics, or physical state. The external environment and the extent to which preventative measures are employed in an attempt to control the biodeterioration processes must also be considered. Environmental factors such as temperature, humidity, pH osmotic pressure, and redox potential are very important in initiating the problem and encouraging its subsequent development. These factors have been reviewed more extensively by Seal and Eggins (1981) and Onions *et al.* (1981). All of these factors will be of special concern to third world countries as their climates tend to be tropical, sub-tropical, or arid, where large changes may occur between day and night temperatures which can cycle the relative humidity between low and high levels, respectively. The monograph produced by the Society of Chemical Industry (1966) is also recommended for an overall review of the subject area.

The established industries in third world countries which encounter biodeterioration problems are those of food production, and the fashioning of natural materials into clothing, housing, tools, and ornaments. The intensification of production so that excess may be exported also brings with it new potential problems associated with storage and transport. The introduction of new industries will bring new problems of a greater magnitude than those of many developed countries because of the lack of experience in operating such industries in more adverse climates.

In the foregoing sections we have thus presented information on biodeterioration problems concerned not only with indigenous industries, but in addition have reviewed other industries, their materials and products, where problems are known to occur. We hope this will enable the relevant agencies to consider this aspect at the outset in the establishment of a new industry to minimize initiation problems which could seriously affect production and markets in the initial years of operation. Whilst we recognize that biodeterioration embraces a wide range of organisms (Biodeteriogens) from birds, rodents, and insects to bacteria, we have been necessarily selective in describing in the main the activities of the micro-organisms. We must conceed, however, that in many Third World countries, the macro-organisms will play a significant part. The recognition and control of this large group is often made easier by their size and a more widely spread understanding of their habits. For this aspect the reader is referred to Allsopp and Seal (1986).

The biodeterioration of materials

In dividing this heading into materials or orientated sections some licence has been exercised in the interpretation of a material so that individual products such as timber and paints can be discussed under their own headings and not in sections headed cellulose or natural oils, respectively. Some materials have been more extensively studied than others and this will be evident from the variation in coverage given to each section.

Foodstuffs

A full appreciation of the biodeterioration of foods cannot be covered in this short overview. We will confine ourselves to defining the problems, stating their extent and offering means for minimizing losses. A more extensive review has been written by Coursey (1983). The biodeterioration of foods is specifically referred to as 'post-harvest decay' and covers the period from harvesting until it reaches the consumer. Two categories of food are harvested—durable and perishable. Durable foods are generally the grains, having a low moisture content (10–15 per cent), low metabolic activity, and a hard texture which reduces damage. These products will keep well and are only adversely affected by external agents of decay. Perishable goods on the other hand have high moisture contents (50–90 per cent), retain high metabolic activity, and decay by a combination of endogenous metabolic factors and the invasion of external agents. Their keeping quality will thus depend on their inherent storage life even if they are not damaged. As it is impossible and indeed, not practical, to separate the influence of biodeteriogens from those inherent effects we normally talk of 'post-harvest loss' which covers both types of deterioration. It is thus important that perishables are stored under conditions which will either slow down metabolic activity or will maintain the produce in its mature state until it is consumed. Damage to the outer protective skin when harvesting or during sorting and packing will also shorten the storage life by allowing the invasion of an indigenous and opportunistic flora of fungi and bacteria which will accelerate the decay process. The control of the external environment by reducing the relative humidity and temperature, and the use of gas atmospheres such as carbon dioxide will retard microbial growth in both perishables and durables allowing long distant transportation or, in the case of grain, long term storage of 2–3 years.

The magnitude of post-harvest losses is very difficult to assess in perishable goods. Estimates for overall losses are put at 25–30 per cent in developing countries. However, even in developed countries losses can be high. In the USA for example, some perishables (e.g. strawberries) may reach 20–25 per cent losses. Loss estimates for different groups of produce in the third world have been assembled from authoritative opinions (Coursey 1983) as follows: roots and tubers, 5–95 per cent; vegetables, 16–100 per cent; and fruits, 14–100 per cent. We may only conclude that significant losses do occur and that it is thus important that losses are minimized.

The control of losses starts in the field where strains may be selected which store better. The harvesting and handling of the produce must minimize damage to its integrity, and storage must take into account the need to prevent inherent deterioration and that brought about by micro-organisms. In this last category the use of both chemicals and environmental control may be advocated. Cooling the produce to reduce its metabolic rate or adequate ventilation to remove moisture or gases (e.g. ethylene from stored bananas), combined with a fungicide to prevent microbial growth are examples of practical control methods.

A type of biodeterioration which is less easily detected at source, or indeed occurs during adverse storage conditions, affects grains which are infected with fungi. This is the production of 'mycotoxins', secondary metabolites produced by species of *Aspergillus* and *Penicillium*. They cause a range of effects in man and animals ranging from convulsions and abortions to carcinomas. They are toxic in small quantities and have been detected in oilseeds, vegetable oils, cereal grains, and nuts. A safe limit of 20 μg aflatoxin/kg of grain has been suggested by the United Nations Protein Advisory Group (Parpia 1982). Based upon this amount a survey in India of foods of the types likely to be contaminated with aflatoxins showed that 68 per cent were above the threshold safety level with all of the groundnut meal samples being in the range 80–100 μg/kg (see Parpia 1982). Losses due to mycotoxins are thus of much concern to the world's population, and their control by correct handling and storage of oil and protein grains cannot be over-emphasized. The detoxification of contaminated grain is a new era for research effort. Hydrogen peroxide has been shown to be successful under controlled conditions for the removal of aflatoxin from a protein concentrate during its manufacture (Parpia 1982).

Wood

Living sapwood in any standing tree, wherever it is growing in the world, is more resistant to microbial attack than heartwood. This is not the case, however, after the tree is felled when the sapwood with its residual cell contents and other food material in the parenchyma is subject to microbial attack. There are several groups of micro-organisms capable of colonizing wood and in some cases degrading it. They include Actinomycetes and other bacteria, moulds, staining fungi, and soft, brown, and white rot fungi (Levy 1969; 1971; Savory 1954; Liese 1970; Carey 1975; Baecker and King 1981). Bacteria are not as troublesome as fungi in the deterioration of wood which is in service, i.e. wood which is not in contact with soil, although Kelly (1983) has recorded bacteria to be present in decaying in-service timber joinery in the UK. Actinomycetes are considered a significant group of organisms attacking wood when in soil contact (Baecker and King 1981). Fungi as causative organisms of decay are recorded not only in the UK and Europe, but also in the third world, Southern Africa (Pizza *et al.* 1984); Thailand (Rananand and Cockroft 1983); Brazil (Cavalcante and Cockroft 1984) and Iran (Niloufari and Cockroft 1984). None of these workers record bacteria as the forerunners of decay sequences.

The superficial decay of wood is known as soft rot in which Ascomycetes and certain Deuteromycetes are responsible for the destruction of cellulose in the wood cell wall (Savory 1954; Carey 1975). Brown and white rot fungi (Basidiomycetes) are responsible for causing a larger degree of degradation. Brown rot fungi attack the polysaccharide of the cell wall leaving the lignin unchanged. The wood becomes dark brown as decay proceeds (Fig. 13.1) with cracking along and across the grain (Carey 1975). White rot fungi are capable of attacking both lignin and cellulose, and may be generally distributed throughout the wood or localized into pockets of decay (Carey 1975).

Fig. 13.1. Severe brown rot in a piece of window joinery.

In some areas of the world, however, decay agencies other than fungal ones are of importance. These include insects in the terrestrial environment, and crustaceans and molluscs in the marine environment.

In Southern Africa there is considerable attack of wood by insects (Pizzi *et al.* 1984). Beetles that attack green or freshly felled timber include the eucalyptus borer (*Phoracantha semipunctata*), whilst seasoned wood is attacked by shot-hole borers (members of the *Bostrichidae*), powder post beetles (*Lyctus brunneus*), furniture beetles (*Anobium punctatum*), and the european 'longhorn' house-borer (*Hylotrupes bajulus*) Fig. 13.2. In Brazil,

Lyctus is the most important genus attacking dry wood and *Xyleborus ferruginea* is the species that is more frequently found in fresh timber (Cavalcante and Cockroft 1984). In Iran, members of the *Anobiidae*, *Bostrichidae*, *Cerambychidae*, and *Lyctidae* are known to damage wood (Niloufari and Cockroft 1984). In Turkey, Sekendiz (1982) records that members of the following genera have wood destroying capabilities: *Xestobium*, *Ernobius*, *Ptilinus*, *Anobium*, *Callidium*, and *Rhyncolus*. In Thailand, the insect pests are placed in two groups, insect pests of damp wood and insect pests of dry wood. A full list of these insects is presented in Rananand and Cockroft (1983).

Fig. 13.2. Beetles which attack seasoned timber.
Left: common furniture beetle (×16)
Centre: death-watch beetle (×10)
Right: house longhorn beetle (×4)
(Photograph: J. Grayson, ICI Paints Division)

Without doubt, termites are the most important group of insects causing the destruction of timber whether it be heartwood or sapwood, softwood, or hardwood (Ocloo 1978). In Thailand they are considered to belong to one of three groups. These groups are as follows:

1. The dampwood termites, which are important ecologically, yet are relatively unimportant economically. These insects are principally members of the *Kalotermitidae*, e.g. *Kalotermes tectonae*.

2. The subterranean termites, these are the most important and widely distributed insects and are made up of members of the family *Rhinotermitidae*, e.g. *Coptotermes premrasmii* and *Termitidae* e.g. *Odontotermes formosanus*.

3. The drywood termites (sometimes called flying termites) which are more restricted in their distribution, but can be destructive in localities where they are found. They are also principally members of the *Kalotermitidae*, e.g. *Cryptotermes thailandis*.

In Southern Africa, wood-attacking termite species have been divided into two groups (Pizzi *et al.* 1984); the drywood termites, e.g. *Cryptotermes brevis*, and the subterranean termites (Figs 13.3 and 13.4), e.g. *Coptotermes formosanus* and *Odontotermes badius*. This latter insect is recorded as the most widely spread wood-destroying species within the borders of the Republic of South Africa. In Brazil, *Cryptotermes brevis*, *Coptotermes*

Fig. 13.3. Termite mound (*Macrotermes* sp.) from the Accra Plains in Ghana. (Photograph: Dr M. Edmunds, Lancashire Polytechnic)

Fig. 13.4. Termite queen (*Macrotermes* sp.) ruler marked in inches (1 in = 2.54 cm). (Photograph: Dr M. Edmunds, Lancashire Polytechnic)

havilandi, and a species of *Kalotermes* are recorded as the most important termites in wood in buildings (Cavalcante and Cockroft 1984). In Turkey, *Reticulitermes lucifungus* has been the subject of a special study (Sekendis 1982). In Iran, flying termites, members of the *Acanthotermes* genus and subterranean termites, including *Amitermes vilis* (Hagen) and *Microcerotermes diversus*, may attack such materials as untreated railway sleepers (Niloufari and Cockroft 1984).

In marine conditions, timber can be attacked by various marine boring molluscs: *Teredo navilis* and *Teredo utriculus* are a problem to Turkish boat manufacturers (Ilhan and Cockroft 1982), and *Teredo navilis*, *Xylotrya capensis*, and *Martesia striata* cause damage to timber in South African waters (Pizzi *et al.* 1984). Marine borers are also active in the Persian Gulf and the Gulf of Oman. For an account of the effectiveness of creosote and copper-chrome-arsenic against teredinid attack on pine and eucalypt timbers in tropical marine waters, the reader should consult Tamblyn *et al.* (1978).

As many of the world's forests are dwindling, there is a global need for wood preservation (Cockroft and Hennington 1983). The need for timber preservation often became necessary as naturally durable timbers became scarce and other suitable, but less durable, species were put into service. This was the case in Thailand as the demand for railway sleepers increased. Modern timber preservation means vacuum/pressure impregnation rather than by soaking/diffusion methods. The impregnation plants that are in operation in Third World countries are of various design and manufacture (Rananand and Cockroft 1983; Pizzi *et al.* 1984; Cavalcante and Cockroft 1984). The need for effective preservation becomes apparent as one recognizes the growing contribution of third world countries to the timber markets within their own ranks and to the world markets (Abdul-Kader 1982; Rananand and Cockroft 1983).

In his account of the development of timber technology in East Africa between 1965 and 1975, Campbell (1978) brings attention to many of the problems that have to be overcome in order to establish a viable timber technology. These problems often stemmed from a general lack of confidence in indigenous species as structural materials. Often it is a lack of information on structural codes, performance specification for timber quality and a lack of standardization generally that can delay the progress of a developing industry of economic importance to a Third World country. The extent to which these problems are being coped with, is discussed in a series of reports produced by the Swedish National Board for Technical Development. These reports should be consulted in order to obtain information on timber resources, consumption patterns, imports and exports, preservation, and perhaps of major importance, information on legislation, standardization, and specifications which have, and are being adopted, by some third world countries. The reports, which are on Wood Preservation, are available for Thailand (Rananand and Cockroft 1983), Iran (Niloufari and Cockroft 1984), Brazil

(Cavalcante and Cockroft 1984), Southern Africa (Pizzi *et al.* 1984) and Turkey (Ilhan and Cockroft 1982). The report of Abdul-Kader (1982) on the Forest Industries of Peninsula Malaysia is most informative as is the report of Cockroft and Henningsson (1983) on the global needs for wood preservation. A few comments and general observations do emerge from the reports concerning preservation and timber utilization.

Creosote has a long history of usage in the Third World and is still in widespread use for the protection of railway sleepers and electricity telegraph poles. Gaining in popularity though are the waterborne preservatives CCA and CCB (copper-chrome-arsenic and copper-chrome-boron). There is some use of these compounds for telegraph poles in Iran (Niloufari and Cockroft 1984), but their main application is in the preservation of constructional timber intended for external and internal use. In Turkey (Ilhan and Cockroft 1982) the use of water-borne preservatives is extensive. The organic solvent preservatives, such as pentachlorophenol, are finding a growing use in Southern Africa with timber intended for the furniture trade or for interior use. Pentachlorophenol is also used extensively for wood treatment in Brazil (Cavalcante and Cockroft 1984). In Southern Africa, mining timbers (*Ecualyptus grandis*) are made fire-retardant rather than treated with preservative since the humid underground conditions, conducive for fungal growth, make preservation measures impracticable.

Fuels and lubricants

Any country with a developed transport system and an engineering industry will use fuels and lubricants. Fuels will be stored prior to use and lubricants will be recirculated in closed or open systems. In both cases there will invariably be an ingress of water and the build-up of contaminant organic material leading to the growth of micro-organisms. The water may form a discrete layer in the bottom of a storage tank or it may be actively mixed with an oil to form an oil-in-water or water-in-oil emulsion. Such emulsions are used as lubricants and coolants in the metal-working industry to aid the machining, or grinding of metal components, the rolling of aluminium, and the drawing of wires. Gasoline fuels used in piston driven internal combustion engines are generally made up of short-chain (less than C_9) volatile alkanes. These are not readily utilized by micro-organisms. Alkanes of chain length $(C_{12}–C_{20})$ are more susceptible to utilization, and this is reflected in the ability of a range of bacteria and fungi to grow at the interfaces of kerosene and water in the fuel tanks of jet aircraft, and diesel and water in gas turbine ship fuel tanks. The problems arising from microbial growth in fuels and lubricants fall into three categories.

1. The formation of troublesome bioslimes which may become detached and block pipework.

2. Losses in the useful properties of an additive.

3. The formation of metabolic products which directly or indirectly contribute to corrosion of metal surfaces.

Fuels

Kerosene is widely used as an aviation fuel in jet engines. For some time it has been known that in the presence of water, which is an invariable consequence of temperature changes during storage linked with solubility of water in oil, fungal growth will occur at the fuel–water and water–metal interfaces (Genner and Hill 1981). Fungal biomass may become detached and drawn into fuel lines or block filters resulting in fuel starvation to the engines. *Cladosporium resinae*, the kerosene fungus (Fig. 13.5), is most commonly

Fig. **13.5.** *Cladosporium resinae*—the kerosene fungus—found growing in aircraft fuel tanks. (×1000)

encountered in these situations although many other genera have been isolated. The problem is normally associated with grounded aircraft and is of particular importance in hot climates where growth of the fungus is accelerated. In flight temperatures may be as low as −40°C which, although suppressing growth, may favour the survival of *C. resinae* over other species in the fuel which are less resistant to the low temperatures during flight.

At the water–metal interface, microbial colonies can result in pitting corrosion as the consequence of a differential aeration cell (see section on Metals). In extreme cases, the pitting may cause complete penetration of the aluminium fuel tank and fuel loss. The production of acidic metabolic products and the utilization of corrosion inhibitors may also help to control and delay corrosion.

Diesel fuel used for driving marine gas turbine engines can readily become contaminated with fungi and bacteria. Unfortunately, the use of sea water as a displacement as the fuel is burnt ensures a constant microbial inoculum and fresh nutrient supply for the micro-organisms. Fuel blockages and corrosion problems have been observed to occur resulting in reduced efficiency, which is of particular concern to naval warships.

The prevention of microbial growths in fuels may be achieved by adopting one or more of the following practices.

1. The regular inspection of aircraft or ship fuel systems.
2. Monitoring the quality of fuel taken on board.
3. Adding biocides (organo-borons) to the fuel in order to suppress the growth of contaminants. Pasteurization has also been suggested for ship fuels (Wychislik and Allsopp 1983).
4. Lining fuel tanks with materials to prevent corrosion. Butadiene nitrile rubber and polyurethanes are currently used in aircraft for this purpose.

Lubricants

Oils used as straight lubricants, where no water is present or at normal operating temperatures above 60°C, are not prone to biodeterioration problems. Two areas of usage, however, have received some attention: metal-working emulsions and lubricating oils used in slow-speed marine diesels. There are a range of emulsions used in the engineering industries for the drilling, cutting, grinding, and rolling of metals. Four classes may be distinguished.

1. Neat oils: straight mineral oils of varying viscosity.
2. 'Soluble' oil emulsions: containing more than 50 per cent by volume of oil in the undiluted form.
3. Semi-synthetic 'soluble' oil emulsions: containing less than 50 per cent by volume of oil—perhaps as little as 5–10 per cent—in the undiluted form.
4. Chemical solutions: containing no oil, but mixtures of synthetic compounds including esters and methyl silicones, which have lubricating and cooling properties.

It is the soluble and chemical solutions which are prone to contamination. A number of additives such as emulsifiers, corrosion inhibitors, extreme pressure additives, and coupling agents act as nutrients for the contaminants, resulting in a change in the properties of the contaminated emulsion. The utilization of the emulsifier, accompanied by a reduction in pH can cause the emulsion to break, the oil separating out as a discrete layer. The development of anaerobic conditions in the sump (storage tank) can lead to the growth of sulphate reducing bacteria which produce sulphide from sulphate in a dissimilatory fashion. The sulphide may combine with any iron present to generate a blackening of the emulsion in a highly reduced environment, or it may be released as hydrogen sulphide, usually first noticed when the emulsion recirculation system is turned on after shut-down at the week-end. Sulphides are, of course, highly corrosive and their effect may be detected on storage of the machined parts.

The conditions in which emulsions are used will often favour the growth and development of a microbial population. The concentrated soluble oil is mixed with water to give a final emulsion concentration of about 2 per cent by volume. This is then continuously circulated around pipework, allowed to drip over machinery where it is reoxygenated, and often maintains a temperature of between 20 and 30°C depending upon the process and the climate. The system is continuously open to atmospheric contamination and to discarded wastes from the work-force. Bioslimes can develop on surfaces and microbes become lodged in the maze of crevices and dead-ends which are characteristic of many circulation systems. A well established biocide industry employing effective formulations is available in Europe and the USA to cope with biodeterioration problems affecting metal-working fluids.

Water can gain acces to the oil in the crankcase of a marine diesel engine from the water of the cooling system. The water will then bring with it a microbial inoculum as well as nutrients (Hill 1984). The oil is maintained at 35–45°C which is suitable for growth of many thermophilic microbial species. The symptoms of an infection are biomass (sludge) build-up, reduction in pH, malodours, and oil emulsification. This results in corrosion problems and the premature wearing of bearings. The use of various oil cleansing systems such as filters, coalescers, and centrifuges reduces contaminating water and particulate matter, and this in turn reduces the likelihood of microbial growth. The use of biocides is necessary only periodically when problems arise.

Rubbers and plastics

Natural and synthetic formulations of rubbers and plastics are extensively used in all parts of the world where they are often exposed to microbial environments in water, and buried or in contact with the soil. Many interacting forces may be involved in a failure problem and it is often difficult to assess the relative importance of micro-organisms in the process. We do, however, have data, some empirical, on the susceptibility of rubbers and plastics to biodeterioration. They contain, in addition to the polymeric backbone, a range of additives to impart stability, flexibility, strength, and aid processing. These must also be considered in the overall assessment. On the whole there is only a relatively short list of susceptible polymers and additive types in commercial use. The biodegradable polymers used in controlled release of drugs and biocides, and as sutures in surgery are not considered in this review. Information on these types has been gathered together by Wise (1984).

The bulk polymers, polyethylene, polystyrene, and polyvinyl chloride, in use today are regarded as recalcitrant molecules and not, in their own right, subject to biodeterioration. Workers have shown though that dimers and short-chain oligomers of styrene are degraded (Higashimura *et al.* 1983) and that unreacted low molecular weight oligomers (less than 500 mol. wt) of ethylene present in polyethylene can result in limited growth on the polymer. However, it requires physical ageing to encourage subsequent microbial utilization which then only occurs at very low rates (figures in the literature vary between 0.3 and 3 per cent weight loss per year; see Stranger-Johannessen 1979). Polymers based upon natural biopolymers, such as proteins and cellulose, tend to be susceptible to degradation and this limits their use to situations where they never or only periodically come into contact with conditions conducive to microbial growth. Thus, products such as cellophane (packaging), cellulose ethers (thickeners in emulsion paints and foodstuffs), casein-formaldehyde (early type of plastic), crepe (latex rubber), and rayon (regenerated cellulose), will all support the growth of micro-organisms under suitable conditions (see Allsopp and Seal 1986).

Natural rubber is a polymer of *cis* 1,4-isoprene and contains between 2 and 3.5 per cent protein. In its purest form (pale crepe) it may still contain significant levels of protein, lipid and carbohydrate, and is capable of absorbing up to 15 per cent moisture. This can result in growth of micro-organisms on products made from crepe (Williams 1982), especially when they are exposed to high humidity conditions. The vulcanization process drastically increases the resistance of rubber to biodeterioration, but if subsequent oxidation takes place microbial growth may occur, resulting in cracking of the surface. Liquid latex is also prone to contamination and has to be protected using chemical preservatives.

There are a number of synthetic rubber formulations in use today including polyisoprenes, styrene-butadiene, neoprene, nitrile, and silicone rubbers. Of these only the polyisoprene and styrene-butadiene have been found to support microbial growth in any significant amounts. The other rubbers appear in both laboratory testing and in the field to be extremely recalcitrant. As we shall see below, it is often the impurities or additives which encourage biodeterioration.

The other main group of polymers worthy of mention are the polyurethanes.

These are a diverse group of polymers based upon the reaction between an isocyanate and hydroxyl-containing compounds such as butan-1,4-diol. In the presence of suitable catalysts, a long chain polymer is produced containing urethane groups which are similar to peptide bonds and subject to hydrolysis. There is much flexibility in the formulation of polyurethanes, and polyesters and polyethers can be inserted into the polymer to give the product elastomeric properties. The presence of ester bonds in a polyester polyurethane makes it more susceptible in the short term to biodeterioration than a polyurethane containing polyether chains (Seal and Pathirana 1982; Pathirana and Seal 1985). Within this generalization, it has been found that the different isocyanates and polyols used can also affect susceptibility (Darby and Kaplan 1968).

Cracking and embrittlement of cables, clothing, and automotive components containing polyester polyurethanes have been observed after relatively short in-service use (2–3 months in warm humid conditions). The storage of some car components in polythene packaging prior to use in the tropics has resulted in severe deterioration.

Many additives are incorporated into plastic formulations including plasticizers, ultra-violet stabilizers, antioxidants, pigments, fillers, processing aids, hydrolysis stabilizers, and fire-retardants. Individually, they are incorporated at low levels and most have been shown to have very little effect on the susceptibility of the formulation to biodeterioration. However, in total they may constitute 50–60 per cent of the total weight and their combined

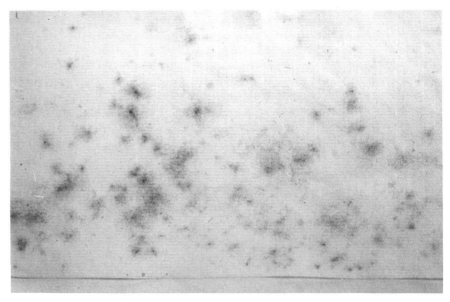

Fig. 13.6. Fungal growth and staining on a plasticized PVC shower curtain.

utilization by micro-organisms may have a significant effect on the properties of the product. The additives most extensively studied have been plasticizers which confer flexibility on an otherwise brittle polymer, and may make up 40 per cent of the total weight of the plastic. Plasticizers are extensively used in polyvinyl chloride where a pliable product is required. They are esters of aliphatic (adipic, sebacic) and aromatic (phthalic) acids and as such may be subject to enzymatic hydrolysis (see Fig. 13.6).

The early literature recorded lists of susceptible and resistant plasticizers (see Berk *et al.* 1957), but later research by Klausmeier (1966) showed that, using suitably isolated soil micro-organisms from enrichment cultures, degradation of previously resistant plasticizers could be observed. Work by Ribbons *et al.* (1984), and Williams and Dale (1983) has further demonstrated that the phthalate plasticizers in particular are not as inert as was previously reported. It is, however, important to consider the physical aspect of the susceptibility of a plasticizer. It must be available at the surface of the plastic and not immobilized within the structure preventing its migration. Low migration rates reduce the rate at which biodeterioration occurs. Short-chain phthalates, such as dibutyl phthalate, are volatile and have been shown to have phytotoxic effects (Hardwick *et al.* 1985). The effect may be extended to micro-organisms under confined conditions.

The presence of other additives such as organic fillers (cellulose based) or processing aids such as lubricants (based on vegetable oils, stearates) will all encourage surface growth which may then extend to other parts of a piece of equipment.

Surface coatings, sealants, and adhesives

These products are formulations of a range of materials, some of which can encourage growth of micro-organisms. Paints contain polymers (acrylates, methacrylates, polyurethanes, and polyvinyl acetate), linseed oil, and cellulose ether thickeners. Sealants may be bitumen or polyurethane based, whilst adhesives may contain casein, starch, or cellulose. We have already encountered most of these materials in previous sections and they are known to promote biodeterioration of the products in which they are incorporated. Water-based formulations of paint and adhesives may further encourage deterioration to occur during storage of the product prior to use (known as 'in-can spoilage'). This results in the loss in viscosity of an emulsion paint due to enzymatic degradation of the cellulose thickener, or the production of an odour in a casein-based adhesive, detected when the lid is first removed. While in service, the paint, sealant, or adhesive film may be subject to colonization by a range of micro-organisms which directly utilize the film causing cracking and loss of usefulness. The passive colonization of algae and fungi may, by the trapping of dirt and water, lead to aesthetic changes or

Fig. 13.7. Surface algal and fungal growth on painted timber fence.

accelerated environmental effects (Fig. 13.7). Water may freeze and crack the surrounding area, and the products of metabolism may accelerate hydrolysis and solubilization of the film (see Allsopp and Seal 1986; Goll and Winters 1974; Upsher 1984.

Pharmaceuticals and cosmetics

Because of the end use of these types of product, there is much control over their manufacture and subsequent storage in order to prevent the colonization and growth of micro-organisms, particularly those species pathogenic to man. Problems do occasionally occur, however, when storage conditions are not adequate, i.e. under warm humid conditions where the product is not in a sterile condition perhaps because its packaging has been damaged, or it is an ointment to be re-used over an indefinite period. Micro-organisms can cause a variety of deterioration effects simply from the appearance of a fungal colony on a topical cream, through the breakdown of an oil-in-water emulsion to the degradation of a drug resulting in a loss in its activity. Antibiotics, such as penicillin, can be inactivated by the β-lactamases of bacteria (Baird 1981), and aspirin has been known to be degraded by *Acinetobacter lwoffi* (Grant 1971). Shampoos can become ropy or slimy due to the presence of bacteria capable of degrading the surfactant, and fungi have been observed growing on toilet soaps. The reviews of Parker (1984) and Baird (1981) deal with the general subject in greater detail.

Metals

The corrosion of metals is an electrochemical phenomenon. Whilst it is probably agreed that micro-organisms contribute to corrosion processes, the extent of the influence which micro-organisms exert on the corrosion process, and the mechanisms by which this occurs, are still under review (Miller 1981; Cragnolino and Tuovinen 1984). The literature cites a number of possible involvements which include the production of corrosive metabolic products such as acids, hydrogen sulphide, and ammonia, the formation of differential aeration electrolytic cells by fungal colonies on metal surfaces, and the assimilation of hydrogen produced at the cathode (termed cathodic depolarization). For a fuller description of these mechanisms the reader is referred to Miller (1981). The effects of these mechanisms are to enhance the rate of corrosion over that which would occur under the same conditions in the absence of micro-organisms. The corrosion results in either perforation

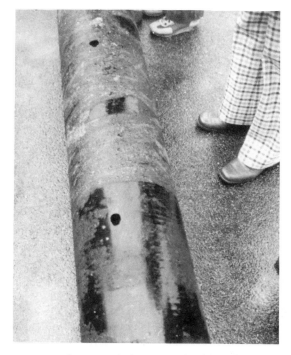

Fig. 13.8. Iron sewerage pipe corroded as a result of low flow rate conditions and anaerobic microbial activity.

(aluminium fuel tanks on aircraft are an example) or the formation of corrosion products (tubercles) which can reduce the diameter of cast iron water pipes and adversely affect flow rates. The acid product of the sulphur-oxidizing bacteria can be sulphuric acid capable of etching and perforating ferrous metals. Its greatest effect has been on concrete storage tanks and sewerage pipes (Fig. 13.8) in hot climates where, in the latter case, the lack of water reduced the volume of water flushed through the sewers each day so that they remained stagnant.

In recent years an economically important group of micro-organisms—the sulphate-reducing bacteria (SRB) (see Postgate's excellent monograph 1984)—has received much attention in connection with its involvement in metal corrosion. This has been due to two factors; the unique ability of the group to reduce sulphate to sulphide in a dissimilatory fashion, and the possession of a hydrogenase. Sulphide is corrosive in forming iron sulphide, which can function as a cathode, whilst hydrogenase is thought to be involved in removing hydrogen produced at the cathode, the presence of which can reduce corrosion. SRB have been implicated in metal corrosion in anaerobic clays, and more recently in oil rigs involved in off-shore oil production where the use of sea-water injection systems introduces contamination into the well resulting in SRB activity. Hydrogen sulphide is produced and the oil becomes sour. Corrosion of pipework and storage areas may follow. Evidence is also accruing on the presence of SRB in the fouling layers of the external oil platform structures. Anaerobic conditions may be established and the SRB then cause pitting corrosion beneath the fouling (Costlow and Tipper 1984; Lewis and Mercer 1984; Southwell et al. 1974).

Other miscellaneous materials

Biodeterioration problems have been recognized in third world countries in unusual cases such as where stone is fashioned (Dukes 1972), or glass is being used for the manufacture of microscope lenses (Nagamuttu 1967). Natural animal products such as leather and wool will also deteriorate under humid conditions. This is of particular concern in the conservation of valuable objects of historic interest. Algae are of particular importance as biodeteriogens in connection with their soiling effects on stonework and buildings (Fig. 13.9) in the terrestrial environment (Grant 1982). Wee and Lee (1980) have reported that high rise buildings in Singapore are badly affected by algal growths. Yong et al. (1972) have also noted that algae will grow on exposed concrete and surfaces coated with cement paints in Singapore. The use of fungicidal paints or periodic cleaning with biocidal washes have been recommended to contain the problem where this is practical (Richardson 1973; Bravery 1981). Glass lenses may be colonized by fungi under high

Fig. 13.9. Brickwork and ornate cornerstone showing epiphytic algal growth.

humidity conditions, leading to etching of the surfaces (Fig. 13.10). Dry storage conditions and the use of volatile fungicides are recommended as control strategies in such cases (Baker 1967).

There are many other materials and products which have been reported to be subject to biodeterioration which we have not mentioned in this brief review. The reader is referred to the published proceedings of the Biodeterioration symposia (Walters and Elphick 1968; Walters and Hueck-van der Plas 1972; Sharpley and Kaplan 1976; Oxley *et al.* 1978; Oxley and Barry 1983) for further details.

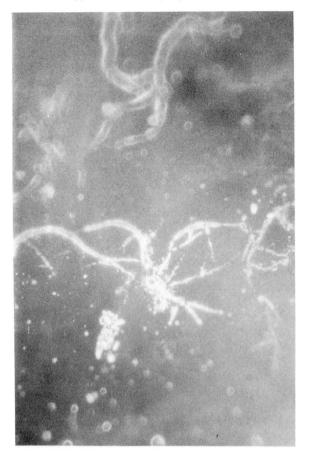

Fig. 13.10. Glass lens etched by the growth of fungi over its surface during storage under humid conditions in the tropics. (×400).

Control methods

The specific control methods cited in the above sections fall into two categories: the use of physical parameters such as temperature, pH, osmotic pressure, gaseous atmospheres, and moisture content/relative humidity; and the incorporation of chemical agents (biocides) to suppress growth or kill the biodeteriogen outright. The choice of method depends upon a number of factors which must be assessed when a control strategy is sought. Biocides are widely used in solvent- and aqueous-based systems where the use of physical agents is not practical. Allsopp and Allsopp (1983) have produced a list of biocides with reference to their use and effective concentration ranges.

Proper control is only possible when the ecology and mechanisms of the biodeterioration problem are fully appreciated. It is all too easy to recommend a method of control in the hope that it will be a panacea for all problems. As this is never the case, it is necessary to recognize the problem and evaluate it before instituting control measures.

References

Abdul-Kader, R. (1982) Current status of forests and forest industries of peninsular Malaysia. *J. Inst. Wood Sci.* 9, 161–7.

Allsopp, C. and Allsopp, D. (1983) An updated survey of commercial products used to protect materials against biodeterioration. *Int. Biodeter. Bull.* 19, 99–146.

Allsopp, D. and Seal, K. J. (1986) *An introduction to biodeterioration*. Edward Arnold, London.

Baecker, A. A. W. and King, B. (1981). Soft rot in wood caused by Streptomyces. *J. Inst. Wood Sci.* 9, 65–71.

Baird, R. M. (1981). Drugs and cosmetics. In *Microbial biodeterioration* (ed. A. H. Rose) pp. 387–429. Academic Press, London.

Baker, P. W. (1967). An evaluation of some fungicides for optical instruments. *Int. Biodeter. Bull.* 3, 59–64.

Berk, S., Ebert, H., and Teitell, L. (1957). Utilisation of plasticisers and related organic compounds by fungi. *Ind. Eng. Chem.* 49, 1115–24.

Bravery, A. F. (1981). Preservation in the Construction Industry. In *Principles and practice of disinfection* (eds A. D. Russel, W. B. Hugo, and G. A. J. Ayliffe) pp. 379–402. Blackwell Scientific Publications, Oxford.

Campbell, P. A. (1978). The development of timber technology in East Africa, 1965–1975. *J. Inst. Wood Sci.* 8, 69–75.

Carey, J. K. (1975). Isolation and characterisation of wood-destroying fungi. In *Microbial aspects of the deterioration of materials* (eds D. W. Lovelock and R. J. Gilbert) pp. 23–37. Academic Press, London.

Cavalcante, M. S. and Cockroft, R. (1984). *Wood preservation in Brazil*. Report no. 445, Swedish National Board for Technical Development.

Cockcroft, R. and Henningsson, B. O. (1983). The global needs of wood preservation. *Int. J. Wood Preserv.* 3, 65–71.

Costlow, J. D. and Tipper, R. C. (1984). *Marine biodeterioration: an interdisciplinary study*. Naval Press, Annapolis, Maryland.

Coursey, D. G. (1983). Post-harvest Losses in Perishable Foods of the Developing World. In *Post-harvest physiology and crop preservation* (ed. M. Lieberman) pp. 485–514. Plenum Press, New York.

Cragnolino, G. and Tuovinen, O. H. (1984). The role of sulphate-reducing and sulphur-oxidising bacteria in the localised corrosion of iron-based alloys—a review. *Int. Biodeter.* 20, 9–26.

Darby, R. T. and Kaplan, A. M. (1968). Fungal susceptibility of polyurethanes. *Appl. Microbiol.* 16, 900–5.

Dukes, W. H. (1972). Conservation of stone: causes of decay. *Architects J.* 156, 422–9.

Eggins, H. O. W. (1983). Biodeterioration, past, present, and future. In *Biodeterioration* (ed. T. A. Oxley and S. Barry) Vol. 5, pp.1–9. J. Wiley & Sons, Chichester.

—— and Oxley, T. A. (1980). Biodeterioration and biodegradation. *Int. Biodeter. Bull.* 16, 53–6.

Genner, C. and Hill, E. C. (1981). Fuels and oils. In *Microbial biodeterioration* (ed. A. H. Rose) pp. 260–306. Academic Press, London.

Goll, M. and Winters, H. (1974). Pseudomonads and their cellulases—mechanism of action and mode of detection. *J. Paint Technol.* 46, 49–52.

Grant, C. (1982). Fouling of terrestrial substrates by algae and implications for control—a review. *Int. Biodeter. Bull.* 18, 57–65.

Grant, D. J. W. (1971). Degradation of acetylsalicylic acid by a strain of *Acinetobacter lwoffi. J. Appl. Bacteriol.* 34, 689–98.

Hardwick, R. C., Cole, R. A., and Fyfield, T. P. (1985). Plastics, phytotoxic vapours, and plant death. *Biologist* 32, 22–4.

Higashimura, T., Sawamoto, M., Hiza, T., Karaiwa, M., Tsuchii, A., and Suzuki, T. (1983). Effect of methyl substitution on microbial degradation of linear styrene dimers by two soil bacteria. *Appl. Environ. Microbiol.* 46, 386–91.

Hill, E. C. (1984). Micro-organisms—numbers, types, significance, detection. In *Monitoring and maintenance of aqueous metal-working fluids* (eds K. W. A. Chater and E. C. Hill) pp. 97–112. J. Wiley & Sons, Chichester.

Hueck, H. J. (1965). The biodeterioration of materials as a part of hylobiology. *Material und Organismen* 1, 5–34.

Ilhan, R. and Cockroft, R. (1982). *Wood preservation in Turkey.* Report no. 294, Styrelsen for Teknisk Utveckling.

Kelly, D. M. T. (1983). Biotic and climatic factors affecting the colonisation of wood-invading fungi. PhD Thesis, Lancashire Polytechnic, Preston, UK.

Klausmeier, R. E. (1966). The effect of extraneous nutrients on the biodeterioration of plastics. In *Microbiological deterioration in the tropics* pp. 232–43. Society of Chemical Industry No. 23, S.C.I., London.

Levy, J. F. (1969). The spectrum of interaction between fungi and wood. *Rec. Ann. Conv. Br. Wood Preservers Ass.* 3, 81–97.

——, (1971). Further basic studies on the interaction of fungi, wood preservatives and wood. *Rec. Ann. Conv. Br. Wood Preservers Ass.* 5, 63–75.

Lewis, J. R. and Mercer, A. D. (1984) *Corrosion and marine growth on off-shore structures.* Ellis Horwood Ltd, Chichester.

Liese, W. (1970). The action of fungi and bacteria during wood deterioration. *Rec. Ann. Conv. Br. Wood Preservers Ass.* 4, 1–14.

Miller, J. D. A. (1981). Metals. In *Microbial biodeterioration* (ed. A. H. Rose) pp. 149–202. Academic Press, London.

Nagamuttu, S. (1967). Moulds on optical glass and control measures. *Int. Biodeter. Bull.* 3, 25–7.

Niloufari, P. and Cockroft, R. (1984). *Wood preservation in Iran.* Report no. 412, National Swedish Board for Technical Development.

Ocloo, J. K. (1978). The natural resistance of the wood of *Terminalia ivorensis*, A. Chev. (Idigbo, Emere) to both fungi and termites. *J. Inst. Wood Sci.* 8, 20–3.

Onions, A. H. S., Allsopp, D., and Eggins, H. O. W. (1981). *Smith's introduction to industrial mycology*, 7th edn. Edward Arnold, London.

Oxley, T. A., Allsopp, D., and Becker, G. (1980). *Biodeterioration. Proc. 4th International Biodeterioration Symposium.* Pitman.

—— and Barry, S. (1983) *Biodeterioration 5. Proceedings of 5th International Biodeterioration Symposium.* J. Wiley & Sons, Chichester.

Parker, M. S. (1984). Microbial biodeterioration of pharmaceutical preparations. *Int. Biodeter.* 20, 151–6.

Parpia, H. A. B. (1982). Some comments on the mycotoxin contamination of agricultural commodities in trade. In *Control of the microbial contamination of foods and feeds in international trade* (eds H. Kurata and C. W. Hesseltine) pp. 203–11. Saikon Publishing Co., Tokyo.

Pathirana, R. A. and Seal, K. J. (1985). Studies on polyurethane deteriorating fungi. Part 3. Physico-mechanical and weight changes during fungal deterioration. *Int. Biodeter.* 21, 41–9.

Pizzi, A., Conradie, W. E., and Cockroft, R. (1984). *Wood preservation in Southern Africa*. Report no. 154, Swedish Wood Preservation Institute.

Postgate, J. R. (1984). *The sulphate-reducing bacteria* 2nd edn. Cambridge University Press, Cambridge.

Rananand, A. and Cockroft, R. (1983). *Wood preservation in Thailand*. Report no. 372, Styrelsen for Teknisk Utveckling.

Ribbons, D. W., Keyser, P., Kunz, D. A., and Taylor, B. F. (1984). Microbial Degradation of Phthalates. In *Microbial degradation of organic compounds* (ed. D. T. Gibson) pp. 371–97. Marcel Dekker Inc., New York.

Richardson, B. A. (1973). Control of biological growths. *Stone Industries* 8, 2–6.

Savory, J. G. (1954). Breakdown of timber by ascomycetes and fungi imperfecti. *Ann. Appl. Biol.* 41, 336–47.

Seal, K. J. and Eggins, H. O. W. (1981). Biodeterioration of Materials. In *Essays in applied microbiology* (eds J. Norris and M. Richmond) pp. 8/1–8/31. J. Wiley & Sons, Chichester.

——, and Pathirana, R. A. (1982). The microbiological susceptibility of polyurethanes—a review. *Int. Biodeter. Bull.* 18, 81–5.

Sekendiz, O. (1982). *Reticulotermes lucifugus, Rossi, and its damages in Turkey*. The International Research Group on Wood Preservation, Document no. IRG/WP/1152, 16th February.

Sharpley, J. M. and Kaplan, A. M. (1976). *Proceedings of the Third International Biodegradation Symposium*. Applied Science Publishers, London.

Society of Chemical Industry (1966). *Microbiological deterioration in the tropics*. S.C.I. Monograph no. 23. Belgrave Sq. London.

Southwell, C. R., Bultman, J. D., and Hummer, C. W. (1974). *Influence of marine organisms on the life of structural steels in seawater*. NRL Report 7672. National Technical Information Service: Springfield, Virginia.

Stranger-Johannessen, M. (1979). Susceptibility of photo-degraded polyethylene to microbiological attack. *J. Appl. Polymer Sci.* 35, 415–21.

Tamblyn, N., Rayner, S., and Levy, C. (1978). Field and marine tests in Papua New Guinea. 1. Performance of creosote and copper-chrome-arsenic preservatives in pine and eucalypt timbers in tropical marine waters. *J. Inst. Wood Sci.* 8, 69–75.

Upsher, F. J. (1984). Fungal colonization of some materials in a hot-wet tropical environment. *Int. Biodeter.* 20, 73–8.

Walters, A. H. and Elphick, J. S. (1968). *Biodeterioration of materials*, Volume 1. Applied Science Publishers, London.

——, and Hueck-van der Plas, E. H. (1972) *Biodeterioration of materials*, Volume 2. Applied Science Publishers, London.

Wee, Y. C. and Lee, K. B. (1980). Proliferation of algae on surfaces of buildings in Singapore. *Int. Biodeter. Bull.* 16, 113–7.

Williams, G. R. (1982). The breakdown of rubber polymers by micro-organisms. *Int. Biodeter. Bull.* 18, 31–6.

——, and Dale, R. (1983). The biodeterioration of the plasticizer diotyl phthalate. *Int. Biodeter.* 19, 37–8.

Wise, D. L. (1984). *Biopolymeric controlled release systems*, Volume 2. CRC Press Inc., Florida.

Wychislik, E. T. and Allsopp, D. (1983). Heat control of microbiol colonisation of shipboard fuel systems. In *Biodeterioration 5* (eds T. A. Oxley and S. Barry) pp. 453–61. J. Wiley & Sons, Chichester.

Yong, F. N., Yeow, C. T., Chua, N. H., and Wong, H. A. (1972). Method for screening and evaluating algicidal and algistatic surface coatings. *J. Singapore Inst. Architects* 53, 13–9.

14

An international network exercise: the MIRCEN programme

E. J. DaSilva and H. Taguchi

1. Introduction

The harnessing of the vast and invisible reservoir of microbial genetic resources in the development of potentially new bio-industrial processes is, increasingly, the prime concern of biotechnology, better known as the applications of microbiology.

The development of new strains as a result of the technique of genetic engineering is expected to bring a techno-economic windfall in the evolution of new processes for the detoxification of wastes and polluted waters, the production of vaccines, the fermentation of local food products, the microbial fixation of nitrogen, the development of protein engineering, and the production of new products from computerized biotechnological processes. In all these processes, without doubt, the preservation and maintenance of mutants, and valuable strains of economic importance are at the very basis of impressive financial investment in long-term basic and applied microbiological research.

The international support of culture collections—the treasure houses of the planet's microbial genetic heritage, can be traced back to the early catalytic support provided by UNESCO, in 1946. Several of today's well-known culture collections have been beneficiaries of such support (DaSilva *et al.* 1977). Moreover, such support was extended also to the promotion of research on the applied use of micro-organisms and to relevant publications such as the *International Bulletin of Bacteriological Nomenclature and Taxonomy*. In 1962, UNESCO's Twelfth General Conference adopted a resolution by the Government of Japan to initiate and intensify research and training activities in microbial biotechnology in view of the growing domestication and use, at that time, of microbial resources in the sector of food, energy, industry, medicine, and agriculture. As a consequence, UNESCO, in 1963, initiated a series of GIAM conferences (Table 14.1) and in subsequent years, undertook joint collaborative activities with the Panel on Microbiology of the International Cell Research Organization (ICRO), the International Association of Microbiological Societies (now known as IUMS), the International Organization for Biotechnology and Bioengineering (IOBB), and the World Federation of Culture Collections (WFCC).

Table 14.1

Global impacts of applied microbiology (GIAM) conferences

Year	Place	Beneficial aspects
1963	Stockholm	Focus on applied microbiology in developing countries
1967	Addis Ababa	Catalyst for research and training in developing countries
1969	Bombay	Institution of national (GIAM-3) fellowship programme
1973	Sao Paulo	Blue-print: Man, microbe, and environmental interaction
1977	Bangkok	South-East Asian network comes of age
1980	Lagos	Emergence of African networks
1985	Helsinki	MIRCENS—mechanisms for international co-operation

The 16th Session of UNESCO's deliberations in 1970 were enriched by a resolution from the Governments of Denmark, Finland, Norway and Iceland calling for the establishment of specialized microbial research centres in developing countries. This foresight by these Governments, in an historic perspective, was the essence of what today is known as the MIRCEN network.

The origins of the MIRCEN network can be traced back to the early days of the UNESCO/ICRO Panel on Microbiology. Established in 1965, the Panel was principally concerned with:

(i) the establishment of an international network for the preservation and exchange of cultures;

(ii) the promotion of the use of micro-organisms as a natural resource; and

(iii) the world-wide training of microbiologists (Porter 1974; Bull and DaSilva 1985).

2. Networks: general principles and examples

Networks are of different types and range, in either an egalitarian, hierarchal, or hybrid manner (Friedman 1974), from loose nets to their formal assembly in bodies, unions, and organizations that are normally non-governmental in character (Table 14.2).

Table 14.2
Types of networks and some characteristics

A. Network types.

Components		
Individuals	Transnational co-operation of individuals for Appropriate/Alternative technologies	TRANET
Unions organizations	Affiliation of scientists within scientific unions geared to specific disciplines	ICSU, ICRO
Institutes	Collaboration of institutes for promotion of international scientific co-operation	IFIAS
Countries	Participation of member countries in net-like systems, functioning within regional or international lines, and interests	U.N. System, OAU, OPEC, CERN, ASEAN, NATO, Warsaw Pact
Agriculture	Institutes specially founded to deal with a specified field/issue	IRRI, CIAT, IITA, ICRISAT, etc.
Health	Participation of institutes from member countries	WHO Referral Centres
Trade	Participation of member countries	UNCTAD

B. Characteristics of decentralization in network progammes

Advantages	Disdvantages
Decisions are made closer to the problems	Increased staffing
Less noise and distortion in the communication channels	Decisions may not always conform to uniformity
Local officers exhibit more responsibility	Feedback flow drastically reduced
Reduces overloading on hierarchical decision makers	
Provides for 'on-the-job' training in project execution	
Enhances organizational morale due to participation as team member	

Other nets involve the linkage of institutes, member countries in the United Nations System, and on a regional basis the pooling together of interests that may be criteria for national and regional progress, and impact in fields ranging from economics, health, self-reliant development, trade, border security, and cultural traits to conservation and utilization of available natural resources. For example, the Network of Community-Oriented Educational Institutions for Health Sciences (often referred to as *Network*) is an international co-operative venture of medical schools throughout the world (Brun 1985). One of its innovations is the new programme of medical training which is built around three anvil concepts: problem-based, student-centred, and community-oriented.

Plucknett and Smith (1984) recently described networking in international agricultural research using as apt examples the Philippines-based International Rice Research Institute (IRRI) and the International Maize and Wheat Improvement Centre (CIMMYT–Centro Internacional de Mejoramiento de Maiz y Trigo) in Mexico. In tracing the genesis of networking in agricultural research, attention was drawn to the role of the International Agricultural Research Centre functioning as catalysts under the aegis of the Consultative Group on International Agricultural Research (CGIAR) and to seven basic principles that are generally valid for all networks.

More recently, representatives from ten developing countries—Cameroon, China, Indonesia, Malaysia, Nigeria, Papua New Guinea, Peru, Philipines, Tanzania, and Thailand, decided to form the Tropical Land Clearing for Sustainable Agriculture Network. This network will focus on 'appropriate technologies for land clearing to allow sustainable agriculture, on ways of re-habilitating degraded lands, and on cropping and soil management systems that will allow continuous agricultural production by small farmers on cleared lands' (IBSRAM 1985).

The newly constituted International Network of Biotechnology, by the Working Group on Technology, Growth and Employment (1984), is comprised of centres, universities, and research institutes, based in Canada, France, the Federal Republic of Germany, Italy, Japan, and the United Kingdom, which participate voluntarily in an international training operation. The network which caters primarily for researchers from the developing countries involves centres and institutions already in existence.

Several Summit Projects geared to collaborative research and development of science and technology were identified by a special working group after the 1980 Versailles Summit. (Anon. 1984). Amongst these, biotechnology was identified as a principal project (Table 14.3).

Most network systems possess a common ancestor. The characteristics of living systems, and albeit, non-living networks, can be traced back through seven interconnecting levels that range from supranational systems through society, organization, group, and organism, to the organ and cell. For each level of evolution of the different networks, there are several variables, some

Table 14.3
Summit projects on biotechnology or with a biotechnological content

Project	Principal participants
Photosynthesis	Japan
Food technology	France, UK
Cultivation of marine products	Canada
Biotechnology	France, UK
Biological sciences	European community

common, others different, and yet which interact in a single or multiple dimension within and across each of the seven levels (Miller 1978).

The patterns, growth, and development of networks vary widely, depending upon a large number of variables. Normally, networks begin with a few like-minded members, whose numbers increase over time. The family, comprised of a family unit and its growth in time, in an apt example. At the organizational level, the rate of increase is different, even for two identical units, on account of the varied responses in adapting to the changing global system, and to the evolving needs of the technical and lay populations existing in the system.

In actual operation, networks flow in a centralized or decentralized state. Centralization involves the allocation of greater financial resources, systemized hierarchical control and, more full-time staffing. Centralization often leads to a non-accommodation of network component viewpoints as the self-serving process is self-blinding in relation to the evolution of internally-developed criteria in the hierarchical environment. An example is the network of a university wherein control of academic affairs is decentralized. Fiscal power, however, is centralized and inhibitory with the result that a certain degree of financial autonomy is obtained through grants from governmental agencies, private foundations or other sources. In exercising such autonomy academic personnel, normally not irresponsible individuals, often decide with minimal bureaucratic constraints, on the mode of expenditure in pursuit of scientific research and development.

Delegation of authority and financial responsibility from higher to lower echelons are the key features of the decentralization aspect of a network operation (Table 14.2). One example of a successful network, which has been cited by non-Asian member states of UNESCO as a model in development, is the Southeast Asian network in microbiology which is funded by the Government of Japan and UNESCO (Network Review 1986).

In sum analysis, decentralization recommends itself since it allows the decision-makers in the top echelons of the system to concentrate on more central issues rather than on matters that do not routinely conflict with defined legal, administrative standards and general policy guidelines.

Feedback is an important characteristic of any network. In centralized systems consistent demands at short intervals may lead to necessary decisions, which although probably right at the time may in the long run be counter-productive. Distortions in feedback arise generally from selective editing and from conformance to prevailing stereotype policy reports and cultural themes, and even verbal nuances (McNamara 1973). Feedback, whether external or internal to the network are either in the form of questionnaires, surveys, polls, or evaluations. Feedback, out of necessity, is rapid in profit-making organizations. In non-commercial organizations and academic networks, feedback is slow on account of the operational elements of decentralization and democracy. Feedback, at all times, remains an important and integral aspect of a network, even though personal biases of network members may be evident.

Nodal points in networks occupy a key position in motivating institutions and individuals to participate in the opertion. Securing and mobilizing funds is another function as such action is necessary to complement 'seed' funding. Another is locating co-operators, either on a regional or international basis. The main characteristics of nodal points functioning as catalysts of change and continuity in network development from concept to operation are curiosity, creativity, competence, and committment. In analysis, most networks involve programatic elements such as: training courses and conferences that build vital 'invisible colleges' geared to critical mass development; fellowships that stimulate nuclei of endogenous scientific growth; the exchange of junior and senior research personnel within and outside network boundaries that allow 'refreshing winds through open research windows'; and the provision of small-scale research grants that consolidate nascent research beach-heads. Promotion of technology in network operations often revolves around:

(i) the improvement of traditional indigenous technologies;
(ii) the acceptance of a scientific 'modern' technology;
(iii) the revival of an old technology;
(iv) the development of a new technology; and
(v) the transfer of technology within regions.

Other desirable features include problem-orientated research based on networks of interacting field projects that are interdisciplinary in nature and complementary in scope and focus (UNESCO/ROSTSEA 1974). This is the rationale behind the various networks of interlinked field projects that countries are engaged in within the framework of UNESCO's global network of Microbiological Resources Centres (MIRCENs).

In summary, network structure is based on existing institutes of advanced study and research in the participating countries of each region. Each regional network links national centres in the discipline under consideration, encourages communication between them, and develops co-operative programmes for improving research and training (DiCastri and Haldey 1979). Such collaboration facilitates the emergence of regional co-operation through regional programmes and activities. National institutions are at the heart of the network and no consideration is given to the establishment of regional centres of excellence.

The network is direct by a regional co-ordinating board, each member being supported by his national authority (as in the case of the South-East Asian network for microbiology). The board meets periodically to determine policy, approve of programmes, and allocate funds keeping in mind the four cornerstones of network structure, i.e.:

(i) reinforcement of national facilities and capacities through regional co-operation;

(ii) conducting regional activities and programmes within regional abilities and capacities;

(iii) development of technical policies and programmes of the network by scientists of the participating institutes;

(iv) involvement of the international scientific community in network planning, function, and evaluation in order to avoid 'isolated' or Utopian growth.

Thus, a sharing of workloads, optimization of resources, exchange of information, collegial decision-making, identification of weak points, team-work approach, and a solidarity of commitment are involved. These features are the hall-marks of the regionalization of the concept of the Microbiological Resources Centres that operate within a global network of microbiological resources centres established in several developing countries (Argentina, Brazil, Egypt, Guatemala, Kenya, Senegal, and Thailand, and interacting with those in Australia, Canada, Japan, Sweden, UK, and USA).

3. The MIRCEN network

Following the United Nations Conference on the Human Environment at Stockholm, 1972, experts from UNEP, UNESCO, and the international scientific community, represented in ICRO, met in 1974 and 1975 at the UNEP Secretariat at Nairobi, to jointly formulate a world-wide programme aid at the preservation of microbial gene pools and making them accessible to developing countries through the establishment of a network of microbiological resources centres (MIRCENs). Subsequently, a UNEP/UNESCO/ICRO project titled 'Development of an Integrated Programme in the Use and Preservation of Microbial Strains for Deployment in Environmental Management' (July 1975–October 1984) was formulated and submitted by UNESCO/ICRO, and approved by UNEP for implementation. The objectives of the MIRCEN network are:

(i) to provide a global infrastructure which would incorporate national, regional, and inter-regional co-operating laboratories geared to the management, distribution, and utilization of the microbial gene pools;

(ii) to reinforce the conservation of micro-organisms, with emphasis on *Rhizobium* gene pools in developing countries with an agrarian base;

(iii) to foster the development of novel technologies native to specific regions;

(iv) to promote the economic and environmental applications of microbiology; and

(v) to serve as focal centres in the network for the training of manpower and diffusion of microbiological knowledge.

The first development in the World network of Microbiological Resources Centres was the establishment of the World Data Centre (WDC) for Microorganisms at the University of Queensland, Brisbane, Australia. The WDC houses a master copy of the World Directory of Collections of Cultures of Microorganisms and serves as a pivotal point for fostering development of culture collections in developing countries.

The total number of culture collections registered with the WDC is 394,

Table 14.4
Research activities of the World Data Centre MIRCEN

A. Facilities available to culture collection curators and researchers

---

The acceptance of data from research workers, matching it against data in the centre, and supplying the output, either unclassified or classified, to the research worker

The location of cultures in various culture collections

The provision of information on the geographical distribution of micro-organisms, host ranges, or the occurrence of organisms with specific properties

The provision of lists of contents of any collection

The provision of lists of all collections maintaining particular organisms

Supplying culture collections with their own collection data

B. Research activities

Collection and updating of strain data

Compilation of data on economically important strains, e.g. *Rhizobium*

Interaction with other MIRCENs in culture collection work

Provision of training facilities within the MIRCEN network

---

from 58 countries. Several facilities are available through the current programmes and activities of the WDC (Table 14.4).

In recent developments to help provide permanency and continuity to the work of the WDC, negotiations have been underway, throughout 1985, to relocate the World Data Center. From July, 1986 the Center will be at the Life Science Research Information Section, RIKEN, Tokyo, Japan.

The MIRCEN at the Karolinska Institute, Stockholm, Sweden, in addition to developing microbiological techniques for the identification of micro-organisms at the WDC, has pioneered the organization of a series of MIRCENET Computer Conferences on biogas production, anaerobic digestion and the bioconversion of lignocellulose (BOSTID 1984).

Computer conferencing is a network system that links geographically scattered individuals together through the use of home or office computers to a remote control computer. Exemplification of the technology of communications is found in the linking of telephone lines to communication satellites which in turn transmit messages to a central computer. Problems of time and travel, slow mails, and incompatible schedules appear to be a thing of the past since sharing of ideas, techniques, and data seem to be continuously on tap.

Table 14.5
Functions of MIRCENET

To help initiate closed computer conferences under defined keys such as microbiology, biological nitrogen-fixation, biogas, networking in culture collections, etc.

To act as an information source for meetings, reviews, identification services, etc.

To provide a platform for discussions on MIRCEN network activities

To provide printouts and records of MIRCENET entries

Apart from attempting to link up the MIRCENs and organizing specialized conferences, MIRCENET has other functions (Table 14.5). The computer conference on the bioconversion of lignocellulosics to fuel, fodder and food, held from May through December 1983, was organized by the International Development Research Centre (IDRC), the MIRCEN in Stockholm, the World Academy of Art and Science, and the International Federation of Institutes for Advanced Study. The aspects of cellulose conversion covered were: upstream processing; processes for food and fodder; processes for liquid and gaseous fuels; and general considerations concerning process technology.

3.1. The BNF MIRCENS

A major problem facing a large number of developing countries is production of more food for their expanding populations. In this quest several developing nations have been expanding their agricultural lands into area which are marginally capable of sustaining productivity, which is invariably limited by the availability of nitrogen fertilizer. By the end of the third quarter of this century, world food production was in fact dependent on a supplemental supply of synthetically fixed nitrogen fertilizer amounting to 40 million tons and costing US$ 8–10 billion.

The production of biofertilizers—cyanobacteria for rice and *Rhizobium* inoculants for leguminous crops—can help greatly in the productivity levels of the plant's soil resource, and in the consequent judicious utilization of petroleum and its hard technologically processed products. The natural age-old phenomena of photosynthesis and biological nitrogen-fixation (BNF) bestow on cyanobacteria and rhizobacteria certain ecological and agricultural advantages. Apart from recycling natural resources in the form of biomass and biological nitrogen, both natural mechanisms promise maximal benefits in augmenting agricultural products in the rural sector. Again, in association with other unicellular photosynthetic algae, the cyanobacteria figure prominently in solar-based integrated algae/fuel/feed/fertilizer production systems with rural poultry and livestock practices. In such systems, the use of algae as biofertilizers provides a cyclic nutrient-supply system with obvious inherent ecological advantages. In 1975, it was estimated that the total nitrogen fixed naturally amounted to 175 million metric tons as compared to 40 million metric tons fixed industrially, and of this amount 35 million tons are fixed by cropped leguminous plants alone.

In this context, the emphasis on the development of biofertilizers or *Rhizobium* inoculant material, particularly in legume-crop areas of the developing countries, appears to be sound. In interaction with other international programmes, modest schemes are already operating through the MIRCENs on a level of regional co-operation in Latin America, East Africa and Southeast Asia and the Pacific, with additional support from FAO and UNEP.

In the area of biological nitrogen fixation five MIRCENs are already operating. These are at the University of Nairobi, Kenya (1977); at Porto Alegre (1978) through two integrated institutions: the Institute of Agronomic Research of the State Department of Agriculture and the Department of Soils of the University of Rio Grande do Sul; at the NifTAL/University of Hawaii/ USAID Project Maui, Hawaii (1981); the Cell Culture and Nitrogen Fixation Laboratory, USDA, Beltsville, USA (1981); and at the Centre National de Recherches Agronomiques, Bambey, Senegal (1982).

Adequate training is one of the most important limiting factors in relation to a more intensive and wider application of the *Rhizobium* technology in the developing countries. In all areas from *Rhizobium* microbiology to legume breeding, legume nutrition and soil fertility, legume extention, etc., it is absolutely necessary to have trained specialists imbued with an awareness of the technology involved and the potentialities of the *Rhizobium*/Legumes symbiosis. Good rhizobia strains are in general available at national or international laboratories, but their use for the production of inoculants and their application of the information is generally ignored in most of the developing countries.

Training has certainly been the most important activity of the MIRCENs towards achieving the established objective of development of research capability and diffusion of the *Rhizobium* technology. Through short courses, intern practical training and graduate degree work a large number of rhizobiologists, agronomists, teachers at agricultural schools, and legume investigators have been trained at the different centers. At the MIRCENs of Porto Alegre, Nairobi, Hawaii and Beltsville a total of more than 300 people have already been trained (Table 14.6), but the number is higher if one takes into account the short courses held at other institutions around the world in the recent past years.

Table 14.6
Number of trainees in *Rhizobium* technology at BNF MIRCENS*

Rhizobium MIRCENS	Type of training			No. of countries reached
	Short courses	Intern	Graduate	
Porto Alegre	94	17	11	15
Nairobi	29		4	11
Hawaii	94	36	18	
Beltsville		5		

* Up to 1983.

Another aspect of the training provided is the advantage of exposing the trainee not only to the *Rhizobium* technology, but also to the problems related to the diffusion and application of the technology. The Hawaii NifTAL MIRCEN has done most in this aspect co-operating in organizing short courses in Kenya, Brazil, Peru, Mexico, Venezuela, Costa Rica, Malaysia, and Thailand. The MIRCEN at Porto Alegre has co-operated in courses in Argentina, Peru, Brazil, Costa Rica, and Indonesia. The MIRCEN at Nairobi has organized courses for participants from a number of African countries such as Nigeria, Ghana, Burkina Faso, Egypt, Sudan, Zambia, the United Republic of Tanzania, Rwanda, and Kenya, with a view to accelerating interaction among workers who had never met before.

The MIRCENs for the East African and West African regions, based at Nairobi and Dakar respectively, address themselves to the field of nitrogen-fixation by legume-rhizobial systems. Broad responsibilities include collection, identification, maintenance, testing, and distribution of rhizobial cultures compatible with crops of the regions. Identification of problems pertinent to the deployment of local rhizobia inoculant technology and promotion of research are other activities. Advice and guidance are provided in the region to individuals and institutions engaged in rhizobiology research.

The MIRCEN in Latin America focuses on nitrogen-fixation with the objective of promoting *Rhizobium* technology. It promotes the identification of leguminous germ-plasm of high symbiotic capacity and soil-limiting factors; the optimal selection of efficient rhizobial strains for soybean, clover, lucerne, lotus, peas, beans, and the cowpea group; optimization of inoculant production for experimental use, for demonstration plots and for small farmers; and the quality control of inoculants for use by private and official laboratories.

The goals of the MIRCENs at Hawaii and Beltsville are to contribute towards alleviating the dependence of developing countries on oil-rich nations for the nitrogenous fertilizers needed to produce basic food commodities. This is accomplished through research to provide a data-base for developing countries to assess the benefits from fuller utilization of legume-based BNF technologies; development and delivery of validated BNF technologies that are appropriate to the needs and circumstances of developing countries; and the provision of support services to facilitate implementation of BNF-based technologies.

The production of inoculants (1500 kg/year) for legume trials and for use by farmers at the IPAGRO Laboratory was started in 1950. Since 1956 a production rate of 400–500 kg per year is maintained. The Nairobi MIRCEN has also a fairly high production of inoculants (1400 packages in 1982) in response to requests from 753 farmers. These laboratory inoculants may also function as standards for comparison with commercial inoculants. The Hawaii MIRCEN has recently developed a technique for the dilution of the broth which can multiply enormously the capacity for the production of the inoculants.

Inoculant production at the East African MIRCEN in Nairobi, Kenya, is now being initiated using strains tested by MIRCEN workers. Inoculants for 11 pasture legumes have been supplied to FAO pasture agronomists at Embu and Kitale. MIRCEN inoculants have been used in the East African bean inoculation trials and soybean demonstration currently being conducted in Kenya.

The IPAGRO laboratory, in Brazil, on behalf of the Federal Government, is now responsible for the quality control of inoculants produced in Brazil. An average of 100 samples are examined per year. On a limited basis this service is also available to institutions in the developing countries. In addition, the

IPAGRO has been producing inoculant material. Average production per year has been around 500 kg. In 1978 the IPAGRO/MIRCEN-Porto Alegre laboratory sent 20 kg of inoculant for soybeans to be used in experiments among small farmers of the FAO Abapo-Izozog Project in Santa Cruz, Bolivia.

In order to produce sufficient rhizobial inoculant material, it is necessary to have good cultures of *Rhizobium*. In this regard, the MIRCENs play a valuable role in maintaining and distributing efficient cultures of *Rhizobium* (Table 14.7). Over 3,000 strains are maintained in the MIRCEN collections collectively.

Table 14.7
Culture collection services of BNF MIRCENs

A. *Rhizobium* culture collections at MIRCENs.

MIRCENS	Number of strains
Bambey	50
Beltsville	938
Hawaii	2000
Nairobi	208
Porto Alegre	650
Total	3846

B. Distribution of cultures of *Rhizobium*

MIRCEN	Number of cultures	Recipient institutions in
Bambey	8	Gambia, Mali, Yemen
Beltsville	508	Zimbabwe, Nigeria, Yugoslavia, India, Spain, Vietnam, Ireland, Malaysia, England, Italy, Canada, Brazil, Mexico, Colombia, South Africa, Senegal, Eqypt, Poland, Argentina, Turkey, W. Germany, Austria, Australia, and New Zealand.
Hawaii	200	Global
Nairobi	95	Uganda, Malawi, Tanzania, Mauritius, Sudan, Congo, Zaire, Rwanda.
Porto Alegre	943	Argentina, Chile, Bolivia, Uruguay, Peru, Ecuador, Colombia, Venezuela, El Salvador, Dominican Rep., Mexico, USA, Trinidad, Brazil.

3.2. The biotechnology MIRCENs

Biotechnology, today, constitutes a frontier area offering a new technological base for the provision of solutions to the problems of all countries—developed and developing. The Japanese—the acknowledged *maître de la biotechnologie classique*—consider the applications of microbiology and the new biotechnology to be the last major technological revolution of this century (Nikkei 1982).

There is no lack of definitions of biotechnology. Whereas several definitions refer to the interaction of the disciplines of biochemistry and microbiology with the engineering sciences, several others are all encompassing in scope. On the other hand, a great majority acknowledge, indisputably, the intrinsic role played by the discipline of microbiology and its applications. Consequently, it is apt to reiterate that biotechnology, to a very great extent, derives its meaning and content from the industrial aspects of microbiology, i.e. fermentation processes (ranging from beer to penicillin), water and waste treatment, aspects of food technology, and a growing range of novel applications ranging from biomedical to the enhancement of oil recovery. In the semantic chaos surrounding the definitions of biotechnology, attention is drawn to the avoidance of extreme views (Teso 1982).

1. The public narrow definition focussing only on genetic engineering even though it is one of many techniques currently deployed.

2. The broad definition that embraces all activities utilizing living matter and particularly the agrofood industry.

Hal'ama *et al.* (1985), in summarizing the discussions of the Third Socialist Symposium on Biotechnology, found the term 'biotechnology' to be broad and imprecise. The term, used in a popular sense, seemed all-embracing and covered every technical aspect of biological processes—the agricultural and food technologies as well as those processes concerning kindergartens, restaurants, hospitals, and cemeteries. To these could be added the newly developed techniques concerning immunology. A reassessment, therefore, seemed more necessary and warranted. In this regard, a professional and realistic view on biotechnology has been prepared primarily for the press and the public (Houwink 1984).

Biotechnology is not an end in itself. It is a resource base for economical incentives, concerns ethical values and is a fount of technological possibilities. On the one hand, biotechnology reinforces prevailing thoughts and views, while, on the other hand, it opens different avenues of research that contribute new dimensions to the growth of human knowledge.

The harnessing of biotechnology for development necessitates a rational approach to the management of time-barred and renewable resources and to the development of appropriate socio-cultural and techno-economic infrastructures that will sustain the progress of the next three decades and more. North–South interaction will, out of necessity, focus on the restructuring of the bioindustrial base, implementation of new strategies of work in the new technologies to feed new markets, re-evaluation of North–South relationships and the evolution of South–South co-operation.

In the area of biotechnology, there are eight MIRCENs in operation. These are at the Thailand Institute of Scientific and Technological Research, Thailand (1976); Ain-Shams University, Cairo, ARE (1976); the Central American Research Institute for Industry, Guatemala (1979); the International Centre of Co-operative Research in Biotechnology, Osaka, Japan (1985); Planta Piloto de Procesos Industriales Microbiologicos, Tucuman, Argentina (1984); University of Maryland, USA (1984); the Institute of Biotechnological Studies, London. UK (1983); and at Ontario, through two co-operating institutions, the Universities of Waterloo and Guelph (1984).

The MIRCEN in the UK is sponsored by the Polytechnic of Central London, University College, London, and the University of Kent at Canterbury. In addition, the first associated laboratory in the network is the Commonwealth Mycological Institute's CMI MIRCEN for mycology.

In the region of South-East Asia, the MIRCEN, with its co-operating laboratories in the Philippines, Indonesia, Singapore, Malaysia, and Hong Kong, and other institutions in Thailand, serves the microbiological community in the collection, preservation, identification, and distribution of microbial germplasm; in the dissemination of information relevant to the cultures and their uses; and in the promotion of research and training activities that are directed towards the needs of the region.

A major task of the MIRCEN was the identification of activities of regional economic development in which micro-organisms play key roles. A survey of culture collections in ASEAN countries revealed that R & D activities were primarily focused on food fermentation with limited work in the area of other industrial fermentations. Regional workshops and training courses are organized periodically at the national and regional levels with a view to strengthening local expertise and regional capability in the utilization of culture collections for economic development. Outstanding examples of co-operation (Table 14.8), with Governments of Australia, Japan, Indonesia, and the Netherlands, and FAO, IAEA, UNU, ESCAP, UNEP, and other prestigious bodies, such as the International Centre of Co-operative Research and Development in Biotechnology, Japan, have been documented elsewhere (Atthasampunna 1984).

In the region of the Arab States, the MIRCEN at Ain-Shams University, Cairo, promotes research and training courses on the conservation and microbial cultures and biotechnologies of interest to the region. With the active co-operation and support of UNESCO's Regional Office for the Arab States, specialized courses have been organized in Sudan, Libya, Morocco, and Iraq.

Through its co-operating MIRCEN laboratory at the University of Khartoum, Sudan, the MIRCEN will have contributed to the establishment

Table 14.8
Some technical activities of the Bangkok MIRCEN

A. Culture collection services

Country requesting cultures	Number of cultures
China	3
France	18
Germany	7
Indonesia	22
Japan	24
Malaysia	13
Philippines	38
Thailand	682

B. Research activities

Project areas	Sponsors
1. Traditional fermented foods: taxonomic study	JSPS, TISTR
2. Polysaccharides of *Rhizobium* strains	JSPS, TISTR
3. Enzyme production in solid state cultivation	JSPS, TISTR
4. Decolorization of molasses pigment by micro-organisms	JSPS, TISTR
5. Production of power alcohol from cassava	Governments of Thailand and Japan
6. Microbial culture collection	ASEAN, Government of Australia

C. Regional and international collaboration

Topic	Sponsors
1. Micro-organisms for lignocellulose	SE Asia Network for Microbiology
2. Bioconversion of lignocelluosic and carbohydrate residues in rural communities	UNU, Governments of Indonesia and the Netherlands
3. Integrated approach to community development	Asian Centre for Population and Community Development

Table 14.8 *(continued)*

Topic	Sponsors
4. Agricultural and agro-industrial residues utilization in the ESCAP region	UNEP, ESCAP, FAO
5. Microbial enzymes	UNESCO, SE Asia Network for Microbiology, Bangkok MIRCEN
6. Research to improve energy production for agriculture, with emphasis on methane and alcohol, aided by nuclear techniques	FAO, IAEA
7. Development and application of technology to renewable and reusable resources	UNEP's Regional Office for Asia and Pacific
8. Productive utilization of organic residues for the production of foods, fuel, and fertilizer	US National Academy of Science
9. Application of science and technology for development	ACAST
10. Microbes for rural development	Government of Japan, ESCAP, ICRO, UNEP, IFIAS, Bangkok MIRCEN

of a culture collection in Sudan specializing in fungal taxonomy. Close collaboration in this regard exists with the Veterinary Research Laboratory at Soba, Sudan.

The co-operating MIRCEN laborating at the Institut Agronomique et Veterinaire Hassan II, Rabat, has made commendable progress through its Departments of Food Technology and Soil Microbiology.

Projects in progress include the following.

1. Food microbiology: isolation and characterization of yeast strains in samples of local butter (smen) at various stages of conservation. Attention is given to the study of the lipolytic activity of the yeast isolates and the factors (temperature, pH, physicochemical nature of the butter) which influence such activities.

2. Soil and agricultural microbiology: isolation of 165 strains of *Rhizobium* from the nodules of two cultivated varieties of chickpea, grown in the various bioclimatic areas of Morocco (from the Saharan to the sub-humid zones). Subsequent work involves evaluating the efficiency of the various strains (nitrogen fixation power, influence of the plant yield) and of determining

their growth characteristics (growth rate, influence of pH, temperature, dessication). The serological properties of the strains are coupled to immunofluorescence studies.

In the region of Central America and the Caribbean, the MIRGEN, in co-operation with the Organization of American States, the Interamerican Development Bank, and several other prestigious agencies, has pioneered the applications of microbiology, process engineering, and fermentation technology in several Member States of Central America and the Caribbean.

Within the MIRCEN framework, 31 co-operating laboratories in Costa Rica, Honduras, El Salvador, Nicaragua, Mexico, Peru, Jamaica, Venezuela, the Dominican Republic, Chile, Ecuador, and Colombia liaise with the MIRCEN at the Instituto Centroamericano de Investigacion y Tecnologia Industrial (ICAITI) in Guatemala.

The MIRCEN network focuses on the establishment of joint collaborative research projects; the exchange of technical personnel; regional training; and the dissemination of scientific information among network institutions (Table 14.9).

Research and training facilities are available in biomass production, biogas,

Table 14.9

Some activities of the biotechnology MIRCEN for Central America and the Caribbean

Year	Activity	Place
1979	Regional course on the utilization of agroindustrial wastes	Guatemala
1981	Travelling biotechnological workshops of MIRCEN fellows	Guatemala, Mexico
1981	Panamerican symposium on fuels and chemicals from biomass	Costa Rica
1981	Production of fuels from sugarcane and agro-industrial residues	Argentina
1982	Panamerican symposium on fuels and chemicals from biomass	Mexico
1982	Solid substrate fermentation workshop	Honduras
1983	Panamerican symposium on fuels and chemicals from biomass	Guatemala
1984	Basic biotechnology – regional course	Nicaragua

liquid fuels, solid waste treatment (aerobic thermophilic solid fermentations), photosynthetic systems in liquid waste treatments, and biopolymer degradation. An innovative feature of the work of this MIRCEN is the release of low-cost scientific publications that help to overcome the barrier of isolation in the region *vis-à-vis* the acquisition of up-to-date scientific and technical literature.

The South American Biotechnology MIRCEN located at Tucuman, Argentina, came onstream in early 1984. Comprised of a regional network with co-operating laboratories in Brazil, Chile, Bolivia, and Peru, this MIRCEN, with similar goals as the Biotechnology MIRCEN for Central America and the Caribbean, has organized regional workshops and training courses in Bolivia, Chile, and Argentina in important fields of economic importance in biotechnology with support from UNESCO and the OAS.

The MIRCENs in the industrialized societies function as an informal bridge with those in the developing countries. In such manner, increased co-operation is promoted between the developed and developing countries. Furthermore, the basis is gradually laid for eventual twinning at a later date. Moreover, as these MIRCENs are engaged in the frontier areas of research, the lead-up time in the transfer of new techniques and knowledge is considerably shortened. For example, the Guelph–Waterloo MIRCEN, with its expertise at the University of Waterloo in biomass conversion technology, microbial biomass protein production, and bioreactor design, is of immense benefit to the work of the MIRCENs at Cairo, Guatemala, and Tucuman.

In similar manner, the MIRCEN at Bangkok has conducted several collaborative research projects with that at the International Centre of Co-operative Research in Biotechnology, Osaka, Japan. This centre conducts the annual Unesco International Postgraduate University Course (of 12 months duration) on Microbiology. Since 1973, it has trained over 150 Asian microbiologists in fields ranging from microbial taxonomy, through fermentation kinetics to biomass residue utilization. It also functions as the Japanese national-point-of-contact for the South-East Asian regional network of microbiology in the UNESCO Programme for Regional Co-operation in the basic sciences.

In the UK, the MIRCEN network is centred upon the Institute for Biotechnological Studies (IBS) sponsored by the Polytechnic of Central London, University College London, and the University of Kent at Canterbury. The first co-operating MIRCEN laboratory in the network is the Commonwealth Mycological Institute's CMI MIRCEN for mycology which has played a significant role in organizing, with Unesco, a number of training courses in the field of biodeterioration biotechnology. For example, a course of 'Microbial Bioconservation in Museums' has dealt with the principles and techniques of entomology, preservation of biomaterials, textile conservation, restoration, restoration of oil paintings, deteriogenic fungi, and museum climatology *vis-à-vis* microbial activities.

The main thrust of the core-MIRCEN at the IBS is the organization of an annual International Postgraduate Diploma Course in Biotechnology for overseas scientists and engineers.

In keeping with the new trends of the expansion of the frontiers of biotechnological research, attention has been given to the potentials of marine biotechnology. A centre of marine biotechnology has been established at the University of Maryland which co-ordinates this activity with other MIRCENs. Work presently underway includes cloning of genes of ecological significance, which permits development of enzymatic processes functioning in the marine environment and, at the same time, allows fundamental elucidation of the evolution of genes and the flow of genes through populations in the marine environment, incuding the deep sea.

The fundamental processes of deep-sea and surface water bacteria are a major line of research. Bacteria have been found from deep-sea trenches to carry plasmids and questions of geological and evolutionary origin of enzymatic systems and relationships between micro-organisms of the deep ocean, surface waters, near-shore, and estuarine environments are under study.

One MIRCEN collaborative study underway is with the Chinese University of Hong Kong and the Shandong College of Oceanography, Qingdao, China.

4. Conclusion

The ultimate success of the MIRCEN concept and the network operations is dependent upon the development of a trained body of manpower capable of applying existing knowledge, and initiating and maintaining further research in the rapidly developing world of biotechnology.

Isolation of researchers, poor library suport, insufficient and deficient dissemination of research findings are amongst the most important constraints in developing countries which hamper the development of the research personnel, their capability, and the application of the technology. The MIRCENs have helped breach the barrier of isolation and advance the frontiers of contemporary research in biotechnology through the ample production of newsletter bulletins, culture catalogues, and research papers. The publication of MIRCEN News annually, the development of MIRCENET, the UNESCO/FAO *Rhizobium* World Catalogue, the *World Directory of Collection of Cultures of Micro-organisms*, and the UNESCO *MIRCEN Journal of Applied Microbiology and Biotechnology* are indications of the gradual emergence of competence and capability of the MIRCENs, and the services they provide on a regional and interregional basis.

Without doubt, the MIRCEN network has made a significant difference to the way microbiology is practised in developing countries. The programme has served to integrate the microbiological and general biotechnological infrastructures of the developed and developing countries by promoting the holding of international conferences in developing countries and helping introduce problems of developing countries into the programmes of conferences held in developed countries. The success of these many vigorous and rapidly-growing activities is due largely to a policy of versatile and flexible cooperation at the working level, with a great many governmental, intergovernmental, and non-governmental organizations participating, all of which result in a constellation of activities, with core funding from UNESCO and other UN agencies providing a high multiplier factor.

In the coming decades, both developed and developing countries will rely heavily on microbial technology to meet problems precipitated in the environment, in the food and industrial sectors, and in socio-economic fields. In this context given the increasing trend towards interdependence, the MIRCEN network of capable microbiological resources centres will have an important role to play on the regional and international scales.

References

Anon. (1984). International Summit Projects. *Look Japan*, 30, 17.

Atthasampunna, P. (1984). Bangkok Microbiological Resource Centre. In *Proc. 4th Int. Conf. Culture Collections* (eds M. Kocur and E. DaSilva) p. 53–62. WFCC, London.

BOSTID (1984). *Computer Conference on Lignocellulose Conversion*. National Academy Press, USA.

Brun, S. (1985). Networking is branching out all over the world. *Counterpart 2*, 27–30.

Bull, A. T. and DaSilva, E. J. (1983). World networks for microbial technology. *Soc. Gen. Microbial Q.* 10, 6–7.

DaSilva, E. J., Burgers, A. C.J., and Olembo, R. J. (1977). UNESCO, UNEP and the International Community of Culture Collections. In *Proc. 3rd Int. Conf. Culture Collections* (eds F. Fernandes and R. Costa-Periere). pp. 107–20. University of Bombay.

Di Castri, F. and Hadley, M. (1979). Research and Training for Ecologically-Sound Development: Problems, Challenges and Strategies. In *Proc. Vth Int. Symp. Tropical Ecology*. (ed. J. Furtado) Kuala Lumpur, Malaysia, 16–21 April. pp. 1229–52.

Friedman, Y. (1974). Groups and Networks. *Int. Ass. Newslett.* No. 1. 274–85.

Hal'ama, D., Blazej, A., and Duda, E. (1984). Biotechnology: Is it Biological or Biochemical Technology? In *Proc. 3rd Symp. Socialist Countries on Biotechnology*, 647–8. Vytlačili Západoslovenste Hacarne, Bratislava.

Houwink, E. H. (1984). *A realistic view on biotechnology*, DECHEMA. Schön and Vetzel, Frankfurt.

International Board for Soil Research and Management, Inc. (1985). *Report of the Inaugural Worshop and Proposal for the implementation of the Tropical Land Clearing for Sustainable Agriculture Network*. Bangkok, Thailand.

McNamara, R. S. (1973). *International Finance Corporation.—Annual Report*. Washington, DC.

Miller, J. G. (1978). *Living Systems*. McGraw-Hill, London.

Network Review (1986). *Ten Years Review of the Activities of the UNESCO Regional Network for Microbiology in Southeast Asia, 1986*. Network headquarters, Department of Biology, Chinese University of Hong Kong, Shatin, Hong Kong.

Nikkei S. S. (1982). *Advanced Technology Background—Biotechnology*, August 18, 2–3.

Plucknett, D. L. and Smith, N. J. H. (1984). Networking in International Agricultural Research. *Science* 225, 989–93.

Porter, J. R. (1974). UNESCO/ICO Panel on Microbiology, *ASM News*, 40, 259–65.

Teso, B. (1982). The promise of biotechnology and some constraints. *OECD Observer*, No. 118, September 1982.

UNESCO/ROSTSEA (1974). *Report—Meeting on Regional Cooperation in the Basic Sciences in Southeast Asia*, 19–25 February 1974. UNESCO/ROSTSEA, Jakarta, Indonesia.

Working Group on Technology, Growth, Employment (1984). *International Network of Biotechnology*. CESTA, Paris.

15

Biotechnology: principles and options for developing countries

J. Lamptey and M. Moo-Young

Summary

Technical breakthroughs in genetic engineering have led researchers to recognize the potential for directing the cellular machinery of living organisms to develop new and improved products, and processes in a wide diversity of applications. These applications could include the production of new drugs, food, and chemicals, the degradation of toxic wastes, and the improvement of production methods for agricultural and other products. In addition, increasing awareness of the enormous potential of novel bioconversion processes utilizing renewable resources has focused attention to a special significance for developing countries in the possibility of alleviating hunger, providing energy, and improving the quality of life in general. There are aspects of biotechnology, however, which may be inappropriate in many developing countries because of local constraints. Future development depends on proper planning and optimal utilization of local talents and resources. Judicious selection is required to determine those biotechnological processes that will provide net-positive socio-economic returns from the investments. Common constraints to most developing countries include limited capital especially of 'hard' currency, and a shortage of skilled labour. Nearly every developing country has plans or contemplates programmes for harnessing the tools of biotechnology for national development. It is, therefore, important that the problems be realistically understood before commitment of scarce funds. In this chapter, emerging trends in biotechnology research are highlighted and possible options for biotechnology development of particular relevance to developing countries are suggested.

Introduction

Recent progress in the development of novel bioconversion processes resulting from technical breakthroughs in recombinant DNA (deoxyribonucleic acid) hybridoma, and bioprocessing techniques, hold important production applications in agriculture, human and animal health, and energy production. Attention is focused on the need to exploit biotechnology worldwide. However, media coverage, especially in the industrialized countries, has created unrealistic expectations of new drugs, new crops, and new fuels. Intense research and development effort are being directed at several aspects of biotechnology (pharmaceutical, agricultural, renewable energy, etc.), but at a more reasonable pace than the general public in many countries has been led to believe (e.g. extensive clinical testing will be required for many new food and healthcare products).

Developing countries have also been caught up in the euphoria surrounding the so-called 'biotechnology revolution'. The basic development goals of most developing countries include increasing agricultural production, improving human and animal health, and making available increased amounts of energy for industrial purposes. For biotechnology to be effective in meeting the needs of most Third World countries, it must be compatible with specific ecological, socio-economical, and sociocultural factors. Compared with the more traditional domestic development programmes, the anticipated results from biotechnology appear to be cost-effective, and within a time-frame and general manpower-requirements of most developing countries (BOSTID 1982).

While some commercial applications of biotechnology are now emerging from current research (in the areas of pharmaceuticals, agriculture, specialty chemicals and food additives, commodity chemicals, and energy production), many basic research issues remain (Baltimore 1982). According to the US Office of Technology Assessment Report on Commercial biotechnology (OTA 1984). 'competitive advantage in areas related to biotechnology may depend as much on developments in bioprocess engineering as on innovations in genetics, immunology, and the other areas of basic science'. Hence, effort should also be directed at the solution of research problems associated with the design and scale-up of bioprocesses for the production of valuable products. Of particular importance are improved bioprocess systems for the cultivation of mammalian and plant cells, and the production of new genetically engineered products. Also, since most of the research carried out so far has been profit-motivated and particularly on problems of interest in developed countries, problems relating to developing countries should be carefully considered before the initiation of any expensive research programmes in biotechnology.

For Third World countries to gain both a realistic appreciation of the usefulness of the new techniques being developed and practical benefits from them, they must evaluate the potential of alternative techniques, screen the results of overseas developments in order to identify those of greatest applicability to local needs and to ensure that the means of their application advances local technological skills and employment opportunities. The priorities of developing countries with regard to biotechnology research can be expected to vary widely. The government of the Philippines, for example, has emphasized research on antibiotics, vaccines, and biomass production, while the Indian government has directed research on genetic engineering, tissue culture, enzyme engineering, and alcohol production. The Brazilian programmes for the large-scale production of ethanol, for use as a gasoline extender and octane booster, to partly alleviate foreign exchange imbalances by reducing reliance on expensive imported petroleum, is well known.

A number of research possibilities relate directly to the needs of developing countries, particularly, modifying agricultural plants to increase resistance to disease (viral, fungal, and bacterial plant pathogens), to tolerate unfavourable environmental conditions, (e.g. drought, frost, soil salinity), to fix nitrogen more efficiently and to produce inexpensive fuels from renewable resources, among others. Of particular importance are those applications which can help to produce substitutes for products or processes which generate environmental pollution hazards, alleviate waste disposal problems or achieve a more efficient utilization of raw materials. In the area of pollution abatement, the development of efficient facilities for the biological treatment of domestic and urban wastes is essential. For example, the production of fuel ethanol, methane fuel biogas, and animal feed single-cell protein (SCP) from renewable waste materials may help alleviate some aspects of these waste disposal problems (Moo-Young and Lamptey 1985). In this chapter, a brief review is given of the underlying principles of biotechnology, and its potential applications, present, and future, in developing countries.

Multidisciplinary nature of biotechnology

Biotechnology, broadly defined, includes any technique that uses living organisms (or parts of organisms) to make or modify products. Biotechnology, therefore, deals with the application of science and engineering to the direct or indirect use of cells from plants or animals, of micro-organisms, in their natural or modified forms, for the production of goods or the provision of services. It is a multifaceted, multidisciplinary activity with high potential for economic development and for improvement of human and animal health. In the past, the 'old' biotechnology (that is, before gene-splicing techniques were developed), has been used extensively in some developing countries, e.g. for the production of silage, compost, and biogas engineering.

Generally, the basic components of biotechnology (organism cultures, enzyme preparations, etc.) are developed by biologists and biochemists. The transfer of biotechnological discoveries requires engineering skills (e.g. applications in useful bioreactor designs are implemented by biochemical/ bioprocess and environmental engineers). For most applications, biochemical engineers are needed for the development, design, operation, control, and analysis of processes in which biological or biochemical phenomena are important.

Figure 15.1 illustrates the multidisciplinary nature of biotechnology with

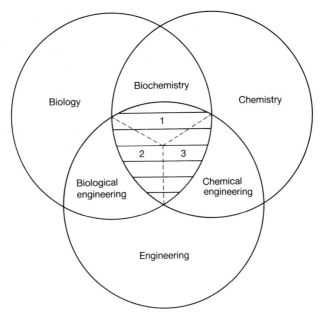

Fig. 15.1. Evolution of biotechnology as a multidisciplinary field. 1. Genetic engineering and molecular biology. 2. Agricultural engineering, biomedical engineering, and biophysics. 3. Biochemical engineering and microbial biochemistry

the principal disciplinary underpinnings being biology, chemistry, and engineering. Intermediate areas of overlap are biochemistry, biological engineering, and chemical engineering which, in turn, have led to three aspects of biotechnology of growing importance: (1) genetic engineering and molecular biology, (2) biomedical engineering and biophysics, and (3) biochemical engineering and microbial biochemistry. Other overlapping areas, such as environmental engineering, fermentation engineering, enzyme engineering, and food and agricultural engineering, are also well known.

The multidisciplinary nature of biotechnology, especially in these days of 'biohype' cannot be over-emphasized. Biotechnology, like any other form of technology (e.g. chemical, nuclear), requires process engineering skills to translate relevant scientific facts into practice. At a molecular level, the so-called genetic engineers may create new life forms or modify existing ones,, but it is at the equipment level where process engineers design and control optimal conditions for the useful and economic functioning of biosystems, be they old or new. The following anecdote illustrates the seriousness of these observations, especially for Third World countries.

At a certain research institution, a biochemist thought he had developed a wonderful new drug and was anxious to produce a large sample of it for product testing. His engineering colleague agreed to prepare such a sample in a pilot plant stirred-tank reactor. Following the recipe, the engineer proceeded mixing various ingredients together in the 500-l steel tank. As the night wore on, the mixture became thicker and thicker, and by midnight had solidified. Panic-stricken, the engineer telephoned the scientist and explained the problem; 'Oh' replied the biochemist, 'I forgot to tell you that sort of thing happened now and again, but I never bothered about it. I would simply break the beaker and take the stuff out'. Obviously the constraints of process engineering can be easily overlooked. Figure 13.2 illustrates other important socio-economic factors to be considered besides basic science and engineering in technology transfer.

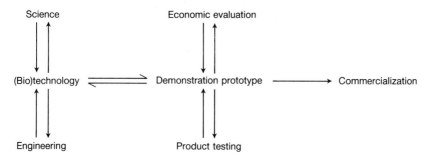

Fig. 15.2. The feed forward/feedback interactions in technology transfer

State-of-the-art of biotechnology

The novel techniques currently being used in biotechnology allow a large amount of control over biological systems; for example, by manipulating the DNA in a cell it is possible to modify its metabolism for the synthesis of special proteins. By this 'genetic engineering', cells may be constructed that make metabolites such as enzymes and other products which they do not normally produce.

A review of the present status of biotechnology would be highly inadequate without mention of the recent remarkable achievements in molecular biology. The discussion will focus on the technical bases and industrial use of recombinant DNA, cell fusion, and bioprocessing techniques. These three 'foundation' techniques find broad applications in areas of pharmaceuticals, plant and animal agriculture, specialty chemicals, environmental applications, commodity chemicals, and energy production.

For additional information on the techniques, the reader is referred to the following references: Bailey and Ollis (1977); Moo-Young and Blanch (1981); Wood et al. (1981); BOSTID (1982); OTA (1984); Bjurstrom (1985).

Recombinant DNA technology

The most widely publicized and one of the most powerful of the novel techniques currently being used in biotechnology is recombinant DNA (rDNA), which includes gene cloning. Current advances in the field of rDNA research have led to techniques which mobilize individual genes, that is, specific sequences of DNA which code the amino acid structure of single proteins and transfer them from a donor to a recipient organism, thus conferring on the recipient the ability to synthesize the gene product. The source of this foreign DNA can be microbial, animal or plant and thus microbial hosts can be directed to produce new products, existing products more efficiently, or large quantities of otherwise scarce products.

Genetic information is stored and transmitted by the information system contained within the DNA. When cells divide each daughter cell receives a complete copy of its parents' DNA, which allows offspring to resemble parent in form and operation. DNA is a linear polymer of nucleotides, which, in turn, are the monomer units of nucleic acids. Deoxyribonucleotides consist of three components: an aromatic base, a pentose sugar, and phosphate groups. The DNA molecule is along double helix consisting of two chains. Each chain is a polymer constructed in part from four nucleotides: adenine(A), guanine(G), cytosine(C), and thymine(T). A gene is an ordered sequence of these letters, and each gene contains the information for the composition of a particular protein and the necessary signals for the production of that protein (Stanier *et al.* 1976; Wood *et al.* 1981).

The structure of the DNA molecule immediately suggests how it can be accurately replicated (Bailey and Ollis 1977). The nucleotide bases are paired in a specific manner to form the rungs of the twisted DNA ladder: adenine in one strand is always paired with thymine in the other, and guanine and cytosine are likewise paired. Only the two base pairs, A–T and G–C, can be accommodated in the double-helical structure (Wood *et al.* 1981). As a consequence, the nucleotide sequence in the other, and the two strands of a DNA duplex are said to be complementary. The pairing provides molecular stabilisation by hydrogen bonding between complementary bases. The entire molecule can thus be described as a linear sequence of nucleotide pairs; the exact order of these pairs constitutes the genetic message which contains all the information necessary to determine the specific structures and functions of the cell (Stanier *et al.* 1976; Bailey and Ollis 1977).

Base pairing also provides a means for duplication of DNA. If the two complementary DNA strands are separated and double helices are constructed from each strand following the base-pairing rules, two new molecules are produced, each identical to the original DNA molecule. Base pairing, therefore, provides a chemical reader for the biological message coded in the DNA nucleotide sequence (Bailey and Ollis 1977).

A bacterial cell can synthesize several thousand different kinds of proteins.

The information required to direct the synthesis of these proteins is encoded by the sequence of nucleotides in the cell's complement of DNA. By the process of replication the chromosome is precisely duplicated, thus assuring that progeny cells receive information enabling them to synthesize the same proteins. The process by which the encoded information of the chromosome directs the coder of polymerization of amino acids into proteins occurs in two steps: transcription and translation (Stanier *et al.* 1976).

In the first step, transcription, the information content of one of the strands of DNA is transcribed into RNA (ribonucleic acid): i.e. the DNA strand serves as a template upon which a single strand of RNA is polymerized. One class of these RNA molecules, termed messenger RNA (mRNA), carries the information encoded in the DNA to the protein-synthesis machinery. In the second step, translation, protein synthesis takes place on ribonucleoprotein particles called ribosomes, which attach themselves to the molecule of mRNA. The information carried by the mRNA molecules is translated into protein molecules by a special class of RNA molecules called transfer RNA (tRNA). The end result of this intricate information transmitting and translating system is a protein molecule. Proteins, in their role as enzymes, directly determine how the cell operates by determining which chemical reactions occur. Thus, the inheritance which a dividing cell leaves to daughter cells must include the information for directing the production of the same protein constituents as found in the parent cell (Bailey and Ollis 1977; Stanier *et al.* 1976; Wood *et al.* 1981; OTA 1984).

The discovery of enzymes called restriction endonucleases led to the development of recombinant DNA techniques which allow a DNA fragment from any source to be joined artificially with procaryotic or eucaryotic DNA molecules that replicate autonomously. Micro-organisms carrying such recombinant DNA molecules can be cloned in large numbers and the exogenous fragment can easily be re-isolated in quantities sufficient for size and sequence analysis (Wood *et al.* 1981).

The code by which genetic information is translated into proteins is the same for all organisms. Thus, because all organisms contain DNA and all organisms interpret that DNA in the same manner, all organisms are in essence, related. This concept forms the basis for the industrial use of DNA. New combinations of genetic information are produced by recombination of DNA sequences. In nearly every instance, a bioprocess using recombinant DNA (rDNA) techniques depends on the expression of DNA from one species in to another. According to the OTA (1984) report, one of the great challenges of rDNA technology is to construct DNA molecules with signals that optimally control the expression of the gene in the new host. The report identified the most difficult part of the cloning process to be the isolation of an appropriate probe. A probe is a sequence of DNA that has the same sequence as the desired gene and has been prepared in such a way that it can be identified after it base pairs with that gene.

Initially, the genes that were first cloned were those that, in certain cells, produced large quantities of relatively pure mRNA. Since the mRNA was complementary to the gene of interest, the mRNA could be used as a probe (Wood *et al*. 1981; OTA 1984). The drawback to ths technique is the fact that most genes do not produce large quantities of mRNA. Recent advances in rDNA research have led to the development of a technique which allows a much greater spectrum of genes to be cloned. With this technique, the amino acid sequence of a protein is determined, and then working backwards through the gene expression scheme, the nucleotide base sequence is determined.

Recombinant DNA techniques have been used principally with bacteria and to a lesser extent yeast. Particular attention is being given to the use of yeasts as host for rDNA cloning because they more closely resemble cells of higher organisms, they can be grown to very high densities, are stable with respect to mutation, produce no toxins, and can secrete proteins.

Apart from their natural function in the protein of organisms via the immune response, antibodies have long been important tools for researchers and clinicians, who use an antibody's specificity to identify particular molecules or cells and to separate them from mixtures. Antibodies also have a major role in diagnosis of a wide variety of diseases. Antibodies that recognize known antigens are used to detect the presence and level of drugs, bacterial and viral products, hormones, and even other antibodies in sensitive assays of blood samples (OTA 1984).

The standard method of obtaining antibodies (for therapeutic, diagnosis, and investigational purposes) is to inject an animal (or human) with the antigen for which an immune response is desired. Blood serum removed from the animal (or human) will then contain a variety of antibodies, each specific to a different part of the injected antigen molecule. Several problems render the use of this technique unattractive. The general technique of extraction is slow and very tedious, and it is very difficult to isolate a specific antibody. The antibodies obtained from several extraction and purification steps are usually only weakly specific, available in limited amounts for any given purpose, and of relatively low activity (Diamond *et al*. 1981; Schulman *et al*. 1978; Davis *et al*. 1982).

In 1975, a new era in immunology was launched with the discovery of the hybridoma technique, a method for creating pure and uniform antibodies against a specific target by Kohler and Milstein (1975, 1976). The technique involves the fusion of myeloma (antibody-producing tumour, i.e. cancer) cells with antibody-producing cells from an immunized donor. The hybrid cell or 'hybridoma' resulting from this fusion has the ability to multiply rapidly and indefinitely in culture, and to produce an antibody of predetermined specificity, known as a 'monoclonal antibody'. The antibodies produced by monoclonal antibody (MAb) technology are homogenous, and the production is predictable and reproducible. The technique permits the analysis of almost any antigenic molecule.

Monoclonal antibody technology has provided a breakthrough in the methods available for the analysis of the antigenic composition of micro-organisms, for rapid diagnosis, and to assist in the development of vaccines. It has enhanced research advances in understanding diseases of humans, animals, and plants at reduced cost. The technique is relatively simple and straightforward to use, and can be readily developed and made available for large scale use in developing countries through co-operation between research centres in developed and developing countries.

The most important factor in the production of hybridomas is the preparation of cells for fusion. Where rodents are used as a source of cells for fusion, immunization protocols must be devised that optimize the proliferative response to antigen in the spleen (Davis *et al.* 1982). A brief outline of the methodology is given here. For more detailed information on the production of MAbs, the reader is referred to the following references (Kohler and Milstein 1975, 1976; Olsson and Kaplan 1980; Baskin 1982; Davis *et al.* 1982).

During the preparation of MAbs, the preferred purified antigen is injected into a mouse, and a few weeks later, the spleen of the mouse is removed. The B lymphocytes (antibody-producing cells) are isolated from the spleen and fused with myeloma cells. The resulting cells are placed in a cell culture medium that allows only the hybridomas to grow. The several hybridomas which result are cloned, and each clone is then tested for the production of the desired antibody. A particular hybridoma clone either may be established in an *in vitro* culture system or may be injected into mice, where the hybridoma grows in the abdominal activity fluid (ascites) from which the antibodies are readily collected (OTA 1984).

Despite the great promise of MAbs, a number of technical problems need to be solved before the full potential of MAb technology can be realized. The potential problems requiring further research and development relate to the following: (i) obtaining MAbs against certain weak antigens remains a difficult task to achieve (Melchers *et al.* 1978; Fox *et al.* 1981); (ii) it has been found that homogeneous antibodies cannot perform some functions, such as forming a precipitate with other antigen–antibody complexes, a necessary function for some diagnostic assays; (iii) the low frequency of fusion, and the stability of the hybridomas and antibodies are potential problems requiring further research (Gefter *et al.* 1977); (iv) some monoclonal antibodies are especially sensitive to small changes in pH, temperature, freezing, and thawing, and can be inactivated during purification (OTA 1984).

Plant cell and tissue culture

The need for novel research techniques that will ultimately lead to increased agricultural productivity is of prime importance for Third World countries.

The use of high-yield crop varieties and efficient production methods have enabled farmers in the United States and Western European countries to increase production markedly in the past several decades, and are of special importance to improved agricultural output in developing countries.

The rapid development of plant cell and tissue culture techniques, and the use of recombinant DNA techniques to crop improvement have generated considerable excitement in agriculture in recent times (Marx 1982; Goodin 1982; Collins 1982). Plant cell and tissue culture techniques offer great potential for the future development of agriculture. Several species of plants can now be cultured and regenerated, allowing many identical plants, including trees, to be grown. Virus-free potato plants, for example, are now obtained from regenerated cell-cultured potatoes, and as a consequence, the yield of these plants has increased substantially (BOSTID 1982).

According to Collins (1982) plant cell and tissue culture techniques provide input to plant breeding efforts at two levels. The first is to make available to breeders, materials such as genetic and cytogenetic variants, haploids, wide-cross hybrids, disease-free stocks, and large numbers of propagules. The second and possibly the most important long-range contribution of plant cell and tissue culture techniques is to provide the bridge between molecular genetic engineering methods and plant breeding. Most of the schemes envisioned involve the transfer of genetic information and require that cells from the recipient plant be cultured. However, the use of improved genetic techniques make possible direct manipulation of pollen or embryos, circumventing the necessity for regeneration.

If single cells or embryos can be cultured, selected, and regenerated, the number of cells in any experiment can be potentially very large, contributing to the overall success of a given experiment. Thus, the potential of plant cell and tissue culture techniques for crop improvement needs to be fully assessed. In spite of the recent notable achievements in plant cell and tissue culture techniques, there are still only a limited number of species that can be easily cultured (Conger 1981). The crop species for which regeneration from cells can be accomplished easily include asparagus, rape seed, cabbage, citrus, sunflower, carrot, cassava, alfalfa, millet, clover, endive, tomatoes, potatoes, and tobacco (Ehleringer 1979; OTA 1984). Current research efforts are being directed at improving the production of corn, wheat, soybeans, and rice at various research centres in both developed and developing countries.

For agricultural applications, it is essential to regenerate whole plants from plant cells. The term 'plant tissue culture' has been loosely applied to all forms of plant cultures grown *in vitro*, ranging from undifferentiated single-unit protoplasts to highly organized and complex organ cultures. In all cases, cell regneration is accomplished by growing the cells under controlled conditions of temperture, light, and with the appropriate growth nutrients, which include various amounts of vitamins, amino acids, inorganic salts, and carbon energy source. Often, complex organic additives such as coconut water, casein hydrolysate, and yeast extract are incorporated for undefined nutritional or hormonal requirements (Collins 1982).

Collins (1982) has reviewed the basic techniques for preparing plant cell tissue cultures. These cultures, which include callus, cell suspension, organ, meristem tip, and protoplast, are summarized below.

Callus culture

During the initial stages of regeneration, masses of undifferentiated cell clusters, known as callus, are formed. Embryonic, seedling, mature vegetative plant parts, and ovules are generally used as initial explant tissue sources.

Cell suspension culture

This technique involves the growth of plant cells and cell aggregates in liquid media. The broth is normally aerated and well mixed. Calli or vegetative tissues are usually used as initiating tissues.

Organ culture

Cultures of plant organs, such as roots, shoots, embryos, anthers, and ovules are usually prepared on solidified agar medium, or on a solid support, in which case a wick is used to provide contact to a liquid medium.

Meristem tip culture

Shoot meristem tips of varying sizes are inoculated into culture on a solidified agar medium followed by transfer of plants to a root-inducing medium.

Protoplast culture

Protoplasts are usually isolated from leaf mesophyll tissues, root tissues, or cell suspension cultures by enzymatic digestion, followed by purification and concentration steps. Cell regeneration is achieved by culturing the protoplasts in a liquid medium. Callus proliferation and plant regeneration are effected by transfer of the cultured protoplasts to an agar medium.

According to Conger (1981) and Collins (1982), one of the major problems encountered in achieving *in vitro* culture capability with all plant species, especially with genotypes of agricultural importance, is the general lack of quantitative information on the physiology, biochemistry, and developmental aspects of plant cells in culture. Successful culturing of most species has generally only been achieved empirically. The specific roles and interaction of growth regulators, including auxins, cytokinins, abscisic acid, and gibberellins, are generally not adequately understood.

Another major problem with *in vitro* culturing is the general ability to regenerate complete plants from single cells and protoplasts. This is especially true with grasses, cereals, and woody species. To be economically attractive, transformed cells, variant cells, and haploid microspores must be capable of regeneration into a complete plant. Restrictions on plant regeneration from cell cultures often include explant source, either embryonic or juvenile, genotype of explant donor, and the limited period for which the culture can be maintained before regenerative potential is lost (Collins 1982). A list of 45 crop plants for which plants have been regenerated from tissue cultures, incuding 25 agronomic species, 11 tree species and nine vegetable species have been reported by Gamborg and Shyluk (1981). Almost none of these examples involved plant regeneration from single cells.

Bioprocessing technology

This section deals with the apparatus level where engineers design and control optimal conditions for the useful functioning of bioprocesses. It is the third of the three constituent basic technologies in biotechnology and is probably the least appreciated at the R&D level, but is often the techno-economic 'bottleneck' in a process implementation (OTA 1984). Therefore, bioprocessing technology is treated in relatively more detail. In these processes living organisms or their components are used to effect desired physical or chemical changes. The term bioprocess is used in preference to the more familiar, but more restricted term 'fermentation' because it includes the broader range of operations involved as pointed out by Moo-Young and Blanch (1981), and OTA (1984). Strictly speaking, a fermentation process refers only to an anaerobic bioprocess.

The use of bioprocesses for the production of valuable products (solvents, chemicals, pharmaceuticals, etc.) is especially attractive due to the high selectivity of the biological agent (e.g. microbe or enzyme) which constitutes an integral part of these processes. There are also compelling reasons for using bioprocesses for the production of certain parts. The most persuasive occurs when products simply cannot be produced effectively by other means, such as is the case for most viral vaccines and complex biologically active molecules (e.g. cells of mammalian origin), and various food/feed-type products (e.g. microbial biomass protein, amino acids, etc). The raw materials used are generally less expensive or more readily available compared to conventional chemical processes.

For most operations, the principal raw materials (carbohydrates) are renewable. In addition, due to their biological nature, these processes are normally carried out under low (ambient) temperature and pressure conditions. Most biochemical processes involve less hazardous operation and reduced environmental impact. Major disadvantages include the low reaction rates and the very dilute end-product concentrations typical of most biological processes. In addition, the production of complex product mixtures results in extensive separation and purification steps during product recovery.

In order to make the new technology economically competitive with technologies already in operation, much effort is being directed at the improvement of the unit processes common to these biochemical operations, particularly in the areas of bioreactor design, and product recovery and purification. For processes based on the use of lignocellulosic biomass materials, intensive research effort is being directed at the development of better pretreatment techniques. The potential of genetic engineering covers entirely new products not made in any other manner, as well as those currently being produced by synthetic routes. The challenge to the biochemical engineer is to translate successfully these processes from laboratory scale to the industrial scale.

As a result of the recent upsurge of interest in biotechnology, advanced approaches to the unit processes in biotechnology have been and are being developed. Concepts such as continuous bioprocessing, vacuum bioprocessing, airlift bioreactors, and immobilized cell/enzyme bioreactors operated in packed and fluidized-bed modes of operation, to name a few, have been recently introduced. In addition hybrid bioprocesses where two or more operations are combined in a single stage such as simultaneous attrition-enzymic saccharification (Ryu and Lee 1983; Lamptey *et al.* 1984) and simultaneous saccharification–fermentation (Lee *et al.* 1983) are becoming increasingly popular.

The general principles involved in the design and operation of bioreactors are briefly reviewed. Principal downstream processing techniques are also presented.

A generalized outline of a typical biochemical process and examples of the processing operations involved are shown in Fig. 15.3. In these operations, the importance of process engineering transport phenomena (mass, heat, and momentum transfer) are evident in all three stages of the process, viz., upstream and downstream processing, where conventional chemical engineering techniques often suffice, in addition to the heart of the process—the bioreactor—which often determines the overall process economics. The need to maintain aseptic conditions is one of the distinguishing features of biochemical processes. It is essential that the bioreactor and its ancillary equipment be sterilizable, and that aseptic conditions can be maintained.

Essentially, the design principles involved in biochemical engineering are similar to those of its parent discipline, chemical engineering. However, additional knowledge of various aspects of the life sciences are often required for the rational application of these principles to process development, equipment design and scale-up. Bioprocesses require a closely controlled

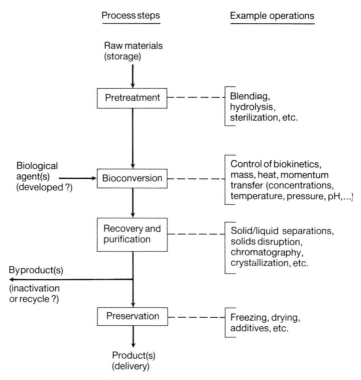

Fig. 15.3. Generalized outline of a typical biotechnological process in the manufacturing industries.

environment for optimal performance, and this necessity markedly influences their design. Special precaution is necessary because of the labile process-sensitive nature of the biocatalysts (immobilized or free suspension cultures of cells or enzymes) and some of the valuable products involved, thus, avoiding protein denaturation, organism inactivation and eventual death, vitamin degradation, etc., especially from heat and/or mechanical shear or impaction. The necessity to maintain a sterile environment has a far-reaching effect on bioprocess equipment design and operation. It is important during the fabrication process to avoid crevices where steam could not reach microorganisms when the vessel is sterilized.

Under controlled conditions, the substrate (raw material) is converted to the product and, when the desired degree of bioconversion has been achieved, byproducts and wastes are separated.

Bioreactor design considerations

A primary task in biochemical engineering is the efficient design, operation, and control of bioreactors. The basic requirement is the optimization of the physical rate processes for mass and heat exchange (between the biocatalysts and their environment) so that the desired intrinsic (inherent) biokinetics of the bioprocess are satisfied or, if this is not possible, the global (gross) process kinetics are maximized. As a prerequisite, adequate conversion yields from the raw materials are required for an economical process. This is especially important in bioprocessing since for several bioprocesses (e.g. ethanol, methane fuel biogas and microbial biomass protein production, etc.), raw materials costs constitute a significant fraction of the total operating costs.

One essential requirement in bioprocessing is a source of carbon for the bioconversion. Carbon sources are required by micro-organisms to supply the energy needed for metabolism and the synthesis of compounds. For some bioprocesses, the carbon source is the substrate used for the bioconversion, as in the production of ethanol from hexose sugars. Carbohydrate feedstocks (the most common substrates) usually in the form of sugars (e.g. from molasses or hydrolysed starch) and starches, triglycerides, and, to a smaller extent, petroleum fractions, serve as carbon sources. After carbon, the next important material required as nutrient substrate is nitrogen (normally in the form of ammonia or ammonium salts) in the ratio of about $C:N = 10:1$ for aerobic conditions and about $C:N = 30:1$ for anaerobic conditions. Other nutrient requirements, used in lesser quantities include phosphorus, potassium and 'trace elements' supplied as mineral salts.

Appropriate stoichiometric relationships are used to calculate the conversion yeilds for various components. In a typical microbial process, an aerobic cell biomass yield is about 50 per cent based on a carbohydrate feedstock and about 75 per cent based on a hydrocarbon one. Anaerobes usually have about one-fifth of these yield values. The cell biomass itself is comprised of about 40–60 per cent crude protein on a dry weight basis, depending on the species.

In addition to cell biomass yield, the biochemical engineer is interested in the conversion factors for the heat of bioconversion, oxygen requirements (if the process is aerobic), etc. Empirical compositions of a particular microbial biomass are often used to arrive at the various stoichiometric equations.

Compared to chemical reactors, bioreactors have special properties which demand caution in the direct application of classical chemical engineering principles. Moo-Young and Blanch (1981), and the OTA (1984) have discussed the potential constraints involved in the design of bioreactors. Some of the conceivable systems and operating constraints involved in the operation of bioreactors are as follows.

1. The use of relatively complex reactant mixtures, some with variable composition, (e.g. lignocellulose), usually results in the formation of complex product mixtures which require extensive (and usually expensive) separation and purification steps.

2. The use of suspended living cells with bulk densities approximately equal to that of their liquid environment tends to minimize interphase mass

and heat transfer rates due to the low relative flow velocities between the dispersed and continuous phases.

3. The use of small microbial cells (in the range of 1–8 μ) coupled with constraint (b) above, usually make it difficult to promote high particle velocities and attain turbulent-flow transfer conditions necessary for optimum mass transfer.

4. The high viscosity and filamentous nature of some mycelial bioprocessing results in bioprocess culture broths which are generally non-Newtonian pseudoplastic. Such non-Newtonian conditions tend to limit desirably high flow dynamics in bioreactors.

5. The formation of relatively large cell aggregates such as mycelia, clumps or pellets in some bioconversions (e.g. fungal bioprocessing) may result in serious intraparticle diffusional resistances leading to oxygen limitations.

6. Careful control of bioreactor operating variable (e.g. solute concentration, pH, temperature, and local pressures) is usually required to avoid damage or destruction of valuable live or labile components.

7. Often, during bioprocessing, the concentrations of reactants (and consequently, products) in aqueous media are usually low, resulting in limited concentration driving forces for mass transfer. In addition, the need to provide and handle large volumes of process water and to dispose of equivalent volumes of high biological oxygen demand wastes, leads to complex and frequently energy-intensive recovery methods for removing small amounts of products from large volumes of water (e.g. distillation of dilute ethanol beers).

8. The use of micro-organisms with slow bioconversion rates results in the use of relatively large bioreactor volumes and consequently long residence times during bioprocessing.

9. The susceptibility of most bioconversion processes, especially during extended continuous operation, to contamination by foreign organisms requires careful consideration during the design and operation of bioreactors.

Solutions to some of these problems through the use of novel and improved biotechnology techniques may make bioprocesses more competitive with conventional chemical syntheses. A combination of novel bioreactor design, operating strategies, recombinant DNA and monoclonal antibody techniques may result in biochemical processes with significantly improved performance economics than they have been in the past. As an illustration of some of the problems imposed by the above constraints, we note that an adequate oxygen supply rate to growing cells is often critical in aerobic bioprocessing (e.g. microbial biomass protein production). The low solubility of oxygen in water represents a significant drawback to efficient bioprocessing since, when the oxygen transfer ability of the bioreactor is exceeded by the ability of the organism to consume oxygen, bioconversion becomes limited. Since oxygen is consumed by the micro-organisms during the bioconversion process, the bioreactor culture broth must be continuously aerated in such a way that the oxygen transfer rate, at least, equals the oxygen consumption rate by the organisms. This is due to the fact that even a temporary depletion of dissolved oxygen could mean irreversible damage to the organisms or a significant reduction in bioconversion rate.

The main bioreactor design parameters which influence the extent of oxygen transfer include: aeration rate, type of bioreactor (mechanically agitated systems versus pneumatically-mixed systems), agitation power input, type of agitator, system pressure, temperature and the rheological properties of the broth. Current techniques for ensuring an adequate oxygen supply include: the use of oxygen-rich gas for aeration, use of high agitation and aeration rates, use of relatively high bioreactor pressures to increase oxygen solubility, and the use of novel bioprocess design and operating techniques (e.g. various designs and the airlift bioreactor). For more details on oxygen transfer in bioreactors see Lamptey 1978; Moo-Young et al. 1978; Moo-Young and Blanch 1981; Kargi and Moo-Young 1985.

Physical rate processes

The kinetics of the physical processes of mass and heat transfer are governed by the hydrodynamic conditions and the physico-chemical properties of the biochemical system. In bioreactors, the interphase transfer rates of supply of nutrients (e.g. sugars, dissolved oxygen) and removal of product metabolites are given by the general expression.

$$Sh = f\,(Re,\ Sc,\ Gr) \text{ for mass transfer} \tag{1}$$

or,

$$Nu = f\,(Re,\ Pr,\ Gr) \text{ for heat transfer} \tag{2}$$

where $Sh = k_L d/D_L$ $Re = dv\varrho/\mu D_L$ $Gr = d^3/\varrho g\Delta\varrho/\mu^2$, $Nu = hd/k$, $Pr = C_p\mu/k$, and k_L = specific mass transfer coefficient, d = effective particle diameter, D_L = mass diffusivity, v = relative particle velocity, ϱ = continuous phase density, μ = continuous phase viscosity, g = gravitational acceleration, h = heat transfer coefficient, k = thermal diffusivity, C_p = heat capacity.

The Sh (Sherwood number) or Nu (Nusselt number) gives the total mass or heat transfer rate relative to the base molecular diffusion; Re (Reynolds number) reflects the intensity of bulk flow ranging from quiescent to turbulent conditions; Sc (Schmidt number) or Pr (Prandtl number) gives the mass or heat diffusivity relative to momentum diffusivity; and Gr (Grashof number) relates the possible effect of gravitational forces relative to viscous ones on the mass or heat transfer rate. The degree of dispersion which governs the interfacial area available for interphase mass and heat transfer also plays an important role in the overall rate processes.

The rate of oxygen transfer in a bioreactor system is given by:

$$OTR = k_L a(C^* - C_L) \tag{3}$$

where OTR = oxygen transfer rate; k_L = oxygen transfer coefficient; a = interfacial area for oxygen transfer; C^* = equilibrium dissolved oxygen concentration, C_L = dissolved oxygen concentration. It is obvious from eqn (3) that one of the easiest ways to increase the oxygen transfer rate is to increase the oxygen concentration driving force.

In general, the oxygen transfer rate is estimated on the basis of the average oxygen consumed with time, by analysing the bioreactor off-gas. The driving force is usually estimated on the basis of the operating conditions of temperature and pressure, which determine the potential oxygen solubility, and from the dissolved oxygen in the culture broth, which is measured experimentally. There are a number of different techniques for measuring *in situ* the dissolved oxygen concentration in bioreactors (Kargi and Moo-Young 1985). The most widely used technique involves the use of an oxygen probe permanently located in the bioreactor. The volumetric oxygen transfer coefficient ($k_L a$) can then be back-calculated, using eqn (3).

For optimum performance of the bioreactor, bioprocess systems are usually maintained within a very narrow operating temperature range by means of efficient cooling systems. The main source of heat that must be removed during bioprocessing is due to the metabolic activity of the micro-organisms. All organisms produce heat as a byproduct of the growth process. The design of cooling systems for bioprocessing is influenced significantly by the scale of operation. A cooling jacket is usually adequate for small bioreactors while some type of internal coil or external heat exchanger is used for large-scale operation.

To promote adequate mass and heat transfer rates in bioreactors and other unit operations (for pretreatment and product recovery steps) in a biochemical plant, processing energy for mixing and agitation is required. Substantial levels of the operational costs are involved in bulk mixing, especially for large scale operation. The pneumatic energy consumption is readily calculated from the gas flowrate and its related pressure drop (compressor power). The mechanical energy consumption is not as simple to evaluate. It can be deduced from a variety of correlations (applicable to mechanically-agitated bioreactors) of the following general form:

$$\frac{P}{N^3 D^5 \varrho} = f\left(\frac{ND^2 P}{\mu}, \frac{N^2 D}{g}\right) \tag{4}$$

where P is the power, N is the impeller rotation rate, D is diameter of the impeller, ϱ is the density of the liquid, μ is the viscosity and g is acceleration due to gravity. In recent years, the energy crisis has encouraged the development of novel configurations in attempts at minimizing mixing costs in bioreactors (Moo-Young *et al.* 1978; Moo-Young and Blanch 1981; Bello *et al.* 1985).

Biokinetics

Various mathematical expressions (mechanistic or empirical) are available for describing quantitatively, cell growth, enzyme hydrolysis, substrate uptake and product formation rates. For example, the simplest and most common cell growth pattern follows an exponential law and in continuous steady-state operations ('chemostat'), the specific growth rate, μ, is given in terms of the maximum specific growth rate, μ_m, the limiting nutrient concentration, S, and a system property known as saturation constant, K_s, as follows:

$$\mu = \frac{\mu_m S}{K_s + S} \tag{5}$$

Various modifications to eqn (5) are used to account for special conditions such as substrate or product inhibition/inactivation or induction/activation (Bailey and Ollis 1977). In addition, there are also various expressions inter-relating product formation to cell growth according to 'growth associated', 'non-growth associated', and 'mixed' kinetics (Aiba *et al.* 1973).

Bioprocess control and instrumentation

The recent upsurge of interest in biotechnology has led to new developments in the area of sensors, data acquisition, and analysis, and general bioprocess control philosophy. Process control of bioprocesses is increasingly aided by computer technology especially via microcomputer/microprocessor facilities. Computercoupled bioreactors are now being used to greatly improved monitoring and control of bioprocesses as well as to analyse data from various sensors. In addition, it is expected that computer interfaces will be used to schedule more efficiently the use of various bioprocess equipment, and to log, store, and analyse data on a continuous basis.

At present, the main difficulty in monitoring plant operation lies in the lack of suitable sensors for certain variables. Existing equipment can readily monitor only temperature, pH, dissolved oxygen concentration, and evolution of gases. Although many other sensors (or probes) have been developed to measure several other variables (e.g. glucose concentration, important metabolite levels), these probes are not rugged enough for long-term repeated steam sterilization or they cannot be used on-line, especially, without undesirable fouling and/or transient responses.

Advanced instrumentation techniques will continue to have increasing use in bioprocess monitoring and control. High performance liquid chromatography (for identification of particular compounds in a mixture) and flow cytometry (for determination of cell size and cell viability) are now being routinely used in various bioprocess operations. Other novel measuring techniques can be expected in the near future following expected breakthroughs from on-going intense research and development efforts.

Bioreactor types

There are numerous types of bioreactors with different operating characteristics. The common classifications are based on various physical, chemical and/or biological criteria as discussed below: The different bioreactor types have been analysed in various texts (Aiba *et al.* 1973; Bailey and Ollis 1977; Atkinson and Mavituna 1983) and only a brief overview of some of the principal types is given here.

1. Based on the mode of flow, there are (a) 'chemostats' (continuous flow steady-state operation), (b) batch, (c) cyclic-batch, and (d) repeated fed-batch systems.

2. Based on the nature of biocatalyst, there are (a) microbial bioreactors, (b) plant or mammalian cell or 'tissue-culture' bioreactors, (c) 'cell-free' enzyme bioreactors.

3. Based on the physical and physicochemical environmental requirements, there are (a) psychrophilic (low temperature), mesophilic and thermophilic (high temperature) bioconversions, (b) acid-tolerant and alkali-tolerant bioconversions (c) aerobic (requiring molecular oxygen) and anaerobic bioreactors, (d) liquid and solid-substrate bioconversions, (e) semi-solid or dense-slurry 'solid-state' bioreactors.

4. Based on the type of mixing energy input, there are (a) mechanically stirred tanks, (b) bubble column bioreactors, (c) air-lift bioreactors with internal or external circulation loops, (d) 'injector' or 'ejector' bioreactors using multiphase pumps for phase contacting, (e) 'fluidized beds' using free suspensions, (f) 'packed-beds' using 'fixed-film' systems, (g) 'immobilized' biocatalyst systems using inert carriers for attachment or entrapment of cells or enzymes. Special operating procedures are required during bioprocessing involving mammalian cells and plant products (OTA 1984; Sahai and Kruth 1985).

In batch processing, the bioreactor is filled with the medium containing the substrate and nutrients. The biocatalyst is added after sterilization, and bioconversion is allowed to occur over a period ranging from a few hours to several days. When the conversion is complete, the contents of the bioreactor are dumped, followed by cleaning of the bioreactor. The product recovered is then purified by various techniques. The turnover time between batches usually accounts for a significant portion of the total processing time.

The agitator in a batch bioreactor (mechanically-agitated batch bioreactor) is usually designed to deliver a range of power inputs through a variable-speed controller, and therefore provides a range of oxygen transfer capability. A given batch bioreactor can be used to produce a variety of products, as in the manufacture of high-value, low-volume bioconversion products (pharmaceuticals) which are sometimes produced on a continuous basis. One major advantage of batch over continuous operation is the maintenance of aspectic conditions. The main disadvantages of batch operation are its extra operating costs (for repeated dumping, cleaning and filling operations) and lower productivity.

The distinction between true batch and the fed-batch operation is related to the way in which the bioreactor is operated. The true batch system has all the substrate and nutrients in the bioreactor, together with the micro-organisms at the beginning of the bioconversion process. The fed-batch system is a batch system coupled with intermittent feeding of nutrients. The principal advantage of fed-batch operation over the true batch system is the capacity to manipulate an extra variable (e.g. feed rate or substrate or product concentration) to overcome inhibition of one kind or another. Fed-batch mode of operation is the most common approach to bioreactor operation in the biochemical process industry today (Bjurstrom 1985).

In continuous steady-state bioprocessing, nutrients are fed to, and spent medium and products are withdrawn from, the bioreactor continuously at equal volumetric rates. The potential advantages over batch operation include significantly higher productivity, greater ease of product recovery due to the lower concentration of biocatalyst in the product stream, lower operating cost due to reuse of biocatalyst and generally improved performance economics. There are three principal types of continuous bioreactors: chemostat, tower and airlift-type bioreactors, and immobilized cell (or enzyme) bioreactors.

The chemostat represents the simplest mode of continuous operation. It is a modified batch bioreactor in which fresh substrate and nutrients are added continuously while the product is also removed continuously at equal volumetric rates. The main disadvantages of the chemostate are the loss of unconverted substrate, nutrients, and biocatalyst due to the continuous withdrawal of the bioreactor broth. The chemostat is usually only used as a laboratory equipment.

Tower bioreactors, broadly defined, are large tubular (generally cylindrical) columns in which agitation is usually provided by the mixing action of a sparged gas. There are no usual facilities for using mechanical agitation to effect mixing.

The draft-tube (internal circulation loop) airlift bioreactor is the largest type of bioreactor being used commercially (Bjurstrom 1985). Le Francois (1955) was the first to patent a bioreactor design based on the 'airlift' concept. In the simplest design, the airlift bioreactor consists of a bubble column with a draft tube submerged in the liquid inside the column. When gas (usually air) is sparged at the base of the draft tube (riser), the apparent density of the liquid in the draft tube is lowered relative to that in the annular space (downcomer). The hydrostatic pressure gradient which is established between the riser and downcomer causes a circulation flow pattern in which the liquid travels up the riser concurrently with the gas-phase and then down the downcomer. Other modifications of the airlift bioreactor include those in which gas is sparged into the annular area, external-circulation loop type (Bello *et al.* 1985) and the bubble-regeneration tower (Lamptey 1978; Moo-Young *et al.* 1978).

The principal advantages of the airlift bioreactor are all related to its geometry, mode of gas sparging and liquid circulation. Most airlift bioreactors used commercially are usually tall (greater than 25 m) which greatly enhances mass transfer at the lower sections of the bioreactor, due to enhanced hydrostatic effects. Multiple gas and substrate feed points to the airlift bioreactor are usually employed in large-scale equipment.

Novel designs of the airlift bioreactor have largely overcome some of the potential drawbacks of the earlier standard designs (Lamptey 1978; Moo-Young *et al.* 1978; Moo-Young and Blanch 1981). Current research is being directed at improving mass transfer in these new designs and at developing improved methods for heat dissipation during bioprocessing.

Immobilized-biocatalyst reactors

The term immobilization is considered as a physical confinement or localization of micro-organisms (whole cells or their enzymes) that permits their economical re-use. The use of immobilized-biocatalyst reactors has been receiving widespread attention in recent years.

Immobilized cells (or enzymes) may display properties quite different from those of freely-suspended cells (or enzymes). Some of these properties may offer advantages to bioprocessing that cannot be obtained with conventional bioreactors. The specific advantages are dependent on the immobilization method employed. Techniques for immobilizing biocatalysts include simple adsorption, chemical (covalent) bonding, cross-linking, entrapment in polymeric matrices, and microencapsulation. The most obvious benefit of immobilization is the capability of recycling or re-using the biocatalyst. Re-use of the biocatalyst provides a means for (a) making a batch bioprocess continuous, and (b) maintaining a high cell density to achieve very high bioconversion rates.

Various biocatalysts (both enzymes and whole cells) have been immobilized on support materials such as ceramic beads, ion exchange resins, wood chips, and various gels with significant success (Lamptey *et al.* 1980, 1985) Robinson *et al.* 1981, 1984). Specific examples of the application of immobilized biocatalysts include the use of a high-rate packed-bed bioreactor containing surface-immobilized yeast on inexpensive wood-chip packing, which is used to continuously ferment carbohydrate solutions (including cellulose hydrolysate) to ethanol (Lamptey 1983). The immobilized-yeast, packed-bed bioreactor operates at 97 per cent conversion, producing ethanol at 12 per cent w/v concentration and at a productivity of 30 g/h. This is the highest concentration and among the highest productivities ever reported for fuel grade ethanol production. Valuable by-products include yeast biomass (useful as protein-rich food/feed) and carbon dioxide (useful for beverages, dry-ice, etc.).

The design and operational procedures for the packed-bed bioreactor (developed by a team of biochemical engineers (M. Moo-Young, J. Lamptey, and C. W. Robinson) at the Department of Chemical Engineering, University of Waterloo) is available for licensing from the Waterloo Centre for Process Development, University of Waterloo (Waterloo, Ontario, Canada).

The production of L-amino acids by the use of immobilized amino acid acylase is an example of an immobilized enzyme bioprocess system. L-amino acids are used extensively in the pharmaceutical industry and in the manufacture of food and animal feed. The Japanese company, Tanabe Seiyaku, has used this novel process (based on initial studies by Ichiro Chibata and his group at Tanabe) for the industrial production of L-alanine, L-methionine, L-phenylalanine, L-tryptophan, and L-valine, for several years. The overall production costs are about 40 per cent lower than obtained with enzymes in free-suspension due to significant savings in labour and enzyme costs. This is in spite of the rather expensive enzyme carrier (DEAE-Sephadex) used in the process.

The most notable commercial application of an immobilized enzyme system is the production of high-fructose corn syrup, catalysed by immobilized glucose isomerase (Klibanov 1983).

The use of immobilized biocatalyst systems for the production of a wide diversity of products, including pharmaceuticals and industrial chemicals and solvents, is especially attractive because of the high potential for using high cell densities in continuous operation. Immobilized-biocatalyst reactors are receiving widespread attention, especially at the laboratory and pilot-plant scales, because they are recognized as being one of the most promising (in addition to pneumatically-mixed bioreactors) areas of bioreactor development. Current research and development efforts in bioprocess design and operation are being directed at improving the design and long-term operational stability of immobilized biocatalyst reactors.

Scale-up

The scale-up of a bioprocess is of prime importance in the commercial development of a new bioprocess. The scale-up translation of a laboratory bioprocess development into industrial practice is often related to the bioreactor design. With regard to the pharmaceutical industry, bioreactor efficiency is generally not considered of prime importance in the overall process economics, because of the low-volume, high product-value nature of the industry. For most bioconversion processes, (ethanol, acetone, butanol, SCP, etc.), however, the efficiency of the bioreactor is a critical factor in the process economics because of the high-volume, low-value of most of these products.

In the past, scale-up criteria were based mainly on maintaining geometric similarity between laboratory and large-scale equipment (usually mechanically-agitated vessels). This technique has serious limitations, especially with respect to the newer bioreactor designs—immobilized cell/enzyme and airlift bioreactors) due to the fact that not all of the important system parameters scale up the same. Various correlations for mass and heat transfer, mixing axial dispersion, and agitation power requirements must be used as the bases to determine the inevitable compromises, suggested by chemical engineering calculations, between geometric and dynamic similarities in actual equipment design and operation.

Downstream processing

Product recovery, concentration and purification are of prime importance in the overall process economics of most bioconversion processes. Often, as much as 60 per cent of the fixed costs of bioprocessing are attributable to the recovery operation in the production of organic acids and amino acids (Aiba *et al.* 1973).

There are a wide range of recovery techniques available, both at the laboratory and industrial levels, for the removal of products, such as whole cells and proteins from the product stream leaving the bioreactor. There are, however, only a limited number of product purification techniques available commercially, the choice of which is dependent on the type and end-use of the particular product. An in-depth discussion of the applicability and limitations of the various recovery and purification techniques available has been given by Aiba *et al.* (1973), Applegate (1984), and Bjurstrom (1985).

A summary of the principal techniques, some of which may be used for both recovery and purification operations, is shown in Tables 15.1 and 15.2. For each technique, the principal methods of operation and typical application are given.

The rationale for the selection of a particular recovery or purification

Table 15.1

Principal product recovery techniques

Recovery technique	Comments
Cell disruption Homogenization	Recovery of intracellular products. Most frequently used large-scale disruption technique. Not suitable for some filamentous organisms. Requires careful temperature control
Bread milling	Industrial method for disrupting especially resistant cells and filamentous organisms
Centrifugation	Widely used in industry for the large scale recovery of bacterial cells, protein precipitates and cell debris from homogenate streams. Widely used in brewing industry to recover yeast cells
Filtration Filter press	Widely used in industry for recovery of microbial cells. Removal of organisms from bioprocess culture broths and recovery of protein precipitates. Slow and cumbersome
Rotary vacuum filter	Widely used industrially for the recovery of yeast and mycelia (especially actinomycetes, e.g. *Streptomyces griseus*) in antibiotics manufacture

Table 15.2

Principal bioprocess purification techniques

Separation technique	Comments
Ultrafiltration	Used for concentration and demineralization of solutions of proteins, sugars, and organic solutes. Used in the food and pharmaceutical industries to recover protein from cheese whey, clarify fruit juices, concentrate and purify various enzymes, and antibiotics.
Reverse osmosis	Recovery of dissolved proteins, ionic salts and small organic molecules; purification of water, and concentration of biological and food processing streams.
Precipitation	One of the oldest methods for recovery and purification of proteins.
Salting out	Citrates, phosphates, sulphates, and chlorides used for recovery, concentration, and purification of proteins.
Organic solvents	Methanol, acetone, ethanol, and isopropanol widely used for protein purification. Ethanol is frequently used in the fractionation of human blood (Cohn's method). Largest volume application of protein precipitation.

Table 15.2 *(continued)*

Separation technique	
Chromatography	Method of choice for purification of biological products.
Ion-exchange	Most widely used chromatographic technique. Effectively separates biomolecules on the basis of net ionic strength and pH. High capacity and flexible technique. Used industrially to purify water and bioprocess culture broths containing small molecules, e.g sugars and amino acids. Purification step in high-fructose corn syrup production.
Affinity chromatography	Promising industrial, highly specific, purification technique. Offers great resolving power for purification of proteins based on biospecific interations. Requires expensive chromatographic media. Suitable for high-value, low-volume applications. Ideal for recovery and purification of products from extremely dilute solutions. Widely used at the laboratory scale for the purification of interferon
Gel-permeation or exclusion chromatography	Widely used industrially in the 1950s. Effects separation on the basis of particle size. Effective for desalting and recovery of solvents from biological products, and fractionation of enzyme mixtures. Not as effective as the other (newer) chromatographic techniques.

technique is influenced by several factors. These include: (i) the product type (extracellular or intracellular), (ii) the physical characteristics of the bioprocess broth (e.g. fluid viscosity, presence of cell debris, filamentous or pelletized organisms), (iii) physical characteristics of the micro-organism (e.g. resistance to disruption, sensitivity of product to heat and shear denaturation), and (iv) location of intracellular product (e.g. bound or unbound; only moderate treatment is required during cell disruption for the recovery of unbound enzymes).

The technique selected should fit within an integrated downstream processing scheme. For example, further downstream processing of cell homogenates, produced by mechanical disruption, is difficult if disintegration is excessive. This is usually due to the formation of significant quantities of submicron particulates (cell debris) and the high viscosity of such homogenates. It is common, in such cases, to incorporate a pretreatment step before cell disruption. Pretreatments used include the precipitation of nucleic acids, or the addition of deoxyribonuclease to the cell suspension before disruption to hydrolyse the nucleic acids as they are released (Melling and Phillips 1965; Bjurstrom 1985).

Current R&D efforts are being directed at improving the design and operational characteristics of membrane filtration systems and improving the packing materials used in chromatography. Various scale-up problems relating to continuous chromatography are also being addressed. At the laboratory scale, immobilized monoclonal antibodies are being used as purification agents for protein products (OTA 1984). Currently, several practical and technical problems are being addressed to enable the large scale use of monoclonal antibodies in downstream processing.

In addition to the major separation and purification techniques discussed in Tables 15.1 and 15.2, crystallization, evaporation, drying, and solvent recovery operations are also used in the final stages of the product recovery process. The principal equipments required for these adjunct operations are similar to those used in the traditional chemical industries.

Process economics

The economic viability of a novel bioprocess development often requires prototype demonstration on a pilot plant scale whereby various operating and systems variables can be optimized, and the bioprocess tested for long-term operational stability and product quality reproductibility. With regard to product quality, fairly large quantities of product sample may be required, especially in the case of new products, for various quality assurance determinations. For example, in the case of the production of new microbial biomass protein (MBP) and pharmaceuticals, animal feeding trials over extended time periods may be required.

A final step in the eventual transfer of the bioprocess into the market-place is a sensitivity analysis of pertinent techno-economic factors, e.g. determination of minimum profitable factory capacity to check on feedstock availability, pay-back time, potential discounted-cash-flow-return on investment (DCFR), and geopolitics as they may relate to product acceptability, tax incentives, government regulations, bank lending rates, etc.

Advances in biotechnology R&D

Progress has been most rapid in the areas of pharmaceuticals and agriculture based on applications of recombinant DNA and hybridoma techniques. Several new products for improvements in human and health care (e.g. human growth hormone and albumin, thymosin alpha-1, tissue-type plasminogen activator, a vaccine for foot-and-mouth disease, etc.) are currently undergoing extensive clinical or animal testing procedures. Although the most notable accomplishments to date have been in the developed countries (mainly USA, Japan, West Germany, UK, Sweden, Switzerland, France), most developing countries can expect to benefit significantly from these developments through judicious planning with regard to technology transfer and direct participation in biotechnology R&D.

A brief discussion of some of the notable achievements from biotechnology R&D is given in this section.

Pharmaceuticals

The initial rationale for the use of novel biotechnological techniques in biomedical research was to reduce production costs. Results from on-going research, using rDNA, MAbs and other novel techniques, indicate that these novel techniques may displace traditional pharmaceutical techniques. These new techniques are making possible what was once inconceivable. In so doing, '. . . by manufacturing human and other proteins (e.g. insulin, interferons, growth hormones) by bioconversion techniques they show promise of revolutionizing the pharmaceutical industry, giving rise to a huge new market for novel, safer and more effective drugs' (Cape 1980). It is expected that through genetic engineering, micro-organisms may be manipulated to meet current bioprocess engineering and plant design requirements.

Interferon is a good example of the well-publicized products of genetic engineering. Interferons, a class of immune regulators, are natural substances (proteins) made and released in minute quantities in higher organisms that regulate the response of cells to viral infections and cancer proliferation. It has been shown to block the replication of many different pathogenic viruses and to exhibit several complicated hormone-like effects on the immune system (Cape 1980). rDNA techniques are currently being used for the production of relatively large quantities of interferon-like proteins for testing as pharmaceutical products for treatment of various viral diseases (herpes genitalis, genital warts, hepatitis, etc.). Research problems related to the stability of the interferons and scale-up of the laboratory bioprocess to large scale operation are being addressed.

In addition to interferon, the large scale production and the effectiveness of various pharmaceuticals are also being actively investigated through the use of novel biotechnological techniques. These include the large-scale production and purification of human insulin (the largest volume peptide hormone used in medicine) for the treatment of diabetics, anti-haemophilic factor, human serum albumin, and gamma globulin. A major obstacle to overcome in the production of these proteins is the relatively low costs of current production methods.

Monoclonal antibody technology and improvements in large-scale cell culture bioprocess techniques and purification methods are currently being used to improve the yields of enzymes (thrombolytic and fibrinolytic) that dissolve blood clots (Klausner 1983). Promising protein compounds actively being developed with rDNA techniques for the treatment of human diseases include human growth regulators (growth proteins), neuroactive peptides (analgesia), and reproductive hormones.

Recombinant DNA, monoclonal antibody techniques, and protoplast fusion are also being used increasingly in the preparation of vaccines for the prevention of human and animal infectious diseases. This area of biotechnology R&D is of prime importance to developing countries since bacterial, viral, and parasitic pathogens are a major cause of human and animal morbidity and mortality in these countries. Immunization is one of the most economical means used to combat infectious diseases in developing countries. Current vaccines consist of the organisms (pathogens) that cause the specific disease being prevented, which have been killed (e.g. pertussis vaccine), treated or attenuated (e.g. modified-live virus measles vaccine) to render them non-virulent. Major problems associated with the use of current vaccines include impotency, drug instability (especially in areas without refrigeration), adverse self-effects, inadequate immunization against all of the various strains of the pathogen, and actual transmission of the disease due to reversion of attenuated organisms to wild type, and inadequate inactivation of pathogens (BOSTID 1982). New inexpensive vaccines (e.g. viral disease vaccines against influenza, herpes, polio etc., as well as parasitic disease vaccines against ascariasis, amoebiasis, malaria, trypanosomiasis, etc.) which are more stable and safer than conventional vaccines are being actively developed by pharmaceutical and biotechnology companies in developed countries.

The development of new and improved drugs is a very expensive proposition. Major problems which remain to be solved include: (i) improvement in basic knowledge with regard to the function of immune regulators and the physiology of cancer; and laboratory culture of pathogens; (ii) establishment of meaningful models of the human disease in animals; (iii) improved drug delivery techniques; and (iv) investigation into the clinical use of the various new products being developed by novel biotechnological techniques to determine their effectiveness and to ensure that they are not toxic in humans.

Agriculture

It is widely believed that agriculture is the field in which biotechnology and will ultimately have the greatest impact in developing countries. Currently, traditional plant breeding and selection techniques coupled with plant cell culture and procedures for cell fusion (*in vitro*) are the principal techniques being used to improve production. Considerable basic research into the genetics and physiology of plant cells is required before the full potential of novel biotechnological techniques can be realized.

In several developing countries, low production outputs have been attributed to the lack of high-yielding crop varieties, plant pathogens, poor agricultural practices, and weak marketing incentives. As a result, intense research programmes into enhanced food production through improved productivity and more intense cropping procedures have been initiated in many countries. The benefits from improved agricultural production practices can be enormous, e.g. the use of high-yield cropping techniques resulted in a two-fold increase in the yield of rice production in the Philippines, from 1960 to 1980, and in wheat production in India over the same period (Swaminathan 1982). Even more dramatic improvements in food production can be expected through the use of novel biotechnological techniques.

Plant agriculture

Novel approaches to crop improvement currently being investigated include increasing crop yields by increasing resistances to disese (viral, fungal, and bacterial plant pathogens) and adverse environmental conditions (e.g. drought, frost, soil salinity) or developing more productive plants, and reducing overall production costs.

Biotechnology is expected to play a major role in improving crop yields and reducing production costs by confering increased resistance to outbreaks of plant pathogens and adverse environmental conditions. According to BOSTID (1982), the major stumbling block has been the inability to produce sufficient quantities of standardized immune reagents, for disease diagnosis and epidemiology, for use on a world-wide basis. The use of monoclonal antibody techniques is expected to overcome this drawback.

It is important that researchers in developing countries find lasting solutions for controlling outbreaks of pathogens which cause significant losses in the production of important cash-crops, and, consequently, in the economic well-being of their individual countries (e.g. infectious diseases in cocoa, rice, maize, cassava, sugarcane, etc.). Countries, such as Brazil, Ghana, and Ivory Coast, have had considerable success in controlling the outbreak of pathogens in cocoa through intense research effort using traditional techniques. However, the development of highly specific antibodies to differentiate between strains of pathogens may make it easier to detect and differentiate between strains of various plant pathogens. Most of the existing disease-resistant genes have been introduced into commercially important lines of interbreeding plant species by traditional plant breeding techniques. In most instances, the interaction between disease-resistant genes and plant metabolism is unknown. Current research effort is being directed at improving our understanding of DNA structure and function, and gene expression in plants. It is expected that the use of such novel biotechnological techniques will lead to a better understanding of the nature of disease-resistance in plants.

Increased resistance to adverse environmental conditions and diseases is being investigated by tissue culture techniques which involve trait improvement in individual cells using stepwise selections under gradually increasing adverse conditions. There is a lack of basic knowledge of the type and function which specific protein molecules play in the control of diseases and adverse environmental factors. There is also a significant amount of work to be done with regard to regeneration of whole plants (e.g. cereal and legume crops) from single cells under actual field testing conditions. Investigations into the mode of action of various chemicals released by some plants that adversely affect neighboring plants is also receiving serious attention (Putman 1983). In addition, plants which produce compounds which are toxic to certain insect species are also being actively investigated (Patrusky 1983).

Improvements in the production of crops of prime importance to developing countries can be effected by using cell and tissue culture techniques and genetic engineering. For rice production, for example, cell and tissue culture may be used for induction and selection of useful mutants at the cellular level (to improve salt tolerance, disease resistance, aluminium toxicity tolerance, lysine and protein content, etc.); reducing breeding time through the use of another and pollen cultures; and in hybrid rice improvement by using protoplast fusion techniques. Genetic engineering can also be used for the incorporation of nitrogen-fixing genes in rice plants (Swaminathan 1982). Intense research efforts on the production of high-lysine and high-protein rice with improved salt tolerance and aluminium toxicity tolerance is being carried out at the International Rice Research Institute in Los Banos, the Philippines.

Intense basic research effort is also being directed at improving the amount of seeds and seed protein formed, and increasing the quality of the stored material. Currently, DNA clones of storage proteins are available from several crop species; soybean, garden bean, corn, wheat and other crops of economic significance to developed countries (OTA 1984). Some preliminary studies in the production of secondary compounds from plants (e.g. agricultural chemicals such as pyrethins, nicotine; drugs such as codeine, morphine, steroids, alkaloids; colourings, and pigments; latex, etc.) are being carried out with cell culture techniques (Bell and Charlwood 1980; Dougall 1981).

Nitrogen is frequently the major limiting factor in achieving high crop productivities. Since commercially available nitrogen fertilizer blends are usually prohibitively expensive in most developing countries, (they have to be imported from the developed countries), crops capable of fixing their own nitrogen have acquired special importance. The production of soybeans using improved biotechnological techniques has become a major source of foreign income in Brazil. In the Sahel regions of Africa, cowpeas (another legume crop) are being widely cultivated both as a source of food and to replenish nitrogen supplies in the soil.

Nitrogen fixation occurs only in certain prokaryotes (cyanobacteria, actinomycetes, bacteria) that contain the enzyme nitrogenase. There has been considerable interest in genetic manipulation of these organisms which interact with plants in nitrogen fixation reactions. Some of these organisms form symbiotic associations with higher plants, e.g. legume-bacteria (*Rhizobium*), non-legume-actinorhiza (*Alnus, Casuarina*) and non-legume-cyanobacteria (*Azolla*) symbioses. During nitrogen-fixation (an energy-intensive bioconversion process) in plant microbe associations, the plant provides both a specialized environment for the microbe and the energy required for the fixation, and receives fixed nitrogen in return (Peters *et al.* 1980). Important physiological features of nitrogen-fixing microbes include the following.

1. In general, cells already supplied with available nitrogen do not typically fix nitrogen.

2. Since nitrogenase is oxygen-sensitive, nitrogen-fixing prokaryotes have mechanisms for limiting oxygen.

3. Amonia must be readily converted to organic nitrogen since it is toxic at high concentrations.

One of the most important plant-microbe symbiotic associations is that between the legumes soybean and alfalfa, with the soil bacteria *Rhizobium*. Improved strains of *Rhizobium* are currently being developed. Researchers at the International Crops Research Institute for the Semi-Arid Tropics (ICRISAT) have been able to increase the rate of nitrogen fixation in legumes by 300–500 per cent using novel techniques (Lewis 1982).

Animal agriculture

The use of novel biotechnological techniques is expected to have a significant impact in the diagnosis, prevention, and control of animal diseases, and in animal nutrition and growth promotion. Monoclonal antibody techniques can be used to diagnose animal diseases while rDNA techniques can be used for the production of improved vaccines and other animal health care products.

The prevention and control of animal diseases are being addressed in a manner similar to human vaccine programmes with rDNA subunit vaccines. Special attention is being given to the control of viral infections that affect animal productivity throughout the world (e.g. food-and-mouth disease (FMD) in livestock, parvovirus in swine and cattle, African swine fever, newcastle disease, etc.). FMD has had a detrimental effect on livestock productivity and export potential in South America, Africa, and the Far East. At least three research groups have cloned the gene that codes for the major FMD viral surface protein (Boothroyd *et al.* 1981; Kleid *et al.* 1981; Kupper *et al*. 1981). Cloning of the genes that code for the surface proteins of viruses of fowl plague, influenza, vesicular stomatitis, herpes simplex, and rabies also has been achieved (Bachrach 1982). The use of MAbs to provide immunity against a variety of viral animal diseases is being explored by several groups throughout the world (BOSTID 1982). In addition, several promising advances in the use of novel biotechnological techniques in fighting bacterial and parasitic infections of animals are being made. The current limitation is a general lack of fundamental knowledge with regard to disease control and drug resistance.

Several laboratories are also exploring various ways to improve animal nutrition and feed-use efficiency. These include the study of gut bacteria that participate in animal digestion, feed additives that could enhance absorption of nutrients, and substances such as growth hormones that may directly stimulate growth and animal productivity (OTA 1984). Commercially available synthetic steroids and natural hormones are currently widely used to promote animal growth (e.g. Synovex, Ralgro, MGA, Compudose, etc.). Growth hormones produced mostly by US biotechnology companies, using rDNA techniques, are currently being tested in humans and animals to determine their effectiveness in stimulating growth. In addition to safety, convenient and cost-effective drug delivery systems and regulatory approval would be required before any new products become commercially available.

Microbial biomass proteins

In this text, the term 'microbial biomass proteins' (MBP) is used instead of the more popular term 'single-cell proteins' (SCP) since it more correctly describes the nature of the various products in question. MBP refers to the dried cells of micro-organisms such as algae, actinomycetes, bacteria, yeasts, moulds, and higher fungi grown in large-scale culture systems for use as protein supplements in human foods or animal feeds.

Micro-organisms have been a component of human foods for several hundred years; e.g. the use of yeast as a leavening agent in bread-making, lactic acid bacteria in the production of fermented milks, cheeses, and sausages; and moulds in making a variety of Oriental and African fermented foods (Litchfield 1978). The first purposeful MBP production originated in Germany during World War I when *Saccharomyces cerevisiae* was grown (on molasses) for consumption as a protein supplement. During World War II, the Germans cultivated *Candida utilis* (Torula yeast) on sulphite waste liquor for use as protein sources for humans and animals (Litchfield 1983).

In recent years, technological improvements in microbial cell production for food and feed include the introduction of continuous processes, the use of airlift bioreactors to avoid mechanical agitation costs, and the development of novel methods for flocculating microbial cells to reduce centrifugation costs

(Lichfield 1980; Moo-Young *et al.* 1979; Moo-Young and Lamptey, 1985). At present, there are several processes based on a variety of micro-organisms, at various stages of commercial development. These include the Waterloo SCP process (University of Waterloo, Canada), Hoechst–Uhde process (Frankfurt, West Germany), the Rank Hovis MacDougall 'New Era Foods' (RHM Research Ltd., High Wycombe, UK), Pekilo process (Finland), the 'Raypro' SCP process (ITT Rayonier, USA), Susa Texaco process (Mexico), the Provesteen process (Phillips Petroleum, USA), Kanegafugi process (Japan), and the Swedish Symba process, among others (Moo-Young *et al.* 1979; Litchfield 1983; Nobile 1985) Solomons 1985; Hitzman 1985).

The Waterloo SCP Process, in particular, could be appropriately applied in large or small-scale operations in most developing countries (Moo-Young *et al.* 1979; Moo-Young and Lamptey 1985). An outline description of this process, which is based on the utilization of waste lignocellulosic materials, is given below. Unlike previous processes which used indirect bioconversion (by growing yeasts on liquid hydrolysates prepared from the solid substrates), the Waterloo SCP Process uses a fungal organism, *Chaetomium cellulolyticum*, for the direct bioconversion of the cellulosic components of agricultural and forestry residues by solid-substrate bioconversion. The non-carbon nutrient supplements are commercial fertilizer-grade chemicals. Optimal bioconversion conditions are in the range: 37°C with a ±10°C tolerance and pH 5 with a ±1.5 tolerance. Specific growth rates of up to 0.25/h can be obtained, the highest for any known direct conversion SCP process. The solid-substrate basis coupled with low pH conditions allow contamination-free operation. Depending on the inherent recalcitrance of the raw material, a mild caustic pretreatment may be required and the pretreatment liquor is concurrently converted. Typically, grain crop residues, such as straw and cornstover, require a 0.25–0.5 per cent w/v NaOH treatment at 121°C for 15 minutes. Certain preprocessed material such as Kraft paper pulpmill sludge and sugarcane bagasse pith require no chemical pretreatment. Depending on the process flow conditions, typical particle size conditions are up to 0.5 cm average diameter for dilute (1–3 per cent w/v) slurry system and up to 5 cm for dense slurry (solid-state) systems.

The process has been successfully tested on a 1200-l bioreactor pilot unit under batch, repeated fed-batch, and continuous (chemostat) conditions for both *C. cellulolyticum* and *Neurospora sitophila*. Preliminary feeding trials on mice, rats, and poultry of products made from wheat straw, cornstover, and papermill pulp sludge, have indicated good protein nutritional value, and no toxic or teratogenic effects. Sensitivity analyses of the process indicate that it is economically feasible for a wide range of industrial and semi-industrial scenarios in several countries, both developed and developing ones (Moo-Young *et al.* 1984; Lamptey *et al.* 1984; Moo-Young and Lamptey 1985). To date, licensing rights for the use and sale of the Waterloo SCP Process has been contracted out to three organizations: Envirocon Limited, Vancouver,

Canada, who have recently built a 2 tonne/day semi-commercial plant; Innotech Inc., Besançon, France; and the provincial Government of Novi Sad, Yugoslavia. Other contracts are under negotiation. In terms of the economics, it is evident that the basis of renewable and waste-residue raw materials allows the Waterloo Process to be more attractive than the existing processes which rely on petroleum resources.

In the United Kingdom, the Ministry of Agriculture, Fisheries, and Food has allowed test market studies on the food use of dried mycelium of *Fusarium graminerum* developed by Rank Hovis McDougall (Solomons 1985), and animal feed use of the ICI SCP Process (dried cells of *M. methylotrophus*) (Litchfield 1983). Several notable developments in the production of SCP from biomass materials are also taking place in several developing countries (e.g. countries in Central America, South-East Asia, Middle East, North and East Africa, etc.).

In addition to short-term toxicological studies in rats, chicks, and piglets, more extensive assessments of carcinogenicity, teratogenicity, and mutagenicity, including multi-generational feeding studies, may be required by most government regulatory agencies before widespread use of SCP products will occur. The Protein Advisory Group of the United Nations has developed guideline for the production and evaluation of SCP products (FO 1970, 1974). The commercial success of any MBP process depends on process economics and feeding performance in broiler chicken, turkey, and piglets, etc., as compared to existing sources of protein feedstuffs such as soybean meal and fish meal.

In human foods, flavour and texture, in addition to nutritional value of MBP products, are important determinants of acceptability (Litchfield 1983). As yet, the commercial production of MBP is still only on a small scale due mainly to economics. In developed countries, the fierce competition with soybean flour in the feed industry has shelved large-scale projects in England, France, Italy, and Japan. In developing countries, the high cost of the standard process and associated risks of obtaining reliable operational data for the biological and nutritional evaluation of the resulting MBP have resulted in negligible commercial developments. The future prospects for large-scale MBP production for animal-feed supplements will be limited to those areas where low-cost substrates, such as waste carbohydrates, are available and conventional protein feedstuffs, such as soybean meal and fish meal, are in short supply. The commercial development of any MBP process will be determined by costs and product quality.

Novel biotechnological techniques can be used to improve the process economics by using improved strains of micro-organisms to achieve better amino acid balance and improve feedstock utilization efficiency. Further improvements in bioprocessing and recovery of MBP products are also necessary.

Energy from renewable resources

Biomass in the form of firewood, agricultural residues, and animal manures, currently contributes significantly to meeting the energy needs of several developed and developing countries. Nearly half the total energy consumed in India, for example, comes from non-commercial fuels (firewood, charcoal, cow-dung, vegetable wastes), and a major portion of these supplies is consumed in rural localities. The most efficient utilization of this source of energy is a programme of major importance in India and other developing countries.

Increases in the real cost of petroleum since 1972, and the possibility of supply shortfalls coupled with the relative price stability of renewable carbohydrate raw materials, have prompted an extensive evaluation of alternative technologies for the production of fuels in several countries. Bioconversion processes in renewable resource utilization of potential or practical interest for developing countries include ethanol and methane fuel biogas (Lamptey 1983; Lamptey et al. 1984; Robinson et al. 1984; Moo-Young et al. 1984; Moo-Young and Lamptey 1985).

Brazil is the world leader in the development of an alcohol fuel program in which ethanol is used as a gasoline (petrol) extender and octane booster. Other countries such as the USA, Canada, Australia, Thailand, Papua New Guinea, the Philippines, Guatemala, and India, while not yet committed to large scale commercial ethanol production schemes, have embarked on feasibility studies and research development programmes.

In planning alternate energy programmes, the diversity of needs and resources among developing countries should be recognied. Brazil, for example, is endowed with fertile land and excellent climate for the large-scale production of a variety of biomass feedstocks (e.g. cassava, sugar cane, etc.), and with its scientific talent is not typical of most developing countries. Pertinent differences in one or more of these factors make it necessary for many developing countries to investigate other avenues (e.g. methane fuel gas production) for applying biotechnology to solving their energy problems. The use of cassava for ethanol production has become particularly attractive for several small farmers in Brazil because it can be intercropped with food crops such as beans, corn, sweet potato, rice, etc. At Brazil's National Research Centre for Cassava and Fruit Crops, scientists are working to develop a new spacing system to allow intercropping called *fileiras duplas* (double rows); the system consists of planting two rows of cassava 60 cm apart. These double rows are then separated by a 2-m space in which other crops are planted. Planting cassava rather than sugar cane for ethanol producton is expected to keep the ecological equilibrium intact. The interested reader is referred to the work of Lamptey et al. (1986) on R&D in energy resources, uses, and technology, and Moo-Young and Lamptey (1985) for further information.

Anaerobic digestion of cellulose-containing wastes, both municipal and agricultural, holds promise as a dual method of methane (energy) generation and waste treatment. Methane production seems attractive to many developing countries because the process requires simple equipment and little control. However, there has been widespread dissatisfaction with operating anaerobic digestion plants because the microbial cultures are easily upset.

Several countries have initiated methane fuel biogas production programmes based on small-scale digestion of manures or organic wastes. China and India have had considerable success in this area. A co-operative system is used in both countries since individual farmers often have insufficient manure or capital, or both, to warrant the construction of a digester. Problems encountered, thus far, include inadequate mixing of digester contents and a lack of clear understanding of the underlying principles of the bioconversion process and the optimal conditions for a more efficient generation of methane.

The prospects for anaerobic digestion are improving because of newer bioreactor designs such as those using packed beds containing immobilized micro-organisms. Details of techno-economic comparisons of single-cell protein, fuel ethanol and methane fuel biogas production from a variety of substrates have been reported by Moo-Young et al. (1984), Lamptey et al. (1984), and Moo-Young and Lamptey (1985). For most of the process scenarios examined, the production of single-cell protein and methane fuel biogas are much more attractive, economicaly, than ethanol production.

Concluding remarks

For most developing countries, the problem of providing enough food is closely related to the problems of energy supplies and population growth (Spitzer 1978). Critical energy supplies can be enough firewood or charcoal for cooking or heating. Energy is critical to increased agricultural productivity. The recent increases in the real cost of petroleum, and consequently in the cost of imported fertilizer, have led to serious imbalances in foreign-exchange capital in most developing countries. In addition, problems of drought and rapid population growth in several developing countries have had a serious negative impact on agriculture. It is generally felt that increased crop productivity is the major key to meeting the aspirations of most of these countries.

It is possible to take advantage of the current advances in biotechnology to increase crop productivity and improve animal and human health. The base of the pyramid lies in education and training (Harrar 1978; Moo-Young 1982). Education at all levels is essential to the development of the manpower required to direct and carry out successful production activities. At present, there is a great gap between the numbers of trained individuals available for service to all facets of the agricultural systems and the needs for such services. Local training, supplemented as needed with developments from overseas, should be emphasized. As the numbers of qualified individuals grow, it then becomes possible to take advantage of the enormous array of biotechnological techniques available, but currently under-utilized.

Most developing countries of the world are blessed with an abundance of biorenewable resources and with environmental conditions which are favourable to bioconversion technologies. In recent years, there has been a strong awareness of the potential benefits to developing countries from more efficient utilization of agricultural and forestry surpluses and residues (e.g. straws, yams, sugar cane, cassava, manures, and other organic resources) for the production of fuel alcohol, methane fuel biogas, microbial biomass proteins, and other useful bioconversion products. To make this dream a reality, several problems relating to socio-political infrastructures, and techno-economic appropriateness are presently being resolved. Recent case histories of education and technology transfer projects sponsored by foreign-aid programmes of UN agencies and certain Western developed countries familiar to the authors (e.g. IDRC, CIDA, OAS, UNEP, UNIDO, UNESCO, UNISTD in Cuba, Mexico, Brazil, Jamaica, Guatemala, India, Nigeria, and Yugoslavia, etc.) could serve to illustrate the basic problems.

Among the option for developing countries are: (i) development of alternative energy sources (fuel ethanol, methane fuel biogas); (ii) improving crop productivity through the use of novel biotechnological techniques; (iii) improving animal and human health through the use of novel genetic engineering and monoclonal antibody techniques; (iv) reversing environmental mismanagement; and (v) evaluating the potential of non-conventional sources of food supplies.

Some recommendations for future research are given below.

Animal and human health

Development of better, safer, and cheaper vaccines for specific diseases using novel genetic engineering techniques and the use of monoclonal antibodies as diagnostic reagents should be given high priority.

Agriculture

1. The use of monoclonal antibody techniques in diagnostic and epidemiological studies involving viral and bacterial plant diseases should be given serious consideration (e.g. in rice, maize, cassava, citrus, etc.).
2. Development of fast-growing, nitrogen-fixing trees.

3. Evaluation of grass species and associated bacteria for their ability to fix nitrogen.

4. Improvements in animal feed and nutrition.

5. Improving the availability of key agricultural necessities, fertilizer, pesticides, and quality seeds.

Energy

1. Use of local raw materials for ethanol production using immobilized cell systems.

2. Examination of processes that yield both fuel and feed.

Particular attention should be given to the choice of research priorities and implementation of biotechnology programs. To avoid later disappointments, and waste of scarce human and material resources, careful thought should be given to the choice of research programmes where new tools will help to accelerate progress and solve national problems. The setting up of regional centres should be considered. It is important to remember that research in the field of biotechnology is a very expensive proposition.

References

Aiba, S., Humphrey, A. E., and Millis, N. (1973). *Biochemical engineering.* Academic Press, New York.

Applegate, L. E. (1984). *Chem. Eng.* June 11, 64–89.

Atkinson, B. and Mavituna, F. (1983). *Biochemical engineering and biotechnology handbook.* The Nature Press, New York.

Bachrach, H. L. (1982). *J. Am. Vet. Med. Ass.* 181, 992.

Bailey, J. E. and Ollis, D. F. (1977). *Biochemical engineering fundamentals.* McGraw-Hill Book Company.

Baltimore, D. (1982). In *BOSTID.* p. 30. National Academy Press, Washington, DC.

Baskin, Y. (1982). *Technol. Rev.* (1982). Oct 19–23.

Bell, A. E. and Charloowd, B. V. (eds) (1980). *Secondary plant products.* Springer-Verlag, New York.

Bello, R. A., Robinson, C. W. and Moo-Young, M. (1985). *Biotechnol. Bioeng.* 27, 369–81.

Bjurstrom, E. (1985). *Chem. Eng.* Feb 18. 126–58.

Boothroyd, J. C., Highfield, P. E., and Cross, G. A. M. (1981). *Nature* 209, 800.

BOSTID (1982). *National Research Council, Priorities in Biotechnology Research for International Development: Proceedings of a Workshop (Washington, DC).* National Academy Press, Washington DC.

Cape, R. E. (1980). In *Proc. 13th Internat. TNO Conference,* Rotterdam, The Netherlands, 18.

Collins, G. B. (1982). In *BOSTID,* 230.

Conger, B. V. (1981). In *Cloning agricultural plants via in vitro techniques* (ed. B. V. Conger) pp. 165–215. CRC Press, Boca Raton, Florida.

Davis, W. C., McGuire, T. C., and Perryman, L. E. (1982). In *BOSTID*, pp. 179–207. National Academy Press, Washington DC.

Diamond, B. A., Yelton, D. E., and Scharff, M. D. (1981). *New Engl. J. Med.* 304, 1344.

Dougall, D. K. (1981). *The biochemistry of plants.* Vol. 7. Academic Press.

Ehleringer, J. R. (1979). *HortSci.* 14, 217.

FAO (United Nations, New York). (1970, 1974). *Protein Advisory Group, Statement No. 4*, Guidelines No. 6 and 7; No. 15.

Fox, P. L., Berenstein, E. H., and Berganian, R. P. (1981). *Eur. J. Immunol.* 11, 431.

Gamborg, O. L. and Shyluk, J. P. (1981). In *Plant tissue culture: methods and applications in agriculture* (ed. T. A. Thorpe), pp. 21–44. Academic Press, New York.

Gefter, M. L., Margulies, D. H., and Scharff, M. D. (1977). *Somat. Cell Genet.* 3, 231.

Goodin, P. (1982). *Agric. Res.* 30, 8.

Harrar, J. G. (1978). In *Proc. 171st ACS National Meeting*, New York pp. 2–4.

Hitzman, D. O. (1985). The 'Provesteen SCP' Process. Paper presented at the Microbial Biomass Proteins Symp. on Nutritional, Safety, and Economic Aspects., Univ. of Waterloo, Waterloo, Canada, June 18.

Kargi, F. and Moo-Young, M. (1981). In *Comprehensive Biotechnology* (ed. M. Moo-Young) Vol. 2. Pergamon Press, Oxford.

Klausner, A. (1983). *Bio/Technology*, July 396–7.

Kleid, D. G., Yansura, D., and Small, B. (1981). *Science* 214, 1125.

Klibanov, A. M. (1983). *Science* 722–7.

Kupper, H., Keller, W., and Kura, C. (1981). *Nature* 289, 555.

Kohler, G. and Milstein, C. (1975). *Nature* 256, 495.

—— and —— (1976). *Eur. J. Immunol.* 6, 511.

Lamptey, J. (1978). MASc. Thesis, University of Waterloo, Ontario, Canada.

—— (1983). PhD, University of Waterloo, Ontario, Canada.

——, M. Moo-Young and Girard, P. (1984). In *Proc. 1984 Int. Chemical Congress of Pacific Basin Societies*, Hawaii, USA. Dec. 1984.

——, ——, and Sullivan, H. G. (1986). *An overview of R&D in energy, resources, uses and technology*. International Development Research Centre and the United Nations University Publications.

——, Robinson, C. W. and Moo-Young, M. (1980). *Biotechnol. Letts.* 2, 541.

——, ——, and —— (1985). *Biotechnol. Letts.* 7, 531.

Le Francois, L., Mariller, C. G., and Mejane, J. V. (1955). *Brevet D'Invention*, France No. 1.102.200.

Lee, J. H., Pagan, R. J., and Rogers, P. L. (1983). *Biotechnol. Bioeng.* 25, 659.

Lewis, C. W. (1982). In *BOSTID*, pp. 147–58. National Academy Press, Washington, DC.

Litchfield, J. H. (1978). *Chemtech*, 8, 218.

—— (1980). *Biosci.* 30, 387.

—— (1983). *Science* 219, 740.

Marx, J. L. (1982). *Science* 216, 1306–7.

Melchers, F., Potter, M., and Warner, N. L. (eds) (1978). *Curr Top. Microbiol. Immunol.* 81.

Melling, J. and Phillips, B. (1965). In *Handbook of enzyme technology* (ed. A.

Wiseman) pp. 55–64. Ellis Horwood Ltd., Chichester.

Moo-Young, M. (1982). In *Proc. Symp. on Bioconversion Processes*, UNIDO Seminar, Novi Sad, Yugoslavia, D.S. Masinac Printers, pp. 243–54.

—— and Blanch, H. W. (1981). *Adv. Biochem. Eng.* 19, 1–69.

——, Daugulis, A. J., Chahal, D. S., and Macdonald, D. G. (1979). *Process Biochem.* 14, 38–40.

—— and Lamptey, J. (1985). *Proc. UNEP Advisory Group Meeting on Applied Microbiology*, Nairobi, Kenya.

——, ——, and Girard, P. (1984). In *Biotechnology advances* (ed. M. Moo-Young) Vol. 2, No. 2 pp. 253–72. Pergamon Press, Oxford.

——, Robinson, C. W., Lamptey, J., and El-Gabbani, D. (1978). In *Proc. Int. Seminar on Momentum; Heat and Mass Transfer in Two-Phase Systems, Dubrovnik, Yugoslavia*. Hemisphere Press.

Nobile, J. (1985). The 'Raypro' Process. Paper presented at the Microbial Biomass Proteins Symp. on Nutritional, Safety, and Economic Aspects. University of Waterloo, Waterloo, Canada, June 18.

Olsson, L. and Kaplan, H. (1980). *Proc. Nat. Acad. Sci.* USA 77, 5429.

OTA (1984). *U.S. Congress, Office of Technology Assessment, Commercial Biotechnology: An International Analysis*.

Patrusky, B. (1983). *Mosaic*, March/April 33–39.

Peters, G. A., Ray, T. B., and Mayne, B. C. (1980). In *Nitrogen fixation: symbiotic associations and cyanobacteria*, (eds W. E. Newton and W. H. Orme-Johnson) Vol. 2. University Park Press, Baltimore.

Putman, A. R. (1983). *Chem. Eng. News*, Apr. 4, 34–45.

——, Moo-Young, M., and Lamptey, J. (1981). *Advances in biotechnology*, (eds M. Moo-Young and C. W. Robinson) Vol. 2. pp. 105–13. Pergamon Press, Oxford.

Robinson, C. W., Lamptey, J., and Moo-Young, M. (1984). In *Bioenergy 84* (eds H. Egneus and A. Ellegard) Vol. 3. Elsevier Applied Science Pub.

Ryu, S. K. and Lee, J. M. (1983). *Biotechnol. Bioeng.* 25, 53.

Sahai, O. and Knuth, M. (1985). In *Biotechnology Progress*, 1–9.

Scharff, M. D., Roberts, S., and Thammana, P. (1978). *Nature* 276, 269.

Solomons, G. L. (1985). Microbial proteins and regulatory clearance for 'New era foods', Paper presented at the Microbial Biomass Proteins Symp. on Nutritional, Safety and Economic Aspects. Univ. of Waterloo, Waterloo, Canada, June 18.

Spitzer, R. R. (1978). In *Proc. 171st ACS National Meeting*, New York pp. 4–8, June.

Stanier, R. Y., Adelberg, E. A., and Ingraham, J. L. (1976). *The microbial world*, 4th edn. Prentice-Hall Inc., Englewood Cliffs, New Jersey.

Swaminathan, M. S. (1982). In *BOSTID*, pp. 38–66. National Academy Press, Washington, DC.

Thomas, C. R. and Dunnill, P. (1980). *Biotechnol. Bioeng.* 21, 2271–302.

Upham, S. K. (1982). *Spudman*, November, 14–9.

Wood, W. B., Wilson, J. H., Benbow, R. M., and Hood, L. E. (1981). *Biochemistry: a problems approach*, 2nd edn. The Benjamin/Cummings Pub. Co., Menlo Park, Calif.

Woodward, J. (ed.). (1985). *Immobilized cells and enyzmes: a practical approach*. IRL Press, Oxford.

16

Establishing a meaningful relationship with your computer

M. I. Krichevsky and Cynthia A. Walczak

1. Introduction

The greatest potential for use and misuse of computers in microbiology is for storage and manipulation of large data-bases. The large data-bases can be as diverse as separation patterns from chromatography and electrophoresis, physiological experiments (enzyme or growth kinetics), phenetic profiles of isolates from ecological or epidemiological surveys, etc. In this discussion, most of the examples will be in relation to isolate descriptions. However, the concepts exemplified may be usefully applied to data from all these diverse data-bases.

The kinds of computer analyses commonly performed for microbiologists include simple searching and report generation from a data-base (either as an end in itself or as a prelude to other analyses), curve fitting and evaluation (e.g. in colorimetry), monitoring and controlling various parameters in a fermenter. cluster analysis and other mathematical techniques useful in numerical taxonomy, computer-aided identification of isolates, growth and development models, nucleic acid and protein sequence matching, risk factor analysis in epidemiology, and so forth.

The limiting factors in management of such large data-bases are the amount of computer memory available, the speed of access to all parts of that memory (i.e. search speeds) and the rate of printing of results. The computational requirements for analyses of most microbiological data are relatively simple. Examples of exceptions requiring more sophisticated mathematics and statistics are models of growth, ecology, physiology of organisms, and multi-dimensional and multi-variate analytic procedures. In most computers having large storage capacity and good search logic, sufficient sophisticated computational ability is also present. Therefore, such computational ability will not be considered further.

As microbiologists (or most neophytes, for that matter) decide to involve themselves with computers, two common patterns of behaviour emerge. The first is to hire a student or a beginning programmer (to save money!), explain the problems to be solved in one or more discussions, and finally, assume that the programmer has some mystical power which allows full comprehension without full communication. When the results are inadequate, the list of the guilty can include the computer, the computer centre, the computer vendor, or the programmer. Creative guilt assessment will include all four. Obviously, the inattentive microbiologist, as the judge, never assesses self-guilt. This pattern usually emerges when the computer in question is operated by some central organization, although it has been known to occur even with personal computers.

The second pattern can occur alone or can follow the first. Here, the microbiologist decides that complete control of the computer is the paramount consideration. So a computer is purchased. Only in this way will absolute control and response to needs be perceived as possible. The usual consequences of this pattern of behaviour are that too small a computer is chosen for the size data-base to be managed (cost saving again) and the microbiologist becomes an amateur programmer.

The microbiologist becoming an amateur programmer can be a very good thing where it results in better communication with the computer professionals. However, when programming becomes an end unto itself, the microbiologist should decide to become a programming professional. Otherwise, the resulting programmes are of dubious quality. As with any profession, programming of high quality requires disciplined training in addition to basic intellect (Krichevsky 1982).

The following discussion will focus on three areas to help the microbiologist function in the wonderful world of computers and minimize the depth of the above pitfalls. The three areas are: (i) organization of the data-base; (ii) programming considerations to facilitate analyses; and (iii) check-points to consider in choosing a computer to use. The emphasis will be on understandable guidelines rather than exhaustive coverage. The points covered are based on personal experience and observation of the problems encountered by others.

2. Data-base organization

In most cases, the data-base will describe collections of physical items (strains, samples) or events (fermenter runs, enzyme assays). In micro-biological surveys, the basic level of description is usually the individual strain or isolate and forms the unit record of the data-base. Within the unit (strain) record are the individual items of information describing the strain. At a higher level of organization, the sample or source from which a group of strains is isolated is described by a combination of the appropriate strain unit records as well as information uniquely associated with the sample or source such as time, place, physical attributes, and history (Fig. 16.1).

The same levels of organization apply to sequential samples from a fermenter and fermenter runs or multiple DNA fragment patterns from plasmids and the host strain, etc.

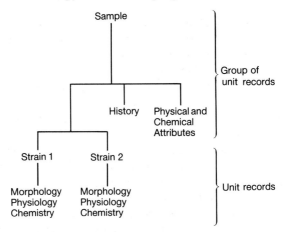

Fig. 16.1. Relationship among data categories recorded in microbiological surveys.

In some circumstances, abbreviated versions of this scheme are used. Ecological or sanitary surveys may only consider the microbial load in the sample, hence no strains. Conversely, central culture collections are almost exclusively strain orientated. In both cases, the unit record, whether sample or strain based, is homogenous in that it contains all the information associated with the record. That is, the information within a single record is non-hierarchically organized.

The obvious first step in computer data management is to decide what information to enter in the computer. The microbiologist must analyse the situation on hand and establish the list of items that will be entered into the data-base.

It is mandatory to build a list system which is flexible from the beginning. Ludicrous, wasteful, and counter-productive examples of computer usage in microbiology have resulted directly from rigid and/or stringent central control of the forms used for recording data. Months and even years pass before change or addition is accomplished. If such inertia is detected at the outset of the system design, the microbiologist should avoid using the computer if at all possible. Otherwise, the laboratory will be working for the computer and performing the real work with the old reliable paper and pencil. This type of rigidity can occur with equally disastrous results when the microbiologist answers to no one other than the manager of the laboratory or personal computer (who may be the microbiologist). The message is the same: if you can't be flexible, use paper, pencil, and eraser.

The list of items should be largely in the form of an open-ended, controlled vocabulary which defines precisely each item to be entered. The list should be maintained in the computer as a separate entity to allow editing, expansion, contraction, and the selection of relevant subsets for individual purposes. However, the initial list will set the tone for what follows. Much scientific and personal frustration can be avoided by application of some simple rules.

Each item should be defined to contain a single attribute. Incorporating multiple features leads to confusing and inefficient retrieval and analysis. That a strain is a Gram-negative rod should be coded separately as positive for being Gram-negative and -positive for having rod-shaped cells. Otherwise, selecting either all Gram-negative strains or all rod-shaped strains is cumbersome at best.

The actual observations (test results) should be entered into the computer, not derived information. For example, clinical microbiologists frequently record only the name (derived information) of an organism upon identification along with the antibiotic resistance pattern found. The biochemical and other phenotypic results are discarded. The resulting problems include the inability to verify the identification, the impossibility of calculation of the frequency of atypical test results, or to later decide if the strain is atypical. Consequently, one cannot detect strain dependent phenomena such as the difference between the introduction of a new antibiotic resistant strain into a hospital versus the acquisition of a new resistance-carrying plasmid into an existing strain. Recording an organism as 'atypical' without recording why it is so judged is informational nonsense.

Coding of the Gram stain information illustrates a number of logical pitfalls in item definition. Historically, the terminology leads to the verbal inversion; the word negative means a positive attribute. Another example of historical inversion is coding of growth in the presence of chemical inhibitors as a positive attribute. Conversely, absence of growth in the presence of antibiotics (sensitivity) is scored as positive.

Gram variable strains can be treated two ways. Both Gram-positive and -negative could be scored as positive or a separate item (Gram-variable) could be scored as such. The latter is the preferred method because of other possible states (e.g. variable staining within a cell, no staining by Gram methods). This latter logic indicates that the Gram staining information should be treated as a mutually exclusive multistate attribute.

In general, multistate items should be entered as if they were separate items. Their multistate nature is invoked after entry for such purposes as error checking (two positive answers for the same strain is an error) and calculation of probabilities (for computer-aided identification programs).

The form in which information is entered can profoundly affect the ease of retrieval and analysis. Fortunately, the overwhelming majority of microbiological data likely to be entered into computers is already recorded as yes/no or numerical observations. By codifying the possible responses and establishing a mechanism for entering yes/no or numerical responses, a 'controlled vocabulary' is established.

However, another important category of strain data is not so easily coded, i.e. data on the sample or source from which a group of strains is isolated and associated information such as time, place, physical attributes, and history (Fig. 16.1).

Such items as names, addresses, comments, literature references, widely varying source information, and other items where it is impractical to anticipate or code most of the possible entries, will always be entered as natural language. The problems with such entries can be minimized only by careful coding and entry. It is tempting to enter familiar words because of naturalness of communication. The very attribute of naturalness creates massive problems. A source of isolation entered as 'dirt surface' versus 'top soil' has both order and synonymy confusion for retrieval by computer. Other common confusions in natural entry are variation in spacing, punctuation differences, upper and lower case letters, spelling (and mis-spelling) differences, etc.

Once the list of items to code has been established, consider the mechanism for data entry. The mechanisms can vary among check lists on paper, menu selection or prompts for computer-aided entry, entering numbers in blank spaces, completing tables where the headings are pre-printed or computer generated. In all cases, the microbiologist is directed (gently, if possible) as to how most of the possible items are to be entered into the computer.

Finally, the list should never be absolute or frozen unless the project is unique and completely self-contained.

The fundamental question to be considered in data-base organization is size. It is imperative for success that careful study be given to both the initial amount of data at hand and the potential growth of the data-base. This growth potential is the source of many unanticipated problems. Is it really sufficient to have only the last year's data immediately available? A desirable question to ask of the data might be: 'In the last decade, what was the distribution of sources of isolation for all strains producing hydrogen sulphide, but otherwise identified as *Escherichia coli*?' This may seem a simple question to some, but it implies a variety of computer capabilities.

Individual items of information must be readily accessible (positive for hydrogen sulphide production and named *Escherichia coli*). The computer system must be capable of accessing and accumulating data across a multi-year span. If the raw data is physically segmented on separate magnetic recording devices such as tapes or discs, a way must exist to select and record elsewhere in the computer system (main memory, a second disc, or a tape) that subset of data that will answer the question at hand. Many smaller computers can perform such a task only by intimately involving the operator for extended periods of time (e.g. to keep track of and change discs).

The exact space occupied by a data-base will be markedly dependent upon its organization in addition to the number of items recorded. Some simple concepts of organization strategies which can greatly decrease the storage requirements follow.

Two strategies for achieving flexibility within a data-base of strain information illustrate the variety of solutions possible. The first is to use a simple table structure (Fig. 16.2).

		Old tests ⟶							New test
	1	2	3	4	5	6	7	8	9
Strains ↓	+	−	+	+	+	−	−	−	/
	−	−	−	+	−	−	−	−	/
	+	+	−	−	−	+	+	+	/
	+	+	−	−	−	−	+	−	/
	+	−	+	+	+	−	−	−	/
New strain	+	+	−	+	−	−	−	+	+

Fig. 16.2. Table structure for storing strain data (+ = positive result, − = negative result, / = blank space due to test not performed)

Each horizontal row contains the data on a single strain. Adding a strain record to the bottom changes the length of all vertical columns by one space each. To add a new feature, a new column is opened on the right side (assuming left to right orientation). A new column must be created but left devoid of information (i.e. marked as blank) for the previous strains. Since these empty spaces occupy the same memory as those containing information, this is wasteful of space in a rapidly changing or highly variable laboratory. Such a simple strategy may be quite adequate for a routine clinical microbiology laboratory wherein the spectrum of tests does not change often. A general purpose culture collection would not find this method efficient or flexible enough.

The second strategy for achieving flexibility allows entry and management of sets of data having possibly differing spectra of microbial information in each set. For example, clinical laboratories will record overlapping, but obviously not identical features for organisms isolated by aerobic versus anaerobic plating methods. Acid production from D-glucose and facultative growth will be recorded in both situations. Obligately aerobic growth need not be tested for the anaerobically isolated strains and vice versa. By logical extension, there will be some mix of common and separate features. Furthermore, the features to be recorded for any particular type of organism may be changed at any time due to availability of new technology.

Each kind of data set is kept separate in the computer and the feature labels are recorded in each data set (Fig. 16.3). Storing the data in sorted order of both the strain designations and the feature lists allows binary or other rapid search techniques to speed data processing. To add or subtract a feature or strain from the data set, the list is modified and the proper rows and columns added or deleted. (To avoid rewriting the files completely each time a change is desired, a technique of location pointers can be used. However, the logic is identical.) The appropriateness of comparison of the data sets is determined by consulting the combined feature lists or temporarily combining the two datasets (Fig. 16.4). In the example given, if feature D is facultative growth, a

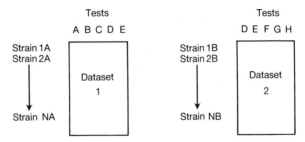

Fig. 16.3. Separation of data sets of two laboratories characterizing strains by partially overlapping test batteries.

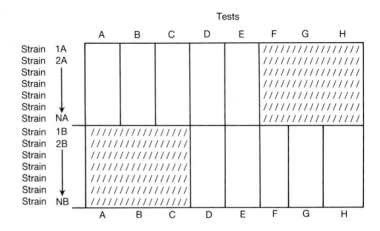

Fig. 16.4. Space requirements for temporarily combining data sets of two laboratories characterizing strains by partially overlapping test batteries.

list of such strains would be simple to produce. Such segmentation of the data into logically organized sets markedly decreases storage requirements and allows consideration of smaller parts of the data in the main (and expensive) memory of the computer.

No matter which data organization is chosen, there are finer details of data management which impact on resource utilization no less than the overall structure. The three types of data coding usually used, binary, numerical, and natural language (termed alphanumeric), can all be entered in ways which will compress the data without loss of meaning. Of these, the most saving of computer memory will be realized in the binary and alphanumeric coding.

In most modern computers, characters (i.e. letters, digits, punctuation) are represented by some form of eight level code. Each level of the code is called a bit (an acronym, for binary digit. The bit represents an individual storage location in the computer memory). The combination of eight bits is called a byte or character. Most often it is convenient (and in small computers, mandatory) to store yes/no information as one answer per byte. Theoretically, such information could be stored as one answer per bit for an eightfold space saving. Because of the often encountered fact of missing data, two bits are required to store such pieces of information. Thus, a four-fold saving is realizable.

Such saving is simple to obtain in computers having the capability of accessing the individual bits of each byte. In computers (usually smaller) that do not allow individual bit access, the storage savings can still be obtained by considering four answers at a time through a coding-decoding procedure. (An analogous coding-decoding system of recording data is used by some commercial identification kit manufacturers. They compress each three test results into a single number, i.e. 'octal' coding, Fig. 16.5.) The needed complications added to the programs to manage this mode must be balanced against the savings in the computer memory (and hence, costly size).

Test A B C	Number
− − −	0
− − +	1
− + −	2
− + +	3
+ − −	4
+ − +	5
+ + −	6
+ + +	7

Fig. 16.5. Translation of three test results (A,B,C) into a single number.

The most data compression is often realized in the alphanumeric (i.e. natural language) data. The most common method of recording alphanumeric information is the use of a fixed length spacing (field), i.e. the familiar technique of filling in as many of the spaces available as needed to contain the desired characters (termed left-justified, Fig. 16.6). The field length is fixed at the maximum number of spaces anticipated for the category of data being recorded. For example, 21 spaces should hold any of the names of bacterial genera (1), the longest name being *Thermoactinopolyspora* and the shortest name is *Mima*. When fixed field format is used, each name will occupy 21 spaces in the computer.

Thermoactinopolyspora
Sporolactobacillus⎯
Streptococcus⎯⎯
Bacillus⎯⎯
Campylobacter⎯⎯
Mima⎯⎯⎯

Fig. 16.6. Storage of genus names in a fixed length field.

Variable length fields can be used so that most of the blank spaces are recovered. A simple format is to end each entry with a special end-of-entry symbol ("|" in Fig. 16.6). If the symbol is unique to the kind of entry (e.g. * = genus, / = specific epithet, @ = source of isolation), the retrieval program can check for both parameters to keep account of item locations in the data-base.

Name length	Strain name	Taxon length	Taxon	Source length	Source
2	A2	14	PSMN Atlantica	9	Station 5
5	BC345	12	PSMN Nautica	4	ATCC

[2]A2[14]PSMN Atlantica[9]Station 5[5]BC345[12]PSMN Nautica[4]ATCC

Fig. 16.7. Variable field length storage of strain data. The data in the table reduces to the bottom line (numbers enclosed in brackets [ ] represent a binary number, which is stored as a single character).

A refinement of this simple strategy speeds up searching as well as compresses the data. A binary number is stored at the beginning of each entry giving the number of characters in the entry (Fig. 16.7). (The binary number is used to a maximum of 256, since it occupies only one byte. That is, one byte = 8 bits; so the largest number to the base 2 is 2^8 or 256.) In addition, a short table of the beginning location of each unit record is maintained. To list the source of isolation of each strain (where the source is the third field) the location table is consulted for the location of the first number of the first field. The number found at that location indicates how many spaces to skip to find the second field size number. That number determines where the desired third field begins. Finally, the third field size number determines how many characters are to be listed for that strain. This process (termed chaining) is repeated for each strain in turn.

The first strategy is simple to program, but requires that each character be inspected to see if it is an entry termination, a slow process. The second strategy is more complex to program but operates considerably faster. Both strategies usually yield compressions of between 25 and 75 per cent and no loss of information.

A data-base wherein both samples and strains must be considered (non-homogenous information) may be organized in a variety of ways. Each has its consequences. Some examples follow.

In the simplest non-hierarchical organization, the information describing a sample would be repeated in every record for those strains isolated from the sample (Table 16.1).

Table 16.1

Non-hierarchical data-base organization for two samples

SAMPLE 1	TIME 1	PLACE 1	STRAIN RECORD 1–1
SAMPLE 1	TIME 1	PLACE 1	STRAIN RECORD 1–2
SAMPLE 1	TIME 1	PLACE 1	STRAIN RECORD 1–3
SAMPLE 1	TIME 1	PLACE 1	STRAIN RECORD ·
SAMPLE 1	TIME 1	PLACE 1	STRAIN RECORD ·
SAMPLE 1	TIME 1	PLACE 1	STRAIN RECORD 1–N
SAMPLE 2	TIME 2	PLACE 2	STRAIN RECORD 2–1
SAMPLE 2	TIME 2	PLACE 2	STRAIN RECORD 2–2
SAMPLE 2	TIME 2	PLACE 2	STRAIN RECORD 2–3
SAMPLE 2	TIME 2	PLACE 2	STRAIN RECORD ·
SAMPLE 2	TIME 2	PLACE 2	STRAIN RECORD ·
SAMPLE 2	TIME 2	PLACE 2	STRAIN RECORD 2–N

The obvious drawback to this organization is the large space overhead used to repeat the same information. The advantage is that every unit record is self-contained. Simple direct sorting of the records is possible, as is direct tabulation. The programming required to arrive at an answer is conceptually straightforward since only rows and columns need to be considered. This kind of data-base organization is quite popular where computer storage is readily available, but programming resource is limited. It is the simplest form of a relational data base (Codd 1982).

A method for avoiding repetitive entry of sample data would be to define

Table 16.2

Hierarchical, position
dependent data-base
organization for two samples

SAMPLE	1
TIME	1
PLACE	1
STRAIN	1–1
STRAIN	1–2
· · ·	
STRAIN	1–N
SAMPLE	2
TIME	2
PLACE	2
STRAIN	2–1
STRAIN	2–2
· · ·	
STRAIN	2–N

two types of unit records, a sample record and a strain record. All strain records pertaining to a particular sample would follow the sample record. Thus, the data organization is hierarchical and position-dependent (Table 16.2). While comparatively parsimonious of space, such organization requires special techniques of information retrieval before sorting or tabulations can be performed. All programs accessing a data-base organized in this fashion must be cognizant of the existence of the two record types whether or not both types are relevant to the task at hand.

A third strategy is to maintain a sample data-base separate from a strain data-base. For those questions limited to samples alone or strains alone, each data-base is utilized exclusively and non-hierarchically. Most transactions with the separate data-bases will be satisfied by considering only one kind of data.

Some transactions will involve both sample and strain data. The transaction may be driven from either direction. A list of associated diseases and the number of patients from whom hydrogen sulphide-producing *Escherichia coli* were isolated would require that links (pointers) to the sample may be contained in the strain record. Conversely, the calculation of ecological diversity in a series of water samples requires pointers to the strains isolated from each of the samples. A simple form of such bidirectional linkage would

Sample code	Sample name	Time	Place
1981–1	Mushrooms CAN 5	1023	Store 22
1981–2	Mushrooms CAN 6	1440	Warehouse

Fig. 16.8. Simple data-base.

Strain designation		Features 1 2 3 4			
1981–1	1	+	–	+	–
1981–1	2	+	+	–	–
1981–2	1	+	–	+	–
1981–2	2	+	–	–	+

Fig. 16.9. Strain data-base.

be to: (i) incorporate the sample designation from Fig. 16.8 in the strain designation, (ii) serially number the strains isolated from the sample (e.g. as in Fig. 16.9, strain 1981–1–2 would be the second strain isolated in 1981 from mushroom can 1981–1), and (iii) enter the number of strains isolated from each sample in its sample record. If the number of isolates is fixed for all samples, step 3 can be omitted. This strategy presumes there is no significance to the order of isolation. Transactions utilizing both types of data-bases would usually require programming to cope with the relationships between the two structures since simple sorting will not suffice to relate all strains to the sample of origin.

The organization of the data-base will have a major impact on the size and type of computer that can cope with the tasks to be performed. However, unless the microbiologist is also the data-base manager, the internal organization of the data-base should not be of concern to the microbiologist as user of the computer. Of great importance to the microbiologist are the programs which allow the desired functions to be performed. On the strength of program construction will acceptance and, hence, success depend.

3. Programming considerations

The computer system and the data management programs must be constructed to give the microbiologist simple and rapid access to any relevant data and analytic programs. Therefore, the user should not be a part of the data management system. If the computer disgorges a message such as: 'Program loaded, remove system disc and insert desired data disc', the system is inadequate. Quite often, the microbiologist will be juggling multiple discs while trying to recall where the desired data were stored. 'Is the streptococcal data on disc S?' No, because S has staphylococci and salmonellae, but no streptococci. They are on the E, for 'enteric', disc with the coliforms!

During an entire interaction with the data, the microbiologist should only have to specify which data are wanted, not their location in the computer. A directory or catalogue of data file names and the nature of their contents should be maintained in the computer. A reasonable system should be large enough and smart enough to find the requested data files and operate on them with no other information than the unique names given to the files. Such file management programs may be supplied with the computer system. If not, they will have to be written *de novo*, a fairly complex task.

Whatever the nature of the computer system or the programs used therein, there is an imperative to good practice in computer usage that is often neglected. Programs must be documented (i.e. described) in two ways, their internal logic (for the programmer) and directions for use (for the microbiologist).

The internal logical or program documentation may be accomplished by being incorporated as comments (i.e. non-executable text) in the body of the program, or as a document separate from the program listing, or some combination of the two.

The program documentation is needed by the programmer for program maintenance and enhancement. An error in program logic may be undetected for long periods of time, sufficient for the logic trail to grow cold. Furthermore, a programmer other than the author may well have to interact with the program. Hopefully, program documentation, while necessary, will not be seen by the microbiologist.

In this last context, beware of obtaining undocumented programs from others unless you are prepared to fight through the logic to modify the program. Changing one thing in a program often will have unforeseen, often subtle consequences elsewhere. Such programs are almost always used at the recipient's risk as there is no reasonable way of supporting them at long distance. Any changes beyond the name of the institution on the report output may turn out to be a major effort.

On a routine basis, directions for use of programs are even more important than program documentation. It is commonplace for users of programs during development to have to consult the programmer on proper use. A program should not be considered ready for routine use before all its options can be exercised and the error messages understood by a neophyte user without asking the programmer for help. Few programs are written to be completely self-explanatory.

Various categories should be included in the user documentation.

1. Instructions for program execution including initiation at the level of turning on the computer or terminal.

2. Complete definition of all terms parochial to computer technology used in the program execution.

3. Complete definition of all abbreviations, mnemonics, codes, etc., used in the program (e.g. does GLY mean acid from glycerol, gas from glycerol, both acid and gas from glycerol, glycerol is utilized, glycerol can be used as sole carbon source or perhaps GLY stands for glycine metabolism in some way).

4. A description of the problem the program is intended to solve and the overall method (termed: the algorithm) used to solve it (e.g. the microbiologist might need to be reminded of the formulae for simple matching and Jaccard's coefficients in a numerical taxonomy program).

The algorithm should be described in enough detail to be followed by a microbiologist or an equivalent literature reference should be given.

5. A complete list of the error messages, likely mistakes to be encountered during program execution, and strategies to overcome the problems.

6. A literal and comprehensive example of program execution including terminal dialogue where appropriate.

Procrastination and resistance to writing program and user documentation have no known bounds. Professional programmers, amateur programmers, and microbiologists, all exhibit such inertial behaviour. Thus, special efforts, often bordering on the heroic, may be required to produce adequate documentation.

A second aspect of accomplishing the cutting of the umbilicus between programmer and program is to write programs that are controlled by external tables or lists of information. For example, the length of fields (natural language items) can be specified in an external table so that changes in field length do not require program changes.

Because of the simplicity of changing individual lines of programs in some computer languages such as BASIC, temptation is great to modify program function by changing the appropriate program line. Such procedures should not be allowed because of the likelihood of inaccurate changes, lack of documentation of such changes, and difficulty in fixing problems that arise, especially when multiple changes are made or when the changes must be reversed for the next time of program execution. Once a program functions successfully, no further changes should be made unless new functions are desired.

Programs are commonly written to be table-driven with respect to the data analysed. That is, the data is submitted as an external table rather than incorporated in the body of the program. The same philosophy is advocated for the labels utilized in the program whether they are considered variable or constant. Using simple codes (preferably numerical) and translation tables to represent labels such as taxa names, features, sources, etc., renders programs more flexible so that the frequency of modification drops dramatically.

A function that must be available to most programs is that of searching the data-base to find data relevant to the task being undertaken. The efficiency of the search strategies used will be a significant factor in overall program efficiency.

The simplest, but least efficient search method is sequential searching. This is the brute force approach of starting at one end of the data-base and searching until the desired data are found. Sequential searching makes no assumptions of order in the data-base. Since this method is the only one feasible in many situations, large computers often have built-in electronic logic facilities for markedly speeding the search process.

Sequential searching may be the most reasonable approach despite its inefficiency. Small data sets and data-base searches with unpredictable search patterns are two examples where sequential searches are the most reasonable.

A small effort invested in analysing the most common uses of the data-base can avoid a great deal of effort and delay in obtaining information. If it is reasonable that the data will be added to infrequently, compared to the frequency of searching (the most common case), then it will be quite advantageous to organize the data-base around the most common search pattern.

For example, in clinical microbiology ordering the records by patient identifier would be more efficient than by date of isolation. Conversely, a study of end products in an artificial rumen would be organized by time, since it is temporal variation that is commonly studied.

Having organized the data-base in this way, one still has to choose a search technique. Some experimentation is worthwhile, especially with small computers.

The two most common search techniques are sequential and binary searching. Sequential searching starts at one end of the data and continues until all matches are found. If the data-base is ordered, the search can be stopped when the search goes beyond the desired point.

Binary searching is the process of searching a list of items which have been presorted into increasing order. Because the list is presorted, one can always tell whether a particular item is above or below any other item in the list. The search pattern is, first, to inspect the middle item in the list. If this item is below the particular item sought, then one need only search the top half of the list. The middle item of the top half is inspected. If it is still below the target item, the top quarter is considered. Thus, each inspection of the list discards one-half the remaining items until the target item is found. Binary search is efficient because the number of items to be inspected is, at most, the logarithm to the base 2 of the number of items in the list.

To illustrate the kind of experiment one might perform, we programmed a sequential search and a binary search in PL/1 and ran them on the IBM 370. We also programmed these algorithms in SAIL, a version of ALGOL and ran them on the PDP–10. The time was noted before the search for an item began and after it was found. This did not include the time to initially read in and store the list of 315 items nor the time to query and read the item to be searched for. There was essentially no difference on the IBM 370. Variations in times were random and we attribute them to being paged in and out of memory in a time sharing environment. Times were on the order of milliseconds. This was also true on the PDP–10. In the PDP–10, increasing the list to 1000 items resulted in barely measurable increased search times which could be ascribed to the extra comparisons required by sequential searching. The maximum increase was 22 ms for an item not in the list at all. The variability is indicated by three identical search requests for an item near the middle of the list resulting in search times of 6, 12, and 13 ms using the same algorithm. The costs for searches in the millisecond range are not major since the cost of reading the data from the disc into the computer core and storing the results back on disc will usually be much greater than the search itself.

Simple binary search is not the only way to search an ordered list. Consider an aphabetized list of abbreviations for the names of bacterial genera accompanied by the full generic names. Such a list would be used in constructing reports where only the abbreviation is used in the data-base itself. Each abbreviation begins with the same letter as the corresponding genus name. Using this last fact, a table is constructed which contains only the 26 letters of the alphabet and a pointer to where in the list of abbreviations each letter begins. (Note, no generic names begin with 'Q') Preliminary use of this table restricts the search of the master list to abbreviations that start with the same first letter. Thus, by first searching an index, only segments of large lists need be searched. Such a procedure also allows the data-base itself to be segmented. (For a full discussion of these strategies, see Krichevsky *et al*. 1980).

It is highly unlikely that repeated searches are random. In fact, according to the '80–20' rule, which has been commonly observed in commercial applications, 80 per cent of the transactions deal with the most active 20 per cent of a file. The same applies to this 20 per cent resulting in 64 per cent of the transactions dealing with the most active 4 per cent of a file. In this case, a very efficient scheme for searching an unordered list is the self-organizing list as described in Knuth (1975). In a self-organizing list, items most frequently used reside near the top of the list, while the least frequently used are near the bottom. There are many ways to come by this arrangement. First, a use count can be kept with each item. They can then be periodically sorted based on this count. Secondly, instead of using memory for counts, an item can be moved to the top of the list when it is accessed. Thirdly (and most efficient), the item accessed can be interchanged with one above it, if it is not already near the beginning of the list.

Thus, there are many reasonable solutions to the very common data processing problem of searching lists.

4. Check-points to consider in choosing a machine and its software

Modern computers of all sizes are sold with a bewildering array of physical components (hardware) and programs (software) to be used to make the computer system 'useful'. Too often the vendor of such systems will be tempted to sell their 'standard' system and 'standard' software. It is the responsibility of the customer to be an 'educated consumer' (S. Syms, personal communication). In this section, we shall provide some points to consider in selecting hardware and software of use to microbiologists.

The check-points will be neither exhaustive nor universally applicable to all situations. However, if the reader can make a reasoned judgement as to their appropriateness, the educated consumer is developing.

4.1. Size of computer

As we have seen, the size of computer as it appears to the user can vary a great deal depending on how it is organized by the software. Another factor is that of single versus multiple users. Whether the multiple users are simultaneous (time-sharing) or serial is a fundamental question. There is much to be gained in economic efficiency by a time-sharing system since the costs of high speed devices can be shared.

The major disadvantage of time-sharing systems is the lack of personal control over their administration. User committees to set policy, no matter how small the computer or the group of users, are necessary for the most effective operation.

Since time-sharing capability now exists to the level of personal computers (thanks to the development of the UNIX operating system) the decision on multiple versus single users applies to almost all sizes of computers.

A variation of time sharing that is becoming popular is 'cluster' systems or 'local area' networks. Here each user has its own workstation which is actually a small to medium size computer, which in turn communicates with the others, and a central computer which operates large memory and output devices. The advantage of such systems is their extreme flexibility and power for the cost.

A major consideration in buying a small computer is its capacity for expansion. The expansion can be in two ways. The internal capacity of the computer and the number of devices that can be directly connected is set out in the manufacturer's specifications. The capability of the computer to talk to larger computers is very important in modern computer applications. This latter consideration has both hardware and software components. The ease of making the physical connections, either through telephone lines or directly, should be investigated. However, the physical aspects are usually the easier to accomplish.

More difficult is the selection of the proper communication programs, which tell the computer how to communicate. The best solution to this problem is to require the vendor to demonstrate the ability of the hardware and software combination to communicate with the desired host computer or network. The user may have to provide the characteristics of the host computer (e.g. transmission speed, codes for characters, parity conventions, access codes, and conventions). The computer jargon for these conventions is 'the handshake'.

If a larger computer is indicated after a needs analysis, the cost of installing the machinery should be determined. Larger computers will require special electrical and ambient temperature control (air conditioning). Provision must be made for connecting electrical signal cables from the computer to the various peripheral devices. This is usually done by installing a raised (false) floor in the computer room and placing the cables under the floor.

Perhaps the most important aspect of proper size documentation is that of speed of access to all the data. The computer should be of a size and have the file management programs to allow access to any part of the data-base without physical intervention of the operator or the user. That is, once the user asks for a piece of information, the process should be rapid and automatic. The changing of discs should be unnecessary for a number of reasons in addition to simple convenience. The potential for error is greatly reduced. Global searches and indexing are made practical. The data can be rearranged for any reason with relatively simple methods. Perhaps the most startling gain is the faster response times, often orders of magnitude faster.

4.2. Specific useful characteristics

Microbiological data analysis is often aided considerably if the computer system (including the software) has some specific characteristics. The lack of these characteristics may not be sufficient to cause rejection of a system, but they may allow a choice between those otherwise similar.

Since much microbiological data are composed of presence or absence of attributes, considerable savings of space and faster search times will be realized if the computer can access the data as individual bits. Most software systems for small computers cannot access individual bits directly. Both the programming languages (e.g. BASIC) and the data-base management systems (e.g. dBase II) are character (byte)-orientated. Simplicity of use rather than efficiency of storage and flexibility motivate this orientation.

Richer languages (e.g. PL/1) and languages closer to machine language (e.g. assembly language, C) allow such bit access, but are time consuming to learn. Unless the microbiologist is willing to spend a great deal of time learning to program at a professional level, these languages are best left to professional, full-time programmers.

As indicated earlier, the ability to store variable length fields and access data stored as such markedly affects the storage capacity required. Data-base management software having this ability allows storage of more data in smaller space. with the rapidly decreasing cost of disc storage, the importance of variable length storage may have more importance in time saving and simpler data file organization than as cost saving.

A hardware feature that can markedly affect search times is built-in search logic. That is, the computer requires only a single machine language instruction to sequentially search an array of memory locations and return the results to the program. This feature is common in large computers, but less so with decreasing size.

Editing is perhaps the most common operation a microbiologist will perform with a computer. Word processing, text editing of programs, updating and correcting of data all utilize editing functions. In a few cases, the editing functions may be contained in data management programs. Much more likely, an editing program provided as part of the operating system of the computer or a word processing program will be used. Therefore, choosing a suitable word processing program is worth some effort.

The more powerful (i.e. capable of complex editing tasks) may have more capability than is required. The more complexity, the more alternatives in the command structure that must be remembered.

Related to this last question is the determination of the number of keystrokes needed to perform simple operations. Consider whether the so-called 'user-friendly' menu-driven word processing program will become quite annoying. The requirement for accessing sequential menus to perform a desired function can slow down the editing process.

A check-point that may be overlooked while being bedazzled by the digital wonderland is that of the availability of local service. Modern computers are highly reliable. However, hardware breakdowns do occur and software can have logic problems. As dependency on the computer grows, so does the need for rapid repair and maintenance.

5. Final thoughts

With the advent of personal computers and the attendant publicity, people have felt a raising of the mysterious curtain obscuring computer usage. The so called 'user friendly' system has led to a commonly held belief that data processing problems are simple to solve.

These raised expectations can be realized with an investment of time and intellectual effort by the user. The difference between success and failure is usually not based on economics. Adequate funding does not insure success. Rather, careful planning is critical to success.

For example, many natural history museums have attempted installing computer-based data management systems. In the USA, estimates vary somewhat, but between two-thirds and three-quarters of the attempts are evaluated to be failures by the user community. Careful planning is commonly the reason given for success.

A few years ago a half-serious comment on deciding how many computers was the correct number for an institution was: 'one computer for every strong-minded individual'. The ready availability of cheap, powerful personal computers makes this comment a very real possibility. Computer network systems have brought the capacity to have personal computer available to the individual.

References

Codd, E. F. (1982). Relational database: a practical foundation for productivity. *Comm. Ass. Computing Machinery* 25, 109–17.

Knuth, D. E. (1975). *The art of computer programming Vol. 3. Sorting and searching.* Addison-Wesley Publ. Co., Reading, Mass, USA.

Krichevsky, M. I. (1982). Coping with computers and computer evangelists. *Ann. Rev. Microbiol.* 36, 311–21.

——, Walczak, C. A., Rogosa, M. R., and Johnson, R. K. (1980). Notes on the interchange of abbreviations and full generic names in computers. *Int. J. Syst. Bacteriol.* 3, 585–93.

17

Bioinformatics and the development of biotechnology

E. L. Foo and L. G. Heden

1. Introduction

The progress of biotechnology has been dramatic over the last three decades. During the 60s the subject was dominated by advances in fermentation technology; in the 70s enzyme engineering was the focus of attention; and now breakthroughs in genetic engineering have inspired a multitude of ambitious programmes and venture capital investments. It is fair to guess that the 90s will be a decade of receptor engineering (synthetic vaccines, directed mutations, quantum biochemistry, molecular genetics, etc.) and of phyto-technology (improved bioproductivity via plant and microbial physiology and genetics, process control, etc.).

The knowledge acquired so far has contributed greatly to the development of biotechnology. First, it was information packaged in the living micro-organisms and cells of culture collections. Then it was the active components of the cells that were needed; those components either had to be isolated and purified, or ordered from catalogues. Now, in order to continue their work which is 'comprises all aspects of the technological exploitation and control of living systems' (Gaden 1984), the bioengineers need access to refined information in the form of molecular spectra, receptor site structures, and nucleotide sequences.

This change in the focus of biotechnology, and the corresponding growth in the information needs of biotechnologists, has been paralleled by three dramatic changes in the associated computer support. In the 1960s the centralized mainframes still dominated information technology. In the 70s decentralized terminals became more common, and a growing number of major libraries and research institutions began to interact via datanets. Now, in the 80s, cheap portable terminals permit individual scientists not only to communicate freely with their colleagues all over the world, but also to search for information in thousands of data banks and databases that form the backbone of bioinformatics.

2. Bioinformatics

2.1. The scope of bioinformatics

The term 'bioinformatics' has been used for the area of interaction between information technology and the life sciences, including biotechnology, and it generally refers to computer-managed information (Heden in press). With the development of means for defining the fundamental intrinsic characteristics of molecules, micro-organisms, and other cells in numerical terms, bioinformatics has developed into an important support activity. It can provide structural and functional information on macromolecules and metabolites, and develops mathematical models illustrating the dynamic interactions within and between cells. It helps the microbial ecologist to understand the digestion processes which occur both inside and outside organisms, and it also indicates how complex phenomena in soil and water might be exploited for human benefit and resource conservation.

Bioinformatics also tells the fermentation engineer where to find the best strains and substrates for a particular purpose, and which 'feedback' strategies are likely to yield the desired industrial product. It provides the artificial intelligence that integrates information from several different instruments into identification labels for bacteria (numerical taxonomy), it provides diagnostic information for doctors, it generates guidelines for biochemists and industrial practitioners of applied microbiology, and it supplies similarly useful information for workers in many other areas of the life sciences.

2.2. Molecular genetics as a source of information

Bioinformatics gained a new dimension when it was understood that all biological processes depend on genetic information stored as linear codes along gigantic chain molecules (DNA). The codes are universal and made up of four basic units paired in a characteristic fashion. Some 300–2000 such pairs make up a functional piece of information, a gene. The number of genes in a cell is determined by its functional requirements: a virus might need some 35 000 pairs of the basic building blocks, a bacterium 35 million, and a human cell perhaps 35 billion. This makes for great differences in the length of the information strands, as well as in the 'packaging' and reading of the messages. In higher, specialized cells, the instructions are normally kept silent except for the little segment needed for a defined function. In fact, in higher animals and plants, the silent regions have increased in size so much that nature has had to devise cutting and splicing procedures that transmit only selected instructions, those which control the specific reactions characterizing the cell in question.

By means of cutting and splicing procedures developed in the laboratory, the genetic engineer can now also exercise control over the ways in which cells react. However, procedures devised to increase genetic diversity in microbial populations and facilitate selection for useful strains, hybrids, and breeds, have long been used. They involve the production of mutants through the destruction, by irradiation or chemicals, of the control mechanisms which the cell has developed over thousands of years. 'New' micro-organisms are thus produced. Some of them can be effective producers of commercially valuable substances, provided that they are supplied with appropriate nutrients in a controlled environment.

Genetic engineers have now also begun to exploit the metabolic machinery of bacteria by making them produce molecules that are foreign to them. This fact has caused much concern, because such micro-organisms can be regarded as 'man-made', even if they might well previously have occurred in nature.

2.3. Biological models for industrial practices

New knowledge about the behaviour of micro-organisms under stress has influenced fermentation technology, as have the results obtained by microbial ecologists studying the stability and versatility of microbial ecosystems. Defined mixed cultures for converting natural products into useful chemicals are on the horizon, and new ways to exploit microbial enzymes are continuously being reported. Those facts underline the need for more comprehensive metabolic mapping of familiar micro-organisms, and for goal-orientated selection and characterization of strains isolated from natural environments.

Even though recent advances in molecular biology have been impressive, much remains to be learned about the production plans which the cell follow when it assembles large protein molecules from a small number of amino acids.

The feedback principles, energy-saving practices, and recycling methods used in modern industry show great similarities with the approaches that nature has developd by trial and error over millions of years. Biodegradation by micro-organisms offers a way to reduce the environmental pollution caused by the use of synthetic chemicals and chemical industries are beginning to understand that nature's method of letting sequential reactions dovetail into each other might eventually help them to save a lot of expensive centrifuges, filters, and flocculation tanks, and also to avoid the use of corrosive chemicals, and high temperatures and pressures. Furthermore, the reduced vulnerability derived from operating on a smaller scale and with a wider spectrum of raw materials might well offset any sacrifices made in processing speed.

2.4. Enzymes as precision tools

The enzymes that break down large molecules were exploited early in industrial processes, but for synthesizing purposes the metabolic processes of whole cells had to be used, first in regular fermentation processes and, more recently, in immobilized systems where the growth of the cells is restricted, but their metabolism largely intact. However, a better understanding of the ways in which cells establish the energy balance of their metabolic processes is now opening new approaches based on defined cell components. These might involve either membrane-bound enzymes or the special molecules (co-enzymes) that transfer electrons between cell locations where energy is either used or stored until needed.

Developments in areas such as recharging of transfer molecules, synthesis of energy-rich compounds (ATP), or the use of enzymatic reactions in abnormal solvent environments, indicate many new opportunities for industrial processes. Cheap substrates, like sugar, can be used to regenerate the co-enzymes that drive the basic life processes, making it possible to modify expensive substrates by metabolic 'hitch-hiking' on properly selected microbial cells. The fact that substances insoluble in water can be attacked by microbial cells, even when the microbes cannot multiply, also opens up a whole new field for biosynthesis. Cholesterol can for instance be oxidized by certain microbial cells even when they are suspended in carbon tetrachloride.

The knowledge gained, for example, about the function of biocatalysts, may eventually lead to the synthesis of enzymes ('synzymes'); and the knowledge of immobilization of enzymes on electrode surfaces can lead to 'biochemical fuel cells'. Such devices might offer new approaches to the decentralized production of electricity from such easily available materials as alcohols, methane, or ammonia and also to the design of various environmental sensors.

2.5. The design and use of molecular probes

Highly specific sensors using enzymes to detect and measure small organic molecules already have medical and industrial applications. They exemplify the type of instrumentation now rapidly broadening the data base of bioinformation.

Cumbersome physical techniques for the study of large molecules have been supplemented with immunological methods using 'monoclonal anti-bodies' which can be used not only to detect and measure very large molecules; they also open up new industrial approaches for 'drug targeting' and immunotherapy. For the biochemist the greatest attraction of such antibodies is, however, that they can be used to separate desired products such as information molecules from crude cell extracts.

The technique for determining the exact sequence of the building blocks that make up a gene has also developed very rapidly in recent years. Thanks to computer science and advanced instrumentation, it is now possible to sequence DNA at a rate of 200 pairs a day. Since this corresponds to a polypeptide made up of 67 amino acids, a gene of molecular weight around 35 000 can be mapped in a few weeks. By back-reading such information, the computer can also derive the structure of the molecule that transmits the instruction from the gene to the protein building site in the cell and also predict the amino acid sequence of the polypeptide that will eventually emerge.

2.6. Benefits and hazards of bioinformatics

Suitably modified synthetic or isolated genes can be attached to an appropriate vector (a ring of DNA called plasmid or a virus), and a particle is thus formed which has the capacity not only to penetrate a microbial cell, but also to force it to read the information provided and to translate it into large quantities of a desired product. Using a technique known as gene amplification, highly productive strains utilizing cheap substrates can for instance cheaply produce hormones for plants, toxins for insects, and antibiotics for human beings and animals.

Genetic engineering also offers great opportunities for saving energy (e.g. biochemically or ecologically protected fermentations to circumvent the need for sterilization) and for simplying 'downstream processing' (autoflocculation, transfer of a bacterial product to a filamentous organism, controlled leakage of desirable products, etc.).

Sequence libraries are gradually developing, e.g. at Georgetown University Centre (2100 sequences, 360 000 amino acids) and the US National Laboratory at Los Alamos (1800 DNA sequences, 2 000 000 nucleotides). They provide predictions about where specific enzymes are likely to cut the DNA chain, indicate how a molecular 'bait' should look in order to 'fish out' desired information strands from disrupted cells, and finally, guide the construction of genes that will be effective in a particular micro-organism.

In view of the practical significance of such libraries, as well as of collections containing cultures of cells which produce monoclonal antibodies, or vectors selectd for their capacity to 'package' and store large DNA fragments, more international co-ordination is needed. The knowledge and materials held in such libraries and collections, which represent our best chance to gain understanding about tumour development, metabolic diseases, and autoimmune reactions, will also benefit industrial microbiology. This is particularly true now that the initial concern about conceivable hazards has subsided (Pergolizzi 1982).

However, genetic engineering must be the subject of continued vigilance, since highly toxic compounds and their derivatives might well be produced as soon as their amino acid sequences can be used as a guide for DNA synthesis. Biological weapons are prohibited by international treaty, but such treaties may of course be disregarded by independent groups (guerilla movements, terrorists, organized crime) so the governments that watch over the Biological Weapons Treaty should monitor the consequences of genetic engineering. The possibility to modify the identification characteristics of certain pathogens or to confine their spread to a defined target area by providing them with specific metabolic requirements for instance needs to be considered.

2.7. The information needs

Two keys are necessary to fully open the treasure chest of genetic diversity represented by the microbial kingdom. One is the hardware of bioinformatics: information about how analyses, biosynthetic procedures, and genetic manipulations are performed, and information about where to find the micro-organisms, vectors, chemicals, and equipment needed. The other key is the software of bioinformatics: the quantitative metabolic information and analytical data that characterize strains, cells, and organic molecules, and the mathematical models which illustrate their interactions. Examples are the computer programs used to analyse the X-ray diffraction patterns revealing the structure of large molecules or the mass spectrometer signals that sort out the composition of complex mixtures of chemicals.

Examples of 'hardware' are the computerized abstracting services and patent files, or the World Data Bank on Micro-organisms (Brisbane). Some of these sources are public and some are private, and they range from generalized services like Medline and Chemical Abstracts, that cover the biomedical and chemical literature, to specialized facilities like the CODATA/ IUIS Hybridoma Data Bank (ATCC 1985; Blaine and Bussard 1985). (See Appendix 1 for acronyms). In the latter case the system development was a collaborative effort of various international organizations, supported by funding agencies in Canada, France, Japan, the United Kingdom, Switzerland, and the United States as well as by the WHO and the CEC.

Much coordination is obviously necessary in the bioinformation field, and major efforts have also been made by several UK agencies and by both governmental (CEC) and non-governmental organizations (ICSU 1985; CODATA 1984). However, access to this information is sometimes difficult, particularly for the working biotechnologists in developing countries. This is why, after discussing some of the various kinds of information sources just mentioned, we conclude this overview by considering the potentials of some other methods to further enhance access to computerized information and data. These are: (i) information exchange initiatives and resource centres; (ii) computer conferencing; and (iii) recent developments in telecommunication.

3. Information exchange initiatives and resource centres

Since a very large number of organizations (Coombs 1985; Crafts-Lighty 1983; Da Silva and Heden 1986; UNESCO 1982; UNIDO 1983, 1984, 1985) fall in this category, only a few of those that MIRCEN/Stockholm is in constant contact with are presented here.

3.1. UNESCO

Unesco has a very extensive international informatics program (UNESCO 1980) that is designed for an exchange of experiences in strategies and policies for informatics. It aims to identify ways and means whereby informatics can contribute to economic, social, and cultural development, to clarify the prerequisites for the elaboration of strategies and policies at the national level, and to draw up a programme for international co-operation and assistance in various areas in the field of informatics. Its educational and training projects extend to several regions and developing countries, with specialized informatics applications, e.g. in education and social sciences, hydrology and seismology, PGI/UNISIST documentation service using the CDS/ISIS software, and various specialized information systems.

3.2. Microbiological Resource Centres (MIRCEN) network

The MIRCEN programme is carried out within the framework of Unesco's regular program activities in cooperation with IUMS, IOBB, WFCC, and other international bodies. It has its roots in the UNEP/UNESCO project on the use and preservation of microbial strains for deployment in environmental management. A world-wide programme for preserving microbial gene pools and making them accessible to developing countries has been launched through the MIRCEN network in environmental, applied microbiological, and biotechnological research. These centres are designed to:

(i) provide the infrastructure for a world network of regional and inter-regional co-operating laboratories geared to the management, distribution and utilization of the microbial gene pool;

(ii) support the conservation of useful strains of micro-organisms, particularly *Rhizobium*;

Table 17.1

UNEP/UNESCO/ICRO Microbiological Resources Centres (MIRCEN)—a network in environmental, applied microbiological and biotechnological research

MIRCEN centres (name of director)	Field of specialization	Other activities
Depts. of Soil Sciences & Botany, Univ. of Nairobi, P.O. Box 30197, Nairobi, Kenya (S. O. Keya)	BNF*	Training courses on micro-biology, fertilizers and bio-productivity, inter-African co-operation, extension services
IPAGRO, Postal 776, 90000 Porto Alegre, Rio Grande do Sul, Brazil. (J. R. Jardim Freire)	BNF	Training courses, extension services
TISTR, 196 Phahonyothin Road, Bangkhen, Bangkok 9, Thailand. (P. Atthasampunna)	Fermentation, food and recycling of organic residues	Culture collections, inter-ASEAN countries co-opera-tion, *Rhizobium*
Ains Shams Univ., Fac. of Agriculture, Shobra-Khaima, Cairo, Egypt. (M. N. Magdoub)	Biotechnology	Training courses on culture collections, food micro-biology, biotechnology, nitrogen fixation
Applied Research Div., ICAITI Ave. La Reforma 4–47 Zona 10, Apdo Postal 1552, Guatemala. (C. A. Rolz)	Biotechnology	Conferences and training courses on bioenergy and bio-chemicals, inter-Caribbean and Central American countries co-operation
NifTAL project, College of Tropical Agriculture and Human Resources, University of Hawaii, PO Box 'O', Paia, Hawaii 96779, USA. (P. Somesagaran)	BNF	Assembly of germplasm resource, antisera service, field testing in developing countries, establishment of regional BNF centres, BNF international service, confer-ences and training courses
Dept of Bacteriology, Karolinska Inst., 104 01 Stockholm, Sweden. (C. -G. Heden)	Biotechnology	Methods development, fermentation technology, computer conferencing, phytotechnology, numerical taxonomy, international net-works, bioinformatics

* BNF = Biological Nitrogen Fixation with emphasis on *Rhizobium*

Table 17.1 *(continued)*

MIRCEN centres (name of director)	Field of specialization	Other activities
Dept of Microbiology, Univ of Queensland, St. Lucia, Queensland 4067, Australia. (V. B. D. Skerman)*	World Data* Centre	World Directories—of collections of cultures of microorganisms, *Rhizobium* culture collections
Centre National de Recherches Agronomiques d'Institut Seneglais de Recherches Agricoles, B.P. 51, Bambey, Senegal. (M. Gueye)	BNF	West-African countries cooperation, management, distribution, and utilization of *Rhizobium* gene pools, appropriate biotechnology
PROIMI, Avenida Belgrano y Pasaje Ceseros, 4000 S.M. de Tucuman, Argentina. (F. Sineriz)	Biotechnology	South American Network for Biotechnology co-operation, biotechnology
Cell Culture & Nitrogen Fixation Lab., Building 011–A, Barc-West, Beltsville, Maryland 20705, USA. (D. F. Weber)	BNF	Culture collection characterization, documentation and preservation of *Rhizobium*, training and conferences on *Rhizobium* technology
ICME, Univ. of Osaka, Suitashi 563, Osaka, Japan. (H. Taguchi)†	Fermentation technology	Training course on microbiology and biotechnology, Asian countries co-operation
University of Kent at Canterbury, Kent (A. T. Bull) Commonwealth Mycological Inst., Ferry Lane, Kew, Surrey, UK. (D. Allsopp)	Biotechnology	Biotechnology biodeterioration, and biodegradation, computerized enquiry service, training courses
Univ. of Waterloo, Waterloo, Ontario & University of Guelph, Ontario N1G 2W1, Canada. (M. Moo-Young and D. Howell)	Fermentation technology and veterinary microbiology	International training courses in biotechnology, computer conferencing
Dept of Microbiology, Univ. of Maryland, College Park Campus, 207742 USA. (R. Colwell)	Marine microbiology	Marine biotechnology, biodegradation, numerical taxonomy

* Now at The Life Science Research Section, RIKENm 2–1 Hirosawa, Wako, Saitama 351–01, Japan (K. Komagata).

† Now known as IcBiotech (H. Okada).

(iii) foster the local development of new 'soft' technologies;

(iv) promote the application of microbiology to strengthen rural economies, particularly in developing countries; and

(v) serve as focal centres for the training of manpower and the diffusion of microbiological know-how.

The current 15 MIRCEN centres, which cover many different geographical areas, all have their own unique capabilities and fields of specialization. They support collaborative research and training, and provide specific service functions like diffusion of information (Table 17.1).

3.3. The International Organization for Biotechnology and Bioengineering

The IOBB is a federation of laboratories and institutes working in the fields of applied microbiology and biotechnology. It has from its inception worked very closely with the UNEP/UNESCO/ICRO Advisory Panel on Microbiology, and lately also with IUPAC. This has made it a link in a communication chain which is now being forged into an Inter-Union Biotechnology Commission by ICSU. Cross-links also exist with the UNU Biotechnology Programme, which aims at training young scientists from developing countries at advanced biotechnology centres. This aim comes close to IOBB's purpose, which is to promote international co-operation, and to inform its member institutes about current trends and opportunities in biotechnology.

3.4. The International Centre for Genetic Engineering and Biotechnology (ICGEB)

In early 1981 a group of experts recommended that UNIDO consider the establishment of an international centre that would help developing countries build the scientific infrastructure needed to gain maximum benefit from recent breakthroughs in genetic engineering and biotechnology. This led to a two-campus facility in Trieste (Italy) and New Delhi (India). The proposed work programme, estimated to cost US $ 40 million, covers the first 15 years of operations. The Trieste facility will probably focus on aspects of industrial microbiology, bioconversion of biomass (fodder, sugars, alcohol fuels, synthetic polymers, etc.), microbial leaching, fermentation, and protein engineering, while the New Delhi component will obviously concentrate R&D and training in agriculture, animal health, and human health (nitrogen fixation, oil microbiology, stress tolerance, and improved nutritional content in plants). A gene bank containing genetic stocks and information is also planned for one of the sites.

Plans to affiliate a number of national and regional centres with this advanced facility highlight the need for very effective communications. This was noted in the preparation of plans for the ICGEB (Heden 1981), and it provided one of the main reasons for the computer conferencing experiments mentioned in section 4.2.

3.5. The Commission of the European Communities (CEC)

The CEC, under its Biotechnology Action Programme, has a Task Force for Biotechnology Information which has initiated studies and exploratory activities such as the establishment of:

(i) the European Data Bank of Nucleic Acid Sequences;

(ii) the European Biotechnology Information and Referral Centre (British Library 1984);

(iii) the computerized information system for EEC Culture Collections of Micro-organisms;

(iv) the European Information System on Enzymes and Enzyme Engineering;

(v) the Technical Workshop on MIRDAB for European Cell and Virus Collections, and

(vi) the inventory of non-vertebrate based approaches and capabilities relevant to toxicological testing.

The European Strategic Programme of R&D in Information Technology covers other dimensions of bioinformatics: modelling, advanced software concepts, human interface, access, and diffusion, as detailed by Cantley (1984, 1985).

3.6. The United Nations University (UNU)

The United Nations University began operations in Tokyo in 1975, and one of its primary aims is to develop and support communication and co-operation to deal with global problems. Other programme areas include: (i) peace and conflict resolution; (ii) the global economy; (iii) energy systems and policy; (iv) resource policy and management; (v) the food-energy nexus; (vi) food, nutrition, biotechnology and poverty; (vii) human and social development; (vii) regional perspectives; (ix) science, technology, and the information society.

Apart from various more conventional forms of information service such as books, newsletters, or video aids (UNU 1985) it co-ordinates various directories, databases, and network newsletters. The International Network of Food Data Systems (INFOODS) at MIT (USA), for example, provides data on the nutrient composition of foods, beverages, and their ingredients on a world-wide basis.

4. Computer conferencing systems

Current advances in computer technology and systems for teleconferencing (e.g. video, computer, audio/graphic, and audio conferencing systems) have now made computer-based message systems (electronic mail) and computer-based conferencing systems (Palme 1984) accessible to a wide public. Computer conferencing, which is a text-based asynchronous mode of communication, has proved particularly useful, and economically competitive when compared with other text-based communication methods such as telegrams and telex. For example, a message costing US$ 20 by telex from Lima to New York might cost around US$ 2 sent by the ITT service which provides a direct connection between its electronic mail system, and the world-wide Telex and telegram networks. Computer conferencing allows rapid and extended discussions among geographically dispersed members, and it of course reduces the problems and costs of travel (Palme 1983; Heden in press). The basic requirements for access are a home computer (even certain electric typewriters or calculators will do) with a modem and a telephone. The rapidly developing links between the various computer-based message systems further facilitate message transmission and allow a user, while traveling for example, to communicate via the nearest system.

4.1. The advantages of computer conferencing

The decentralized, asynchronous computer conference offers several advantages, both as a complement and an alternative to more traditional congresses.

1. The number of participants is limited only by the conference organizer, and costs are low enough to permit the participation of groups with limited or not travel funds and small operational budgets.

2. Many participants, in small face-to-face groups, can generate and contribute ideas with much more freedom than the international symposium allows. A psychological barrier is removed if the chore of message typing is limited to a selected individual (the rapporteur).

3. The conference structure is flexible, and overload can be avoided by including sub-groups and associated telephone conferences.

4. The need for interpreters is reduced and opportunities for participation are increased, since regional workshops can use local languages while text transmissions are in one common language.

5. The challenge of rising travel costs can be met by dispersed conferences with regional nodes.

6. International symposia can be made more effective by preparatory collaboration (selection of programme items, presentation of papers, etc.) via the computer conferences.

4.2. Computer conferencing activities in biotechnology

Computer-based conferencing systems (Table 17.2) have been extensively used by computer enthusiasts, but only recently have biologists started to use them. MIRCEN Stockholm, in co-operation with other organizations (IDRC, WAAS, IFIAS, NAS, UNU), started to evaluate its application in fields such as biological nitrogen fixation, anerobic digestion, bioenergy, and bioconversion of lignocellulose only a few years ago.

The computer conference on 'Bioconversion of Lignocellulose for Fuel, Fodder, and Food Needed for Rural Development in Poor Countries', which ran from May until December 1983, was probably the first major international computer conference which has been devoted to a central development issue. The technical aims were to assess the potential of the computer conferencing technique to facilitate scientific research and information exchange, and to lay the groundwork for future applications on a global basis. Two computer host systems (EIES and QZCOM) were used. The asynchronous computer conference ended with a dispersed 4-day workshop involving small groups of scientists gathered in a number of 'nodes' at Stockholm, Ottawa, Manila, Tokyo and elsewhere (Balson 1984; NAS 1984).

To follow up the Bioconversion Computer Conference, MIRCEN Stockholm then started two computer conference series in the QZCOM computer conferencing system. The first, MIRCENET, was intended to serve the 15 MIRCEN centres; it generated two computer conferences ('Mircenet Newsletter' to provide news on events related to biotechnology and 'Mircenet Comcon GIAMVII' for the preparation of computer conferencing activities during GIAM VII. The second series, Anaerobic Digestion MIRCEN, concentrates specifically on all aspects of methanogenic fermentations, biogas technology, and landfills. Its purpose was to link individuals and organizations for information exchange and to encourage on-line and off-line discussions, as well as to facilitate electronic discussions of the papers and posters presented at face-to-face meetings, e.g. at the International BioEnergy '84 Conference in Gothenburg, Sweden, June 1984, and the 3rd EC Conference on Energy from Biomass, Venice, March 1985. There are now two permanent computer conferences, 'Anaerobic Digestion (Mircen) Technical', and 'Anaerobic Digestion (Mircen) Cocktail', and several other short-term conferences (Foo 1984). MIRCEN Stockholm is now working on additional ways to use the telenetwork on anaerobic digestion, via QZCOM and COSY, to facilitate electronic discussions and to serve as an electronic extension of symposia, workshops, training courses, and even computer seminars. Special emphasis is given to the 'dispersed conferences' concept, which demonstrated its usefulness in the 4-day workshop mentioned earlier. Such co-ordinated face-to-face meetings have the advantages as summarized in Table 17.3.

In preparation is also a 'small information exchange club' called the 'Phytotechnology Project'; its background is that controlled environment

Table 17.2

Vendors for computer conference systems

System	Availability	Connect cost	Processor
MATRIX Transaction Exchange International Tele/Conferencing Co. Suite 405, 1877 Broadway Boulder, CO 80302	Telenet may be licensed	Set-up fee plus $6.75–19.75/hr 5-year license: $10 000	DEC System
PARTICIPATE Participation Systems Inc. 50 Cross Street Winchester, MA 01890	The Source or Dialcom	Initial fee – $100 plus $7.75–25.75/hr $10/month minimum Initial fee – none $18.50–22.50/hr $100/month minimum	
CONFER II Advertelev Communications Systems, 2067 Ascot Ann Arbor, MI 48103	Telenet Autonet direct dial may be licensed	$19–20/hr licence cost depends on implementation	Amdahl
EIES Electronic Information Exchange System New Jersey Institute of Technology 323 High Street, Newark, NJ 07102	Telenet Uninet direct dial	$75/month group accounts $200/month plus $10/month per user plus $8/hr	Perkin-Elmer 3230
COM Stockholm University Computer Center Box 27322 102 54 Stockholm Sweden	May be licensed	$4–40/hr	DEC 10 DEC 20
AUGMENT Tymshare 20705 Valley Green Drive Cupertino, CA 95014	Tymnet may be licensed	$14–18/hr licensing cost depends on implementation	DEC 20 with TOPS 20 operating system

Source: NAS (1984). Computer Conference in Lignocellulose Conversion—Staff Summary Report. National Academy of Sciences, Washington, D.C.

Table 17.3

Advantages of asynchronous computer conferencing

Reduced need to travel; saves time and energy when many geographically dispersed individuals need to carry out joint research projects, to plan conferences, etc.

Rapid communication technique compared with postal mail; no time lost in trying to contact others by phone; mail can be received from any location especially when the receiver is travelling

Freedom to control time and frequency of communication means that nongoing work is not disrupted by travel or time-zone constraints

Access to private files and reprints, as well as to local co-workers; improves the message content, and the precision is improved by editing before transmission

No comments missed and 'clocked' transcripts available as priority records or as widely distributed 'instantaneous publications'

Reduced risk for monopolised, prejudiced or emotional discussions and greater opportunities for junior scientists to participate and those with difficulties in conversation with foreign language

Printed texts more easily understood, particularly when foreign languages are involved or low-quality voice-lines must be used

The road is paved towards Computer-Assisted International Team Research (CAITR), co-operative modelling, setting of standards, joint authorship, follow-up, and augmentation of face-to-face meetings, etc.

Voting with immediate display, anonymous or private messages, computer-aided sorting and other aids are available to chairmen/gatekeepers/programme managers/rapporteurs, and others.

---

agriculture has special relevance for the development of the arid zones in the world. However, the project also encompasses the type of 'microcosm' studies that will be needed to test some 'large-scale release' biotechnologies to be sure that they are quite fail-safe (McGarity 1985; Alexander 1985). Those applications (the use of genetically modified micro-organisms to reduce frost damage to crops, to destroy oil and toxic substances, to improve mineral recovery from low-grade ores, to improve tertiary oil recovery, etc.) are potentially so important that an international exchange of test protocols, probably including 'biochemical fingerprinting' (Illeni 1981; Kühn 1985) may well become an important future element of bioinformatics.

A standard 'test-rig' (Heden in press) for evaluating various combinations of physical and physiological control principles, including also microbiological rhizosphere manipulations, is now being developed as a basis for this telecommunications network.

4.3. Research co-operation and project co-ordination

Inspired by their experience with the 'bioconversion' computer conference mentioned above, a number of international organizations have started their own networking experiments. UNU is in the process of launching a computer

Table 17.4

The Consultative Group on International Agricultural Research (CGIAR)
and its centres

Name of centre	Main function	Telecommunication facilities
1. Centro Internacional de Agricultura Tropical (CIAT), Apartado Aero 6713, Cali, Colombia	Concentrates on poor urbans and rural food consumers through R&D in food production techniques and development of technologies	International direct dialing line available telex; DAPAQ-TELECOM and access to CMBS, DIALOG information services
2. Centro Internacional de Mejoramiento de Maiz y Trigo (CIMMYT), Londres 40, Mexico 06600, D.F., Mexico	World-wide improvement of maize, wheat, barley and triticale; serves as the world's major repository for maize germplasm	International direct dialling available; TELEPAC; has DEC VAX-11/780 as main computer
3. Centro Intercional de la Papa (CIP), Apartado 5969, Lima, Peru	Improvement of potato in high-altitudes and tropics; world collection of potato germplasm (15 000 genotypes)	International direct line available; telex
4. International Board for Plant Genetic Resources (IBPGR), FAO, Via delle terme di Caracalla, Rome 00100, Italy	To preserve germplasms of world's crops and promote network of genetic resource centres to improve the collection, conservation, documentation and use of plant germplasms; world authority on germplasm collection, storage and data compilation	ITALCABLE, ITALPAC; IBM 370 mainfame and a PRIME 750

Table 17.4 *(continued)*

Name of centre	Main function	Telecommunication facilities
5. International Center for Agricultural Research in the Dry Areas (ICARDA), PO Box 5466, Aleppo, Syria	To increase agricultural productivity in West Asia and North Africa and to serve as regional centre for improvement of barely, lentils and broad beans	No international data network access; VAX 11/780 mainframe with 24 ASCII terminals mostly in work stations
6. International Crop Research Institute for the Semi-Arid Tropics (ICRISAT), ICRISAT Patancheru PO, Andra Pradesh 502–324, India	Improvement of major food crops and management of soils and water; an economics program that conducts needs assessment and investigates factors that affect acceptance of improved technologies	International calls through operator; DEC VAX 11/780 with 87 ASCII terminals, international data network not available
7. International Food Policy Research Institute (IFPRI), 1776 Massachusetts Avenue NW, Washington DC 20036, USA	Identifies and analyses strategies for meeting world food needs, reduce hunger and malnutrition, food distribution, trade, trends, and statistics	Six lines for data communication
8. International Institute of Tropical Agriculture (IITA), PO Box 5320, Ibadan, Nigeria	Concentrates on lowland agriculture with major efforts on roots and tubers, cereals and grain legumes; improved traditional farming systems	International calls can take 3–4 days via operator; no international data network access
9. International Livestock Center for Africa (ILCA), PO Box 5689, Addis Ababa, Ethiopia	To improve livestock production and marketing systems, training of specialists, and relevant African livestock industry documentation	Telex; international calls through operator; no international data network access
10. International Laboratory for Research on Animal Diseases (ILRAD), PO Box 30709, Nairobi, Kenya	To develop control measures for two major livestock diseases—trypanosomiasis and theileriosis	International calls through operator; no international data network access; MEMORY 7500 as main computer

Table 17.4 *(continued)*

Name of centre	Main function	Telecommunication facilities
11. International Rice Research Institute (IRRI), PO Box 933, Manila, Philippines	Improvement of the quality and quantity of rice; world germplasm collection	Two data communication lines; IBM 4331, WANG 2200 as main frame computer
12. International Service for National Agricultural Research (ISNAR), Oranje Buitensingel 6, 2511 VE Hague, The Netherlands	To help nations to specify goals and policies for agri-cultural development, to improve national capacities to organize and manage research, communication of results, funding, human expertise, etc.	One data communication line; WANG OIS-140 as main computer
13. West Africa Rice Development Association (WARDA), PO Box 1019, Monrovia, Liberia	Regional self-sufficient in rice in West Africa; all matters related to rice– disease control, post-harvest aspects, storage, processing, marketing, export	International direct line available; no international data network access; WANG 2200 MVP with seven terminals
14. CGIAR Secretariat (CG SEC), 1818 H Street, NW. Washington DC 20433, USA	Has several main functions of a Secretariat—ensure funding, adequate resources, policy issues, implementation, informa-tion support, etc.	Also has access to World Bank's VAX 780s
15. TAC Secretariat (TAC SEC), FAO, Rome 00100, Italy	Analogous to CG SEC including identification and analysis of scientific policy issues, recruitment of experts, documentation, reviewing programs and budgets, etc.	Uses FAO's facilities

conference to support *Brucella* vaccine research in Latin America. IFIAS and WAAS now use the technique for project co-ordination and conference preparations. IDRC, one of the major actors in the bioconversion conference, is now well advanced in using data links to facilitate the operations of its regional offices. This illustrates that developing countries can now often be reached via electronic communication facilities, a fact which is also indicated by the successful electronic networking of the majority of the laboratories of CGIAR (Balson 1985; Telematics International 1984) (Table 17.4a–6). WAITRO is also considering a test study by its members involved in anaerobic digestion, to evaluate computer conferencing techniques for members active in other areas of biotechnological research and development.

5. Recent developments in telecommunication technology

The developing countries are lagging behind in exploiting the opportunities offered by improved telecommunication (Manassah 1982) to support their infrastructure in applied microbiology and biotechnology. However, a great many now have satellite links which their scientists might well use internationally, provided that they manage to persuade their PTTs to let them sidestep bad telephone lines, and can find an aid agency that will pick up the hard currency communication bill.

Neither is easy, because PTTs prefer to do business with airlines, banks, and multi-national corporations, and they often take a dim view of small groups of scientists that want to make use of their premises and of their precious time. With few exceptions aid agencies, on the other hand, are still notoriously suspicious of 'high technology', and much prefer to provide funds for air travel and sophisticated short-term training abroad, than for supporting their trainees with vital information once they return home.

However, decision makers in industrialized and developing countries alike are now starting to realize that the flow of information is as essential for the building of a scientific infrastructure as the blood which flows through the umbilical cord to an unborn baby.

Also, some non-profit organizations are making great progress in providing information support to developing countries. Through the pioneering efforts of VITA, AMSAT, and Interpares, a new way to provide developing countries with cheap text communication has recently been demonstrated (BOSTID 1985). An inexpensive satellite in polar orbit receives, stores and re-transmits messages, reappearing over the same spots every day, ready to communicate with simple ground stations (costing about US$ 1600 per unit) that consists of a transmitter/receiver, a crossed dipole antenna, a portable computer, and a terminal node controller. The dedicated system, PACSAT, could be in operation as early as 1987 (Y. Pal, unpubl. obs., 1982; Ramani and Miller 1982; Valantin and Balson 1985).

In many places we now also find a new generation of decision makers, who have lost the 'mainframe awe' of their predecessors and who are used to working in an environment of telephones, telexes, and telefaxes. They are also starting to show concern about 'information poverty' as one of the causes for the poor efficiency of many scientific and technical endeavors in developing countries.

These decision makers also know that some lap-size computers can be used for data collection (surveys in forestry, fisheries, farming, health, etc.) far away from all power and telephone lines. Not only does this save time, but transcription errors are also avoided when the information is brought home for analysis and incorporation into some master bioinformatics data base.

Obviously, many factors now interact in ways which make it reasonable to assume that bioinformatics will become a vitally important factor for biotechnology not only in the industrial, but also in developing parts of the world.

References

Alexander, M. (1985). Ecological consequences-reducing the uncertainties. *Issues Sci. Technol.* 1, 57.

ATCC (1985). Hybridoma data bank. *Am. Type Culture Collection Q. Newslett.* 5, 1–2.

Balson, D. (1984). *International computer-based conference on biotechnology: a case study.* IDRC, Canada.

—— (1985). *CGNET: A data transfer network for the CGIAR.* Paper presented at the 7th IAALD World Congress. June 2–6.

Blaine, L. D. and Bussard, A. (1985). The hybridoma data bank: a multidisciplinary resource for microbiology. In *Book of Abstracts for 7th International Conference on Global Impacts of Applied Microbiology*, p. 159.

BOSTID (1985). Satellite provides communities links. Editorial in *BOSTID Developments*. 5.

British Library (1984). *Background papers for a seminar on biotechnology information.* Science Reference Library, London.

Cantley, M. F. (1984). Bio-informatics in Europe: foundations and visions. *Swiss Biotech.* 2, 7–14.

CODATA (1984). *CODATA Newslett.* Oct. No. 30.

Coombs, J. (1985). *The international biotechnology directory.* Macmillan, London.

Crafts-Lighty, A. (1983). *Information sources in biotechnology.* Macmillan, London.

DaSilva, E. J. and Heden, C.-G. (1986). The role of international organizations in biotechnology. In *Comprehensive biotechnology* (eds C. W. Robinson, and J. A. Howell) Pergamon Press, U.K.

Foo, E. L. (1984). *Anaerobic digestion Mircen.* Volumes 1 and 2. Published by MIRCEN-Stockholm, Karolinska Institute, 10401 Stockholm.

Gaden, E. L., Jr (1984). Biotechnology for development: promise and problems. *BOSTID Developments, Newslett. Sci. Technol. Int. Development*, US Nat. Res. Coun. 4, 1.

Heden, C.-G. (1981). *The potential impact of microbiology on developing countries.* UNIDO/IS.261.

ICSU (1985). *International Council of Scientific Unions Yearbook 1985.* ICSU, 51 Boulevard de Montmorency 75016 Paris.

Ileni, T. (1981). Laser light scattering on bacterial colonies as a possible tool for rapid identification. In *Scattering techniques applied to supramolecular and non-equilibrium systems*. (eds H. Chen *et al.*) pp. 887–92. Plenum, New York.

Kühn, I. (1985). Biochemical fingerprinting of *Escherichia coli*: a simple method for epidemiological investigations. *J. Microbiol. Methods* 3, 159–70.

Manassah, J. T. (1982). *Innovations in telecommunications*. Parts A and B. Academic Press, New York.

McGarity, T. O. (1985). Regulating biotechnology. *Iss. Sci. Technol.* 1, 39.

NAS (1984). *Computer Conference on Lignocellulose Conversion—Staff Summary Report*. Nat. Acad. Sci. Washington, DC.

Palme, J. (1983). *The COM and PortaCom computer conferencing systems*. Stockholm University Computing Centre, Box 27322, 10254 Stockholm.

—— (1984). *Survey of computer-based message systems*. QZ Computer Center, Box 27322, S 102 54 Stockholm, Sweden.

Pergolizzi, R. G. (ed.) (1982). *Genetic engineering and biotechnology source book*. McGraw-Hill, Washington DC.

Ramani, S. and Miller, R. (1982). A new type of communication satellite needed for computer-based messaging. In *Pathways to the information society* (ed. M. B. Williams). North-Holland Publishing Co.

Reiniger, P. and Cantley, M. F. (1985). Bioinformatics in Europe. In *Book of Abstracts for 7th International Conference on Global Impacts of Applied Microbiology*. p. 23.

Telematics International (1984). *CGNET: A data transfer network for the Consultative Group on International Agricultural Research*. Unpublished Report to the CGIAR and IDRC.

UNESCO (1980). *Informatics: a vital factor in development*. Informatics Section, SC/SER, UNESCO, Paris.

—— (1982). *International Directory of new and renewable energy information sources and research centers*, 1st ed. UNESCO.

UNIDO (1983). *Directory of industrial and technological research institutes: industrial conversion of biomass*. UNIDO/IS.372.

—— (1984). *Directory of international non-governmental organisations in consultative status with UNIDO*. Published by UNIDO, Austria.

—— (1985). *Directory of industrial information services and systems in developing countries*. UNIDO/IS.205/REV.1.

UNU (1985). *The United Nations University today. Introduction and basic facts*. UNU, Tokyo.

Valantin, R. and Balson, D. (1986). Choosing information technologies: Case Studies in Computer-based communications. In *New information technologies and development* (ed. Aklilu Lemma) pp. 55–8. Centre for Science and Technology and Development, United Nations Secretariat, New York.

7. Appendix 1: Acronyms

AMSAT: Radio Amateur Satellite Corporation
ATCC: American Type Culture Collection
BOSTID: Board on Science and Technology for International Development
CDS/ISIS: Computerized Documentation System/Integrated Set of Information Systems (of Unesco)

CEC: Commission of the European Communities
CGIAR: Consultative Group on International Agricultural Research
CODATA: Committee for Data on Science and Technology
QZCOM: Computerized Conferencing System (Stockholm University)
EIES: Electronic Information Exchange System
FAO: UN Food and Agriculture Organization
GIAM VII: 7th International Conference on the Global Impact of Applied
Microbiology (12–16th August, Helsinki, Finland)
ICGEB: International Center for Genetic Engineering and Biotechnology
ICRO: International Cell Research Organization
ICSU: International Council of Scientific Unions
IDRC: International Development Research Centre
IFIAS: International Federation of Institutes for Advanced Studies
IOBB: International Organization for Biotechnology and Bioengineering
IUIS: International Union of Immunological Societies
IUMS: International Union of Microbiological Societies
IUPAC: International Union of Pure and Applied Chemistry
MIRCEN: UNEP/UNESCO/ICRO Microbiological Resources Center
MIRDAB: Microbiological Resources Databank
MIT: Massachusetts Institute of Technology
NAS: National Academy of Sciences, Washington D.C.
PACSAT: Packet Satellite Communication
PGI: General Information Programme (of UNESCO)
PTTs: National Post-Telephone-Telegraph Administrations
UN: United Nations
UNEP: UN Environment Programme
UNESCO: UN Educational, Scientific, and Cultural Organization
UNIDO: UN Industrial Development Organization
UNISIST: Programme of International Co-operation in Scientific and Technological
Information (of UNESCO)
UNU: United Nations University
VITA: Volunteers in Technical Assistance
WAAS: World Academy of Art and Science
WAITRO: World Association of Industrial and Technical Research Organizations
WFCC: World Federation for Culture Collections
WHO: World Health Organization

18

Patent developments in microbial technology and genetic engineering

Sisko Knuth* and H. G. Gyllenberg

1. Introduction

Inventions have been patented ever since the 17th century. The British statute of Monopolies originates from 1624 and in the USA the first patent law dates back to 1790 (Wegner 1979). Similar rules were introduced in France during the Revolution. The roots of the patent system are in the materialistic world of machines and equipment. However, with new technological developments the patent laws have been extended to cover new areas. As stated by Armitage (1977): 'From the crafts of the 17th century and the mechanical devices of the industrial revolution, patents have adapted to the chemical, industrial developments of the 19th century and to the electric, electronic, plastics, agricultural, and nuclear industries of the 20th century'. For example, the US Constitution (Article 1, section 8) states that 'The Congress shall have power to promote the progress of science and useful arts, by securing for a limited time to authors and inventors the exclusive right to their respective writings and discoveries'.

Applying old industrial property laws to new fields of technology is difficult. When it comes to computer technology we know that software is not patentable (Bent 1982; Schulte 1981), whereas the hardware is. Patenting in some new fields of technology, like microelectronics, is not considered profitable, because technological development is so rapid that keeping inventions as trade secrets is the only way to maintain an advantage over the competitors (Dickson 1980).

What form will the patenting of the newest developments in biotechnology take? Or is patenting generally the best method of protecting inventions in the field of, for example, gene technology? Patents have been frequently applied for in this field, but until now only a few patents have been granted.

---

* New address: Research Laboratories ALKO Ltd., POB 350, SF–00101, Helsinki, Finland.

2. Patent system

2.1. Patents and conditions of patentability

A patent means an agreement between the state and the inventor. The State, for a limited time, grants the inventor the right to exclude others from using his invention, unless given permission by the inventor (patentee). In return the inventor must make a full disclosure of his invention in the patent specification, which becomes public after a certain period of time. The meaning of granting patents is two-fold: to promote technology with the help of new inventions, and to prevent inventions remaining secret. Granting a patent for a certain invention stimulates competitors to search for alternative solutions to the same problem or even to improve the invention (Crespi 1982; Vossius 1981, 1982).

According to the laws applied in most countries, the conditions for patentability are novelty, inventiveness or non-obviousness, and utility or industrial applicability. Besides this the invention must be reproducible. Although the conditions are the same, the patent authorities and courts have applied them differently in different countries. In several countries absolute novelty is demanded. This means that the prior publication of an invention elsewhere, and even outside the country where the patent is being applied for is a bar to novelty (Crespi 1982), though in the USA the publishing of an invention a year before the actual filing date of the application does not preclude patentability. A grace period like this is also found in the WIPO Model Law for Developing Countries on Inventions (Hüni and Buss 1982).

A patentable invention must be inventive or non-obvious. Inventiveness is evaluated with regard to the state of the art, and the so-called knowledge of the average expert. The European Patent Convention (Article 56) defines: 'An invention shall be considered as involving an inventive step if, having regard to the state of the art, it is not obvious to a person skilled in the art', and the US Patent Act (Section 103) (Sinnott 1983) includes the demand that the invention must not have been obvious at the time the invention was made to a person having ordinary skill in the art. An invention must show some form of practical utility. In almost all patent laws, a prerequisite for an invention is industrial character or applicability. Article 57 of the European Patent Convention defines that the invention can be in any kind of industry including agriculture. The demand for practical utility is particularly strong in the USA (Rosenberg 1980), although it need not be of an intrinsic industrial character.

2.2. Applying for a patent

To obtain a patent for an invention, the inventor has to file a patent application at the patent office. In various countries this becomes public 18 months after the filing date, although in the USA and in the Soviet Union, usually only granted patents or inventor's certificates (in the Soviet Union) are made public (USSR Chamber of Commerce and Industry, Patent Department 1981). Patent protection for an invention is granted only for a limited time which, in the countries that are members of the European Patents Convention (EPC), is 20 years from the filing date of the patent application (Schulte 1981) and, in the USA, it is 17 years from the date of issue of the patent (Sinnott 1983).

Priority date means the first date of filing a patent application. The signature states of the Paris Convention (International Convention for the Protection of Industrial Property)—until now 93 countries have ratified this agreement—(Anon. 1984*a*) accord to foreign nationals the same rights as they accord to their own nationals. After filing an application in one of these countries, a person can apply for a patent for the same invention, or part of it, in the other countries and still enjoy the benefits of the filing date of the earlier application (Crespi 1982).

A patent document (or patent specification) is divided into two parts; the patent description (or specification) and the claims. The disclosure of the invention is given in the patent description, the scope of the protection required by the applicant is defined in the claims. A patent application contains one or more claims, and protective coverage varies from the first (or main) claim, to subclaims that are more limited.

It is possible to apply for a national patent in one or more countries, or for a regional patent. It must be remembered that a patent is in force only in the country where it has been granted for the given invention. Patents are worth application in countries where the competitors and markets are. The Patent Co-operation Treaty (PCT) is an international agreement, ratified by 33 countries, and deals with international preliminary investigation on the novelty and patentability of inventions. After a preliminary investigation, the applicant decides whether he wants to continue the application on a national or regional basis (Anon. 1984*b*; World Intellectual Property Organization 1978).

A European patent can be applied for in the states that are members of the European Patent Convention (EPC), which at the moment includes Austria, Belgium, France, Federal Republic of Germany, Great Britain, Italy, Liechtenstein, Luxembourg, Netherlands, Sweden, and Switzerland. According to that convention, an applicant can obtain a patent in one or more of the member countries by filing an application at the European Patent Office (European Patent Office 1983).

The member states of the African Intellectual Property Organization (OAPI) are Benin, Cameroon, Central African Republic, Chad, Congo, Gabon, Ivory Coast, Mauritania, Niger, Senegal, Togo, and Upper Volta (Anon. 1984*c*). For these states, a regional patent protection is available (Anon. 1983).

2.3. Patenting in different countries

Although the conditions for patentability are the same in many countries, each country has its own way of dealing with patent applications. In some countries the patent office examines only whether the application is formally acceptable (registration or non-examining system). In most industrialized countries the grounds for examination are based on whether the invention meets the conditions of patentability (examining system). Each country may also have different concepts as to what can be patented, how the claims are interpreted, how the patent or patent application shall be published, duration of patent protection, maintenance of the granted patent, etc. (Vossius 1981). According to Crespi (1982) there are six different patent systems: the British patent law, the US patent law, the German patent law, the Euro-Latin patent system, the Japanese, and the European patent law. Cooper (1982) distinguishes between statutory protection in the Eastern Bloc, the Western Bloc and in the Third World.

2.4. Patentable and non-patentable inventions

Section 101 of the US Patent Act states: 'Whoever invents or discovers any new and useful process, machine, manufacture, or composition of matter, or any new or useful improvement thereof may obtain a patent therefor subject to the conditions and requirements of this title' (Sinnott 1983). So the US statute does not actually exclude any category of inventions from patentability. A more precise definition can be found in court rulings. In the case of Parker versus Flook (1978), which dealt with a patent application for a computer program, the US Supreme Court stated in 1978: '. . . Phenomena of nature, though just discovered, mental processes and abstract intellectual concepts are not patentable, as they are the basic tools of scientific and technological work . . .' (Cooper 1982; Vossius 1982).

Article 52 of the European Patent Convention defines as non-patentable discoveries, scientific theories, mathematical methods, and computer programs. Things such as surgical and therapeutic treatments, or diagnostic methods that are used on people or animals are outside the scope of patentability, because they are not considered to be susceptible to industrial application. Article 53 of the convention also defines as non-patentable plant and animal varieties, and the biological means of producing them. On the other hand, microbiological processes and the products thereof belong to patentable inventions. National patent laws in some countries defines the following as non-patentable: substances produced by chemical methods, medicines, foods, and mere mixtures of foods or medicines (Crespi 1982).

Biotechnology can be defined as the technological utilization of living organisms or otherwise active materials, and substances of biological origin

(Gyllenberg 1982). Traditional biotechnology, such as different fermentation methods and the production of antibiotics, have been fitted inside the patent law in a similar manner as chemical inventions, the inference being the utilization of certain chemical reactions taking place within the cell (Hüni and Buss 1982). Problems have arisen only when patent protection has been sought for new micro-organisms produced by selection, mutation, and gene manipulation.

3. Patenting in biotechnology

3.1. Patenting micro-organisms as products

Now and then patent applicants have tried to get a micro-organism protected as a product, especially when a new mutant has been isolated or developed, but in several countries patent offices have expressed the opinion that a micro-organism is not patentable if it is living (Halluin 1982), or the patent has not been granted because the micro-organism is considered to be a natural product, or its production has not been considered to be repeatable (von Pechmann 1972). The patent practice in various countries was influenced by the US patent office changing its policy as a result of judicial proceedings in the cases of Bergy and Chakrabarty. The case of Bergy (1977) claimed a pure culture of *Streptomyces vellosus* isolated from nature which produced lincomycin. The case of Chakrabarty (1978) included the claim 'to a bacterium from the genus *Pseudomonas* containing at least two stable energy generating plasmids each having a separate hydrocarbon degradative pathway'. On the 16th of June 1980, the US Supreme Court decided by five votes to four that Dr Chakrabarty's 'man-made plasmid-injected *Pseudomonas*' was patentable. According to the court decision 'there is no longer a patent law distinction between living and non-living compositions of matter as long as they are man-made'. There was no Supreme Court decision in the case of Bergy because it was withdrawn before a decision was reached (Halluin 1982).

3.2. The problem of natural products and patenting 'life'

In connection with the cases of Bergy and Chakrabarty there was a lot of discussion about the problem of natural products and the patenting of higher forms of life (Cooper 1982). Novelty is one of the criteria for the patentability of inventions, and a compound or a micro-organism that may have existed for thousands of years in nature cannot be considered novel in this sense. The idea that a patent may be rejected on the grounds that it is a natural product stems from the fact that the public cannot be deprived of the rights that it is already enjoying (Saliwanchik 1982). If a patent is sought for a natural product, the existence of which the public has not even known, granting the patent seems justified and the public has not lost any of its rights. The inventor is the first to discover that the natural product has biological activity, isolated and purified the product, or defined its chemical structure. Recently in the USA, patents have been issued for biologically pure cultures of micro-organisms even though a final court decision was never reached in the case of Bergy (Cooper 1982). In the German Federal Republic, court decisions have shown that, although an invention exists in some form in nature, this is still no bar to its patentability (Anon. 1978a, b, c).

When Chakrabarty was issued a patent for his man-made micro-organism it was feared that the case would lead to the patenting of higher organisms as well. This has proved groundless since it would be difficult or impossible to describe an invention concerning a higher life form so that it could be repeatable (Wegner 1979) or present an invention as having technical usefulness (Vossius 1982). The same problem has been dealt with in the German Federal Republic with respect to the cases of Rote Taube (Anon. 1969) and Bäckerhefe (Anon. 1975). Some countries have allowed the legal protection of new plant varieties by plant patents or plant variety rights (Crespi 1982). The European Patent Convention (Article 53), however, defines that '. . . patents shall not be granted in respect of . . . plant or animal varieties . . .'

3.3. Product and process patents

Basically, there are two categories of patents, product and process patents, according to what the patent claims define as the subject matter of the invention. Among product patents there are 'substances', 'compositions of matter' and 'devices', whilst among process patents there are 'processes of preparation', 'working methods', and 'uses' (Vossius 1982). Protection for an invention can be sought in different claim combinations in the same application; for the product, for the process of preparing the product, and for the use of the product. A patent application must, however, meet the demand of unity (only one inventive idea can be accepted in one application). Product protection is generally considered the most efficient form of protection. There are two types of product claims. Product *per se*—a claim that gives a product absolute protection regardless of how it has been made or what it is used for. Product-by-process claims only protect a product that has been made by a certain process (Crespi 1982). Claims like this are admissible if a product cannot be identified by distinctive characteristic parameters. Hüni and Buss (1982) list over 30 countries where it is possible to obtain protection for a micro-organism as a product though, in some of the countries, indirectly as a product of a microbiological process.

3.4. Describing micro-organisms in a patent application

Article 83 of the European Patent Convention defines: 'The European patent application must disclose the invention in a manner sufficiently clear and complete for it to be carried out by a person skilled in the art'. This so called 'disclosure requirement' is common to all patent systems. Section 112 of the US Patent Act provides that the written description of an invention is '. . . in such full, clear, concise and exact terms as to enable any person skilled in the art to which it pertains, or with which it is most nearly connected, to make and use the same'. This claim is the basis for the so-called 'enabling disclosure' in the USA. Section 112 also implies presenting the best mode of carrying out the invention. Generally speaking, an invention must be described in such detail that an average expert can, on the basis of the description, use the invention without having to actually invent anything himself, although some experiments may be necessary. The manner in which a micro-organism must be described in a patent application depends on whether it is known and well-characterized micro-organism, which is easily available, or an unknown micro-organism which has recently been isolated or newly produced by mutation or genetic engineering. If a micro-organism is well-known and characterized, a general description of its characteristics and a reference to literature is enough (Biggart 1982; Saliwanchik 1977).

If a micro-organism is recently isolated, new or man-made, it is recommended that it is characterized as to its genus and species (or what the micro-organism most resembles), how the micro-organism was obtained, where it was isolated from, what mutagenesis has been performed on it, or the genetic techniques used to modify it. In addition, such features as morphology, metabolic characteristics, antibiotic resistance, tolerance to heavy metals, growth conditions like temperatures, requirements for vitamins, etc., are also used (Biggart 1982; Halluin 1982). The above-mentioned way of characterizing new micro-organisms also concerns applications where protection is not sought for the micro-organism *per se*, but for the methods in which they are used.

There are two problems encountered in describing micro-organisms, namely the difficulty of taxonomic description and the need to fulfill the requirement for repeatability. A species of a bacterium has been described as follows: 'A species of a bacterium is the type culture or specimen together with all other cultures or specimens regarded by an investigator as sufficiently like that type (or sufficiently closely related to it) to be grouped with it' (Buchanan 1948). Since amongst the criteria for the patentability of an invention, there are listed such things as novelty, inventiveness, or non-obviousness, inventors seek to show in their patent applications that their micro-organism is new even when it is not, and the vague definition of a species offers an excellent opportunity to do this (Gyllenberg 1982).

It must also be noted that the methods and criteria that are applied to microbial taxonomy have developed rapidly in the past 10 years and nowadays micro-organisms are described with almost completely different taxonomic terms than 15 or 20 years ago (Gyllenberg 1982). Thus, comparing micro-organisms described in new patent publications to ones presented in older publications is a difficult task. Micro-organisms produced by gene techniques are often described in patent publications with the help of a new plasmid or vector that has been transferred into it. This is motivated because novelty is often focused on the fact that new genetic material has been introduced into a micro-organism (Crespi 1982).

Although the taxonomic description of a micro-organism is given in as detailed a form as possible and the method for obtaining the organism has been described, that still does not guarantee that 'one skilled in the art' could use the invention. In other words, the criterion for reproduction is not fulfilled. The described micro-organism must be available; for this reason it has become practice that the micro-organism should be deposited in a culture collection and is an integral part of the patent publication.

3.5. Deposition of micro-organisms

The first deposit in a patentable process was made in 1949 when the strain *Streptomyces aureofaciens* which produced chlortetracycline was deposited at Northern Regional Research Laboratory, Peoria, USA (Cooper, 1982). Court cases that have had an effect on developing the deposit system have been Argoudelis (1970) and Feldman versus Aunstrup (1975) in the USA, and the cases of Levorin (Anon. 1974), Bäckerhefe (Anon. 1975) and Mikroorganismen (Anon. 1977) in the German Federal Republic. The European Patent Convention included in its Implementing Regulations special rules 28 and 28a. If an invention includes the use of a micro-organism (Rule 28) '. . . which is not available to the public and which cannot be described . . . in such a manner as to enable the invention to be carried out by a person skilled in the art . . .' the micro-organism must be deposited in a culture collection at the time when the patent is filed. The details concerning the identification of the strain must be revealed in the application, and the deposit must be made available to anyone from the first date of publication (18 months after the priority date).

Applicants were not, however, satisfied with rule 28 since other inventors only need to publish information about their invention and are not required to deposit it. If a micro-organism has been deposited in a culture collection it is available to third parties who would then have the physical means of utilizing the invention. In 1980 rule 28 was revised so that during the time between the publication of the application and the granting of the patent the strain should only be available to independent experts, at the discretion of the inventor, and should not be passed on to a third party. Rule 28a concerns the redeposition of a micro-organism. In the USA and in Japan, a microbial deposit is only available after the patent has been granted (Crespi 1982).

In 1977 the Budapest Treaty on the International Recognition of the Deposit of Microorganisms for the Purposes of Patent Procedure was signed and came into force in 1980. By the beginning of 1984, the treaty was ratified by 13 states (Anon. 1984d). According to the Budapest Treaty, deposition of a micro-organism in a culture collection approved by the treaty means that the deposit is valid in all those countries that have ratified the treaty. If the culture collection happens to be in the applicants home country then this makes deposition easier than having to send the deposit abroad where there may be difficulties with export and import licenses. It must be noted that, according to the Budapest Treaty, micro-organisms include bacteria, fungi, yeasts, viruses, animal and plant cell lines, protozoa, and algae which can be deposited (Hüni and Buss 1982). Hence, the concept of a 'micro-organism' encompasses a wide range of things and does not comply with the term used in scientific circles.

The German court rulings have been particular about the demand for reproducibility. If protection is sought for a strain *per se*, then a repeatable production method must be made known, a mere deposit in a collection is not enough (Vossius 1982). If the micro-organism has been isolated, e.g. from the soil, it is difficult to describe a repeatable method of isolation, and thus in Germany it is not possible to get product protection for strains of micro-organisms that have been isolated from nature.

3.6. Scope of protection

An inventor tries to define his invention so that protection will be as wide and beneficial as possible. When dealing with microbiological inventions, the more details there are in the claim then the narrower the patent protection is for that invention. In wording the claim, careful consideration must be given as to whether the invention can be used in the area of the defined protection. If it cannot be used, then the patent application can be rejected or the patent already granted revoked.

Woodruff et al. (1970) have discussed the limitation of a claim to a particular microbial species. They suggest that some new inventions are so important and widely applicable that limiting protection to one species gives insufficient protection. They argue that the deposited strains are only 'examples' of those organisms that can be successfully employed in the invention and the inventor must describe in the patent specification how micro-organisms that correspond to the one used in the invention can be found in nature. On the other hand, it can be shortsighted to limit patent protection to strains that have traditionally been included in a certain taxon if the strains have many untypical characteristics. According to Woodruff *et al.* (1970) an experimental definition of the species for each patent could be used to overcome the problem of interspecies isolates. Hayhurst (1971) is of the opinion that the applicant, in trying to define a micro-organism exactly by trusting criteria that are currently in vogue, may include in his claim micro-organisms that are useless.

The applicants often try to broaden the scope of patent protection to include natural and man-made mutants obtained from the micro-organism described in the application. This is logical from a biological point of view because 'a descendant should be grouped with its ancestor' (Gyllenberg 1982). Patent authorities, however, take a negative view against the protection of mutants. A patent only protects that which has been verified in the description of the invention. It is difficult to predict the usefulness of natural or man-made mutants, and the same applies to micro-organisms that have been produced by gene technology.

4. Patenting inventions in gene technology

4.1. Objects of patent protection

The patent protection of genetically-engineered micro-organisms has already been referred to, and after the Chakrabarty case it is possible to obtain patents for these in several countries. A bigger problem, however, is presented by the new tools of gene technology: plasmids, vectors, re-combinant-DNA, and DNA-fragments. In principle DNA can, from a chemical point of view, be regarded as an organic polymer just as the polymers in the plastic or textile industry. The essential difference is that DNA is a polymer made up of four bases, adenine, guanine, cytosine, and thymine, the exact positions of which are extremely important for the functioning of the DNA molecule. The other essential difference is the fact that DNA is a polymer that can be reproduced in a cell or enzymatically in a test tube. An organic polymer from the plastics industry is on the other hand a 'dead' chemical compound where there is regular repetition of a certain molecule, e.g. polyvinylchloride (Jackson 1981).

Do these differences influence the patenting of a DNA molecule? Jackson offers this interesting comment '. . . it is going to be very difficult to define when a DNA molecule, an inanimate object, stops and a micro-organism, an animate object, begins'. Living micro-organisms have already been patented so what should prevent the patenting of its component parts.

In gene technology patent applications protection has been sought for a DNA-fragment (most often a gene), a plasmid, a vector, recombinant-DNA, and a micro-organism as a product, and also for proceses for producing them

(Suetina *et al.* 1982). Patent authorities have, however, preferred to grant process protection rather than product protection, and moreover for a micro-organism itself rather than the new products of gene technology.

4.2. Characterizing inventions in gene technology

One of the problems of patenting inventions in gene technology is how to characterize them in patent publications. In the descriptive part of the publication, the invention should be described so clearly that a person skilled in the art can use the invention on the basis of the description. The patent's scope of protection is defined in the claims of the publication.

When patents are applied for in a new field of technology it is difficult to know what the person skilled in the art knows about the subject, and it is also difficult to speculate beforehand the way in which the courts of law would react to the wording and definitions within the patent. In the first published patent applications and granted patents concerning gene technology, termin-ology has been confused, characterization of the inventions variable, and the protection sought has been rather wide (Knuth *et al.* 1984). Examples of wide protection are Cohen and Boyer's US-patent number 4237224 (Cohen and Boyer 1980) and European patent applications numbers 78300596 (Itakura and Riggs 1979), 783000597 (Riggs 1979) and 783000598 (Itakura 1979). As has already been observed, 'extension of protection beyond the content of the specification may lead to rejection of the application, or to revocation of an already granted patent.

A plasmid, vector, recombinant-DNA, and a DNA-fragment have mainly been characterized in patent publications by using 12 different characteristics (Knuth *et al.* 1984): name (designation), constituent parts (functional sub-units), name of structural gene (or encoded product), restriction enzyme cleavage sites, molecular weight/length, production or isolation method, host range, function, order of genes, origin of DNA insert, nucleotide sequence of DNA insert, regulatory region. Also physical map and genetic map have been presented quite often. In most cases, each object has been characterized by 6 or 7 different characteristics, but there are also patent publications where in the claims only the designation is given for the object of protection. The base sequence of a DNA-fragment is fairly often presented in the claims. For example, in the US-patent number 4393201 (Curtis and Wunner 1983), the base sequence is the only characteristic listed in the claims. However, presenting the base sequence without variations may guarantee an invention only a very narrow protection. At the other extreme the DNA-fragment can be defined according to its function. There are no internationally approved recommendations as to how the objects of gene technology should be characterized in patent publications, and this may cause delays in the procedure. Only when patent attorneys and patent authorities have become acquainted with this new field of technology the applications may become better formulated and the handling of patents be speeded up.

4.3. Deposits

The problem connected with the disclosure of the tools of gene technology concerns the deposition. Should the deposition of plasmids, vectors, recombinant-DNA, and DNA-fragments be demanded or not? The matter has been discussed in several articles, but opinions are divided. According to one opinion, DNA should be regarded as a chemical compound that can be defined so exactly that no deposition is necessary. On the basis of a description, plasmids, vectors, recombinant-DNA, and DNA-fragments can be produced in a reproducible manner (Bent 1982). The other viewpoint states that the literal description is insufficient because when plasmids or vectors are being produced there may be small variations in their structure, which can cause decisive changes in their function (Cooper 1982; Halluin 1982). In the USA, the best mode of carrying out an invention must be presented and, according to Saliwanchik (1982), a plasmid should be deposited for the sake of safety, because '. . . the fact remains that the best mode of carrying out the invention is the plasmid itself'.

A deposit must be made on the filing date of the patent application (Schulte 1981; Halluin 1982). A deposit made later on will not be accepted. If a deposit is not made the disclosure of the invention can be considered insufficient and no patent will be granted. On the other hand, if a deposit has been made and no patent granted, the deposit will nevertheless become public in most countries when the application becomes public. This would mean that the inventor is making his own 'miniature factory' available to the public when he still may not receive protection for it.

4.4. Conditions of patentability

In principle, the inventions of gene technology are patentable if they fulfill the conditions of patentability, which are the demands for novelty, inventiveness, utility, and reproducibility. The conditions are considered with regard to the prior art at the filing (or priority) date and the knowledge of the average expert. When it comes to inventions in the field of gene technology the claim will certainly be made that they are not novel. Applications will also be rejected because in the opinion of the patent authorities they do not meet the demand for inventiveness. Useful answers as to the limits of patent protection within gene technology will be at hands only in the future when there occur prejudicing court rulings regarding these matters.

5. Future of patenting biotechnology

Alvin Toffler (1980) in his, perhaps exaggerated, visions of the future society and the future technology, lists biotechnology, including genetic engineering, amongst the 'revolutionary' technologies which will characterize the approaching information society. Although all of Toffler's views may not be realized in all of their dimensions (at least not in the near future), his points are well-worth serious attention. Amongst many other questions, it must be asked, if the inventions of the information society can be evaluated and protected by practices and laws whose roots emanate from the dawn of the so-called first technological revolution.

Cooper (1984) suggests '. . . the development of a patent law written with biological inventions in mind'. In this context Cooper makes reference to the US Plant Variety Protection Act of 1970, and notes that in that act, the traditional patent requirements for novelty, utility, and non-obviousness have been replaced by requirements for distinctness, uniformity, and stability. Cooper concludes that these concepts may be applied not only to plant varieties, but also to animal varieties, cell lines, and micro-organisms.

Goldstein (1984), on the other hand, has considered the possibility of protecting genetic engineering inventions by copyright law. The Copyright Act of 1976 in the USA extended the copyright protection to include computer programs as 'literary works'. Genetic sequences could be seen as analogous to computer programs and accordingly could be protected under copyright law. However, Goldstein (1984) notices that copyright protection might be useful for well defined gene sequence combinations in a mature stage of the art. At the early stages, where broad protection could be obtained, patents would be by far superior.

As far as micro-organisms *per se* are concerned, procaryotes and eucaryotes represent different approaches. Procaryotic taxonomy is obviously approaching deep reconsiderations, particularly with regards to the species concept. The species concept as applied to bacteria varies to meet different purposes (e.g. Gyllenberg 1985). Accordingly, the concept of a bacterial species has hardly any legal implications, and is therefore of limited value as a basis for patent applications. In the future, patent specifications and claims based on a description of 'new species', are no more significant than a distinct description of the organism in terms of, for example DNA-homology to previously known strains, or using other markers indicating bacterial identity.

Prokaryote taxonomy will certainly proceed from phenetic observations to the basic genetic structures including plasmids. In the future, 'patent taxonomy', i.e. the distinction of organisms for patent application purposes, will consequently deal with basic DNA-structures. Possibly the development of a completely new prokaryote taxonomy will be enhanced by the achievements within the field of microbiological patents.

It is also claimed that the very rapid developments within the field of genetic engineering will provoke a movement towards a continuous hiding of scientific progress instead of the publication of innovations by patent applications. In this respect one has to consider that the patent system does not only serve the purpose of protecting inventions, but also to provoke potential competitors to reveal the level of their know-how in a particular area.

Possibly, the future development in the field of biotechnology will take the direction where the really big technological breakthroughs are kept as secret as possible, but where the patent system is used to explore the existing know-how in potentially promising areas. This development is of course confusing since it changes the whole philosophy of patenting from a means of obtaining protection for a given invention to a means of collecting information about developments within some particular technological area.

Patenting in biotechnology is an attractive intellectual adventure where the final norms and rules cannot be anticipated, and the legal approach to biotechnology is a challenging aspect of the future.

References

Anon. (1969). Rote Taube. *Gewerblicher Rechtsschutz und Urheberrecht* 1969, 672–6.
—— (1974). Levorin. *Gewerblicher Rechtsschutz und Urheberrecht* 1974, 392–3.
—— (1975). Bäckerhefe. *Gewerblicher Rechtsschutz und Urheberrecht* 1975, 430–4.
—— (1977). Mikroorganismen. *Gewerblicher Rechtsschutz und Urheberrecht* 1977, 30–2.
—— (1978*a*). Naturstoffe. *Gewerblicher Rechtsschutz und Urheberrecht* 1978, 238–41.
—— (1978*b*). Menthonthiole. *Gewerblicher Rechtsschutz und Urheberrecht* 1978, 702–5.
—— (1978*c*). *Lactobacillus bavaricus. Gewerblicher Rechtsschutz und Urheberrecht* 1978, 586–8.
—— (1983). International Treaties: Africal Intellectual Property Organization (OAPI). *Official J. Eur. Patent Office* 1983, 68–72.
—— (1984*a*). Paris Convention. Membership 1.1.1984. *Indust. Property* 23, 6–8.
—— (1984*b*). Patent Cooperation Treaty. Membership 1.1.1984. *Indust. Property* 23, 14.
—— (1984*c*). African Intellectual Property Organization. Membership 1.1.1984. *Indust. Property* 23, 28.
—— (1984*d*). Budapest Treaty. Membership 1.1.1984. *Indust. Property* 23, 16.
Argoudelis (1970). In re Argoudelis. 434 Federal Reporter, Second Series 1390, 168 United States Patent Quarterly 99 (Court of Customs and Patent Appeals 1970). (Ref. Halluin 1982)
Armitage, E. (1977). British Patent Law. 200 Years of English and American Patent, Trademark, and Copyright Law 9. (Ref. Cooper 1982).
Buchanan, R. E. (1948). How bacteria are named and identified. In *Bergey's manual of determinative bacteriology* (ed. R. S. Breed, E. G. D. Murray, and A. P. Hitchens), 6th edn. The Williams & Wilkins Company. Baltimore.

Bent, S. A. (1982). Patent protection for DNA molecules. *J. Patent Office Soc.* 64, 60–86.

Bergy (1977). In re Bergy. 573 Federal Reporter, Second Series 1031, 195 United States Patent Quarterly 344 (Court of Customs and Patent Appeals 1977). (Ref. Halluin 1982).

Biggart, W. A. (1982). Patentability in the United States of microorganisms, products produced by microorganisms and microorganism mutational and genetic modification techniques. J. Law Technol. 22, 113–36.

Chakrabarty (1978). In re Charkrabarty. 571 Federal Reporter, Second Series 40, 197 United States Patent Quarterly 72 (Court of Customs and Patent Appeals 1978). (Ref. Halluin 1982).

Cohen, S. N. and Boyer, H. W. (1980). US-patent 4237224 issued at 2.12.1980.

Cooper, I. P. (1982). *Biotechnology and the law*. Clark Boardman Company Ltd, New York.

—— (1984). Do we need a special patent law for biological inventions. *Biotechnology* 2, 179, 192.

Crespi, R. S. (1982). *Patenting in the biological sciences*. John Wiley & Sons. Chichester/New York.

Curtis, P. J. and Wunner, W. H. (1983). US-patent 4393201, issued at 12.7.1983.

Dickson, D. (1980). Patenting living organisms—how to beat the bugrustlers. *Nature* 283, 128–9.

European Patent Office (1983) *How to get a European patent. Guide for applicants* (6th edn). Munich.

Feldman versus Aunstrup (1975). 517 Federal Reporter, Second Series 1351, 186 United States Patent Quarterly 108 (Court of Customs and Patent Appeals 1975) (Ref. Halluin 1982).

Goldstein, J. A. (1984). Copyrightability of genetic works. *Biotechnol.* 2, 138–42.

Gyllenberg, H. (1982). Man-made microbes and man-made law. *Acta Biotechnologica* 2, 207–12.

—— (1985). Progress and expectations in biology. In *The identification of progress in learning* (ed. T. Hägerstrand). Cambridge University Press, Cambridge.

Halluin, A. P. (1982). Patenting the results of genetic engineering research: an overview. In *Banbury report 10. Patenting of life forms* (eds D. W. Plant, N. J. Reimers, and N. D. Zinder), pp. 67–126. Cold Spring Harbor Laboratory, New York.

Hayhurst, W. L. (1971). Patents relating to industrial microorganisms in the english speaking countries. *Ind. Property* 10, 189–98.

Hüni, A. and Buss, V. (1982). Plant protection in the field of genetic engineering. *Indust. Property* 21, 356–68.

Itakura, K. (1979). European patent application 78300598, published at 16.5.1979 (publication number 0001931).

—— and Riggs, D. (1979). European patent application 78300596, published at 16.5.1979 (publication number 0001929.)

Jackson, D. A. (1981). Patenting of genes: What will the ground rules be? In *Patentability of microorganisms: issues and questions* (eds R. F. Acker and M. Schaechter), pp. 23–7. American Society for Microbiology, Washington DC.

Knuth, S. H., Belyakova, L. G., Velkov, V. V., Dementyev, V. N., Lommi, H. I., Harper, R. A., and Gyllenberg, H. G. (1984). Characteristics of genetic material with reference to patent documents. *Acta Biotechnol.* 4, 195–208.

Parker versus Flook (1978). 198 United States Patent Quarterly 1893 (Ref. Vossius 1982).

Riggs, A. D. (1979). European patent application 78300597, published at 16.5.1979 (publication number 0001930).

Rosenberg, P. D. (1980). *Patent law fundamentals*, 2nd edn. Clark Boardman Company Ltd, New York.

Saliwanchik, R. (1977). Patentable distinctions among microorganisms. *Dev. Indust. Microbiol.* 18, 327–31.

——, (1982). *Legal protection of microbiological and genetic engineering inventions.* Addison-Wesley. London.

Schulte, R. (1981). *Patentgesetz. Heymanns Taschenkommentare zum gewerblichen Rechtschutz. 3. Auflage.* Carl Heymanns Verlag KG. Cologne, Berlin, Bonn, Munich.

Sinnott, J. P. (1983). *World patent law and practice. Patent statutes, regulations and treaties,* Volume 2B. Matthew Bender and Company. New York.

Suetina, R. L., Belyakova, L. G., and Velkov, V. V. (1982). *Biotechnology and recombinant DNA.* Scientific Center of Biological Research of the Academy of Sciences of the USSR in Pushchino. Pushchino.

Toffler, A. (1980). *The third wave.* William Morrow and Company Inc., New York.

USSR Chamber of Commerce and Industry, Patent Department (1981). *Patenting foreign inventions in the USSR. Practical guide.* Moscow.

von Pechmann, E. F. (1972). Über nationale und internationale Probleme des Schutzes mikrobiologischer Erfindungen. *Gewerblicher Rechtsschutz und Urheberrecht* 1972, 51–102.

Vossius, V. (1981). Patent protection for biological inventions. In *Biotechnology. A comprehensive treatise in 8 volumes. Microbial fundamentals* (ed H.-J. Rehm and G. Reed) Vol 1, pp. 435–52. Verlag Chemie, Weinheim Deerfield Beach, Florida/ Basel.

——, (1982). The patenting of life forms under the European Patent Convention and German Patent Law: Patentable inventions in the field of genetic manipulations. In *Banbury report 10. Patenting of life forms* (eds D. W. Plant, N. J. Reimers, and N. D. Zinder) pp. 67–126. Cold Spring Harbor Laboratory, New York.

Wegner, H. C. (1979). Patenting the products of genetic engineering. *Biotechnol. Lett.* 1, 145–50.

Woodruff, H. B., Currie, S. A., Hallada, T. C., and Putter, I. (1970). Importance of the description of the producing organisms in obtaining patent protection for fermentation processes. In *Abstract Book of the 1st international symposium of genetics of industrial microorganisms,* pp. 403–22. Academy of Sciencies of Czechoslovakia, Prague.

World Intellectual Property Organization (1978). *Patent Cooperation Treaty, Applicants Guide. General Information for users of the Patent Cooperation Treaty.* Edition with replacement sheets separately listed. December 1978.

Index